SALVAGE

STATISTICAL MECHANICS AND FIELD THEORY

Statistical Mechanics and Field Theory

Lectures given at the 1971 Haifa Summer School,
with additional contributions from participants
at the 1971 Europhysics Conference

Edited by

R. N. SEN and C. WEIL

Department of Physics
Technion—Israel Institute of Technology
Haifa, Israel

HALSTED PRESS

A Division of JOHN WILEY & SONS, INC., New York

ISRAEL UNIVERSITIES PRESS

Jerusalem · London

ISRAEL UNIVERSITIES PRESS
is a publishing division of
ISRAEL PROGRAM FOR SCIENTIFIC TRANSLATIONS LTD., Jerusalem

© 1972
by ISRAEL PROGRAM FOR SCIENTIFIC TRANSLATIONS LTD.

Sole distributors for the Western Hemisphere

HALSTED PRESS, a division of
JOHN WILEY & SONS, INC., NEW YORK

Sole distributors for Europe, Africa, and the Middle East

JOHN WILEY & SONS LTD., CHICHESTER, ENGLAND

Distributors for the rest of the world

KETER PUBLISHING HOUSE, JERUSALEM

ISBN 0 7065 1263 4   IUP/Jerusalem–London
ISBN 0 470 77595 5   Halsted Press/New York
Library of Congress Catalog Card Number 72-4108
IPST catalog number 25011

Technical editor: Ora Hand

This book has been composed, printed and bound by Keter Press, Jerusalem, Israel, 1972

Contents

FOREWORD

In July-August 1971, a summer school, a Europhysics Conference and an informal workshop on Statistical Mechanics and Field Theory were held at the Technion—Israel Institute of Technology. It was decided to publish the lectures given at the summer school. We took the opportunity provided by the conference and the workshop to solicit some additional expository articles from certain individuals. The present volume is the result.

We hope that this collection captures some of the flavor of a very plesant and stimulating meeting, in which mathematical and non-mathematical physicists made a conscious effort to communicate with each other. The mathematically advanced reader may find that some of the articles move at a leisurely rather than exhilarating pace. The more traditional nonmathematical physicist may at times be out of breath. However, with a little goodwill he may discover in these pages that learning by logic stimulates, and *saves a vast amount of time. Most of the articles collected here were not written for the expert. "Perhaps for this reason, experts will enjoy them."*

We would like to acknowledge the financial support, extended through the Technion—Israel Institute of Technology, of the Yad Avi ha-Yishuv, of the Technion itself, the Israel Academy of Sciences and Humanities, the Bar-Ilan University and the Tel Aviv University. We also take the opportunity to thank Madeleine Levy for looking after the organizational details with efficiency, and charm.

Particular thanks are extended to the summer school lecturers who shouldered the heaviest burden. Much of the success of the Europhysics Conference was due to the care with which they prepared the ground. Finally, to all participants who journeyed from near and far to this meeting:

לשנה הבאה בירושלים!*

R.N. SEN

* "Next year in Jerusalem!"

C. WEIL

STATISTICAL MECHANICS AND FIELD THEORY
R. N. Sen and C. Weil, Editors

Introduction to Operator Algebras

HUZIHIRO ARAKI

Research Institute for Mathematical Sciences, Kyoto (Japan)

Abstract

Some basic concepts in the theory of operator algebras such as the GNS construction, the Gelfand isomorphism, etc. and some advanced topics such as KMS states, modular automorphism and decomposition theory are illustrated by examples of finite dimensional cases. As a further illustration of physical application, the determination of KMS states for a free Bose gas is described.

In this short series of lectures, I intend to give a brief introduction to some aspects of the theory of W^*- and C^*-algebras and its application to physics by way of simple examples.

Both W^*- and C^*-algebras are algebras of operators on a Hilbert space and hence it is not necessarily surprising that they provide us with a natural mathematical setup for some class of problems in quantum physics, including quantum field theory and quantum statistical mechanics. It turns out that some aspects of classical statistical mechanics can also be formulated in the language of (commutative) C^*-algebras.

Crudely speaking, there are two distinct aspects in the theory of operator algebras, namely the algebraic and the topological aspects. For the algebraic aspects, an algebra of finite matrices is a good place where one can see the essence of an argument, while for the topological aspects, a commutative algebra of functions on some compact space is often a convenient example where one can see the essential features. For more difficult problems, the two aspects are actually twisted together and the simple examples mentioned above do not give a correct picture. Despite this, I shall pick up algebras of finite matrices to illustrate the KMS condition as an example of the algebraic aspect (although the really difficult part is topological) and the decomposition of states as an example of the topological aspect. To give an example of applications, I shall illustrate the free Bose gas in connection with these two general theories.

1

1. PRELIMINARIES ON A SINGLE OPERATOR

In the following, $\mathfrak{H}$ denotes a Hilbert space, which includes as a special case the n-dimensional Hilbert space $\mathfrak{H}_n$, $n = 1, 2, \ldots$. The inner product of two elements Ψ and Φ of $\mathfrak{H}$ is denoted by (Ψ, Φ). We adopt the physicist's convention: $(\Psi, c\Phi) = c(\Psi, \Phi)$ for any complex number c. (Mathematician's convention $(\Psi, c\Phi) = \bar{c}(\Psi, \Phi)$.)

A linear operator A on $\mathfrak{H}$ can be described in terms of its matrix elements $A_{ij} = (e_i, Ae_j)$ relative to any fixed complete orthonormal basis $e_i \in \mathfrak{H}$.

The norm $\|A\|$ is defined by

$$\|A\| = \sup_{\Psi \in \mathfrak{H}, \Psi \neq 0} \|A\Psi\|/\|\Psi\|, \tag{1.1}$$

where $\|\Phi\| = (\Phi, \Phi)^{1/2}$. If $\|A\| < \infty$, A is said to be bounded. We shall limit our attention most of the time to bounded A.

The adjoint A^* of A is defined by $A^*\Psi = \chi$ if

$$(\chi, \Phi) = (\Psi, A\Phi), \qquad \chi \in \mathfrak{H} \tag{1.2}$$

for all Φ. If $A^* = A$, A is called selfadjoint.

We shall list simple properties of the norm and the adjoint, with some indications of their proofs.

From the positivity of $(\Psi + c\Phi, \Psi + c\Phi) \geq 0$ for all c, we obtain the Schwarz inequality $|(\Psi, \Phi)| \leq \|\Psi\| \, \|\Phi\|$. Hence

$$\|\Psi\| = \sup_{\Phi \in \mathfrak{H}, \Phi \neq 0} |(\Phi, \Psi)|/\|\Phi\|, \tag{1.3}$$

where the sup is reached by $\Phi = \Psi$ (if $\Psi \neq 0$). Therefore

$$\|A\| = \sup |(\Phi, A\Psi)|/\{\|\Phi\| \, \|\Psi\|\} \tag{1.4}$$

from which we obtain

$$\|A^*\| = \|A\|. \tag{1.5}$$

From (1.1), $\|A\Psi\| \leq \|A\| \, \|\Psi\|$ for all Ψ. Therefore, $\|A_1(A_2\Psi)\| \leq \|A_1\| \, \|A_2\Psi\| \leq \|A_1\| \, \|A_2\| \, \|\Psi\|$ and hence

$$\|A_1 A_2\| \leq \|A_1\| \, \|A_2\|. \tag{1.6}$$

As a special case, we have $\|A^*A\| \leq \|A^*\| \, \|A\| = \|A\|^2$. On the other hand, $\|A\Psi\|^2 = (\Psi, A^*A\Psi) \leq \|A^*A\| \, \|\Psi\|^2$ and hence $\|A\|^2 \leq \|A^*A\|$. Therefore,

$$\|A^*A\| = \|A\|^2. \tag{1.7}$$

A bounded linear operator A^{-1} satisfying $A^{-1}A = AA^{-1} = 1$ is called a bounded inverse of A, which is unique if it exists. The set of complex numbers c such that $c - A$ has a bounded inverse is called the resolvent set of A and its complement is called the spectrum of A.

Example

Let A be a selfadjoint operator on $\mathfrak{H}_n$. Let $\mathfrak{H}_A(\lambda)$ be the set of vectors Ψ in $\mathfrak{H}_n$ satisfying $A\Psi = \lambda\Psi$. Then an arbitrary vector $\Phi \in \mathfrak{H}$ has a unique decomposition $\Phi = \sum_\lambda \Phi(\lambda)$ where $\Phi(\lambda) \in \mathfrak{H}_A(\lambda)$ and the sum runs over those λ such that $\mathfrak{H}_A(\lambda)$ contains a non-zero vector of $\mathfrak{H}_n$. The operator $E_A(\lambda)$ defined by $E_A(\lambda)\Phi = \Phi(\lambda)$ is an orthogonal projection (i.e. $E_A(\lambda)^2 = E_A(\lambda) = E_A(\lambda)^*$) and $A = \sum \lambda E_A(\lambda)$. Then

$$(c - A)^{-1} = \sum (c - \lambda)^{-1} E_A(\lambda) \tag{1.8}$$

exists if c does not coincide with any eigenvalue λ. On the other hand, if $(\lambda - A)\Psi = 0$ for $\Psi \neq 0$, then $(\lambda - A)^{-1}$ obviously cannot exist. Therefore, the spectrum of A is the set of all eigenvalues of A. In particular, if $c > |\lambda|$ for all λ, then

$$(c - A)^{-1} = \sum_{n=0}^{\infty} c^{-n-1} A^n \tag{1.9}$$

as is seen from (1.8). (This is true for any A if $\|A\| < c$.)

Due to the orthogonality $E_A(\lambda) E_A(\lambda') = 0$ for $\lambda \neq \lambda'$, we have

$$\|A\Psi\|^2 = \sum |\lambda|^2 \|E_A(\lambda)\Psi\|^2, \qquad \|\Psi\|^2 = \sum \|E_A(\lambda)\Psi\|^2. \tag{1.10}$$

Therefore $\|A\| = \sup|\lambda|$. Thus we have the following algebraic characterization of $\|A\|$ for an arbitrary A:

$\|A\|$ is the positive square root of the supremum of real c such that $c - A^*A$ does not have a bounded inverse.

This characterization holds in any C^*-algebra.

2. C^*- AND W^*-ALGEBRAS

Example

Let $\mathscr{B}(\mathfrak{H})$ denote the set of all bounded linear operators on $\mathfrak{H}$. Specifically, consider $\mathscr{B}(\mathfrak{H}_n)$. It has the following operations defined in it: $A_1 + A_2$, cA, $A_1 A_2$, A^*. In terms of matrix elements, we have

$$(A_1 + A_2)_{ij} = (A_1)_{ij} + (A_2)_{ij} \tag{2.1}$$

$$(cA)_{ij} = cA_{ij} \tag{2.2}$$

$$(A_1 A_2)_{ij} = \sum_k (A_1)_{ik} (A_2)_{kj} \tag{2.3}$$

$$(A^*)_{ij} = A_{ji}. \tag{2.4}$$

It is an algebra (over the complex field) in which the *-operation satisfies the following properties:

$$(A_1 + A_2)^* = (A_1)^* + (A_2)^* \tag{2.5}$$

$$(cA)^* = \bar{c}A^* \tag{2.6}$$

$$(A_1 A_2)^* = A_2^* A_1^* \tag{2.7}$$

$$(A^*)^* = A. \tag{2.8}$$

Such an algebra is called a *-algebra.

Definition:

A *-algebra $\mathfrak{A}$ of operators on $\mathfrak{H}$ is called a concrete C^*-algebra if it is closed under the norm, namely if $A_n \in \mathfrak{A}$, $n = 1, 2, \ldots$ and $\lim_{n \to \infty} \|A_n - A\| = 0$ for $A \in \mathcal{B}(\mathfrak{H})$ imply $A \in \mathfrak{A}$.

Definition:

A *-algebra $\mathfrak{A}$ equipped with the norm is called an abstract C^*-algebra (or B^*-algebra) if $\mathfrak{A}$ is complete with respect to the norm and the following properties are satisfied:

$$\|A_1 + A_2\| \leq \|A_1\| + \|A_2\|, \tag{2.9}$$

$$\|cA\| = |c| \, \|A\|, \tag{2.10}$$

$$\|A_1 A_2\| = \|A_1\| \, \|A_2\|, \tag{2.11}$$

$$\|A^*A\| = \|A\|^2. \tag{2.12}$$

(From these, it follows that $\|A^*\| = \|A\|$.)

From what we have seen in Section 1, any concrete C^*-algebra is an abstract C^*-algebra. The converse also holds, in the sense that any abstract C^*-algebra has a faithful representation on a Hilbert space, where we use the following terminology:

Definition:

For each element A of a *-algebra $\mathfrak{A}$, let $\pi(A)$ be a bounded linear operator on $\mathfrak{H}$. It is called a *-representation if

$$\pi(A_1 + A_2) = \pi(A_1) + \pi(A_2), \tag{2.13}$$

$$\pi(cA) = c\pi(A), \tag{2.14}$$

$$\pi(A_1 A_2) = \pi(A_1)\pi(A_2), \tag{2.15}$$

$$\pi(A^*) = \pi(A)^*. \tag{2.16}$$

It is called faithful if $\pi(A) = 0$ implies $A = 0$.

From the algebraic characterization of $\|A\|$ in Section 1, we see that $\|\pi(A)\| \leqq \|A\|$ and the equality $\|\pi(A)\| = \|A\|$ holds if the representation is faithful.

Thus the norm of a *-algebra which makes it a C^*-algebra is unique if it exists.

If we use the so-called weak topology instead of the norm topology in the definition of a concrete C^*-algebra, we have

Definition:

A *-algebra R of operators on $\mathfrak{H}$ is called a von Neumann algebra (or W^*-algebra) if it is weakly closed and if the identity operator 1 is in R. Here R is weakly closed if for every generalized sequence A_α in R, satisfying $\lim(\Psi, A_\alpha\Phi) = (\Psi, A\Phi)$ for all Ψ and Φ in $\mathfrak{H}$ and some $A \in \mathcal{B}(\mathfrak{H})$, A belongs to R.

For the finite dimensional case, a *-algebra on $\mathfrak{H}_n$ is at the same time a C^*-algebra and a W^*-algebra. In the general case, a W^*-algebra is a concrete C^*-algebra, but the converse is not necessarily true. For example, consider the Hilbert space of all square integrable functions on the real line denoted by $L_2(R)$ (or $L_2(R, dx)$). The multiplication by a bounded function is an element of $\mathcal{B}(L_2(R))$. The set of all such operators is a commutative W^*-algebra, denoted by $L_\infty(R)$. Its subset $C(R)$, consisting of the multiplication by a bounded function which is continuous, is a C^*-algebra but is not a W^*-algebra. In either case, the norm of the operator is the (essential) supremum of the absolute value of the function.

An important characterization of a W^*-algebra is given by the double commutation theorem of von Neumann.

Definition:

For any set R of operators on $\mathfrak{H}$, R' denotes the set of $A \in \mathcal{B}(\mathfrak{H})$ such that $[B, A] = 0$ and $[B^*, A] = 0$ for all $B \in R$.

Then $(R')' \equiv R''$ is the smallest W^*-algebra on $\mathfrak{H}$ containing R, i.e., elements of R'' can be approximated (in the weak topology) arbitrarily closely by polynomials of 1, $B_i \in R$, $i = 1, \ldots, N$ and B_i^*, $i = 1, \ldots, N$, $N = 1, 2, \ldots$.

3. STATES OF A C^*-ALGEBRA

Let $\mathfrak{A}$ be a concrete C^*-algebra on $\mathfrak{H}$ and $\Omega \in \mathfrak{H}$ be a unit vector. The expectation value

$$\omega_\Omega(A) \equiv (\Omega, A\Omega) \tag{3.1}$$

satisfies

$$\omega_\Omega(A_1 + A_2) = \omega_\Omega(A_1) + \omega_\Omega(A_2) \tag{3.2}$$
$$\omega_\Omega(cA) = c\omega_\Omega(A) \tag{3.3}$$

(linearity)

$$\omega_\Omega(A^*A) \, (= \|A\Omega\|^2) \geqq 0 \quad \text{(positivity)} \tag{3.4}$$

$$\|\omega_\Omega\| \equiv \sup_{A \in \mathfrak{A}, A \neq 0} |\omega_\Omega(A)| / \|A\| = \|\Omega\| = 1 \quad \text{(normalization)}. \tag{3.5}$$

Any normalized positive linear functional on $\mathfrak{A}$ is called a state. If $1 \in \mathfrak{A}$, then the normalization condition can be replaced by $\omega(1) = 1$. ω_Ω is called a vector state by Ω.

A vector Ω in $\mathfrak{H}$ is called a cyclic vector of $\mathfrak{A}$ if $\mathfrak{A}\Omega$ is dense in $\mathfrak{H}$ where $\mathfrak{A}\Omega$ denotes the set of $A\Omega$, $A \in \mathfrak{A}$.

An important theorem due to Gelfand, Naimark and Segal gives the converse, namely for any state ω, there exists a Hilbert space $\mathfrak{H}_\omega$, a *-representation π_ω of $\mathfrak{A}$ on $\mathfrak{H}_\omega$ and a unit cyclic vector Ω_ω of $\mathfrak{A}$ in $\mathfrak{H}_\omega$ such that

$$\omega(A) = (\Omega_\omega, \pi_\omega(A)\Omega_\omega) \tag{3.6}$$

for all $A \in \mathfrak{A}$. Such a triplet $\mathfrak{H}_\omega, \pi_\omega, \Omega_\omega$ is uniquely determined by ω (up to unitary equivalence). We shall describe the construction of the triplet canonically associated with ω through (3.6) by using the simplest example given in Section 2: $\mathfrak{A} = \mathscr{B}(\mathfrak{H}_n)$.

The easiest way to understand the situation is to start from the conclusion of the theorem. Since Ω_ω is to be cyclic, the set of vectors $\pi_\omega(A)\Omega_\omega$ is to be dense in $\mathfrak{H}_\omega$, and it is exactly $\mathfrak{H}_\omega$ if $\mathfrak{A} = \mathscr{B}(\mathfrak{H}_n)$. Hence, to construct $\mathfrak{H}_\omega$, we consider the set of symbols $\pi_\omega(A)\Omega_\omega$ depending on $A \in \mathfrak{A}$, define its addition $(\pi_\omega(A_1)\Omega_\omega) + (\pi_\omega(A_2)\Omega_\omega) = \pi_\omega(A_1 + A_2)\Omega_\omega$, the multiplication by a complex number c: $c(\pi_\omega(A_1)\Omega_\omega) = (\pi_\omega(cA_1)\Omega_\omega)$ and the inner product

$$(\pi_\omega(A_1)\Omega_\omega, \pi_\omega(A_2)\Omega_\omega) = \omega(A_1^* A_2). \tag{3.7}$$

We can determine the right-hand side of (3.7) by computing the left-hand side using the equality $\pi_\omega(A_1)^* \pi_\omega(A_2) = \pi_\omega(A_1^* A_2)$. We then consider those symbols with zero norm to be zero, and (the completion of) the quotient space so obtained happens to satisfy all the requirements for $\mathfrak{H}_\omega$, where $\Omega_\omega = \pi_\omega(1)\Omega_\omega$ and $\pi_\omega(A)$ is defined by $\pi_\omega(A)(\pi_\omega(B)\Omega_\omega) = \pi_\omega(AB)\Omega_\omega$.

In the next section, we shall see more concretely the structure of $\mathfrak{H}_\omega$, π_ω and Ω_ω for the special case of $\mathfrak{A} = \mathscr{B}(\mathfrak{H}_n)$.

4. MODULAR OPERATORS

We concentrate our attention on the simple example $\mathfrak{A} = \mathscr{B}(\mathfrak{H}_n)$. The trace of a matrix A is defined by

$$\mathrm{Tr}\, A = \sum_{i=1}^{n} A_{ii}. \tag{4.1}$$

If we divide by the dimension n, we obtain a state called the central state:

$$\tau(A) \equiv \frac{1}{n}\mathrm{Tr}(A). \tag{4.2}$$

Since $\mathrm{Tr}\, AB = \sum_i \left(\sum_k A_{ik} B_{ki} \right) = \mathrm{Tr}\, BA$, we have

$$\tau(AB) = \tau(BA). \tag{4.3}$$

A state of $\mathscr{B}(\mathfrak{H}_n)$ satisfying (4.3) is actually unique: Let e^{ij} be a matrix with all matrix elements 0 except the (ij) matrix element which is 1 $((e^{ij})_{kl} = \delta_{ik}\delta_{jl})$. Then $A = \sum A_{ij} e^{ij}$. First consider A^0 such that $\mathrm{Tr}\, A^0 = 0$. Then

$$A^0 = \sum_{i \neq j} A^0_{ij} [e^{i1}, e^{1j}] + \sum_{i=1}^{n-1} (A^0_{11} + \ldots + A^0_{ii}) [e^{i,i+1}, e^{i+1,i}]. \tag{4.4}$$

If τ' satisfies $\tau'([A_1, A_2]) = 0$ for all A_1 and A_2, then $\tau'(A^0) = 0$. Since $A^0 \equiv A - \tau(A)1$ has the property $\mathrm{Tr}\, A^0 = 0$, we have $\tau'(A) = \tau'(\tau(A)\,1) = \tau(A)$, which proves the uniqueness.

This also shows that the trace is independent of the choice of the orthonormal basis. For a selfadjoint $A = \sum \lambda E_A(\lambda)$, we may take a complete orthonormal basis consisting of eigenvectors of A and obtain $\tau(A) = (1/n) \sum \lambda \dim \mathfrak{H}_A(\lambda)$, where $\dim \mathfrak{H}_A(\lambda)$ is the multiplicity of an eigenvalue λ. For A^*A, all its eigenvalues are nonnegative and hence $\tau(A^*A) = 0$ only if $A = 0$. Hence $\pi_\tau(A)\Omega_\tau = 0$ only if $A = 0$, which implies that the correspondence $A \in \mathscr{B}(\mathfrak{H}_n) \leftrightarrow \pi_\tau(A)\Omega_\tau \in \mathfrak{H}_\tau$ is one-to-one. We denote by A_Ψ the unique element of $\mathscr{B}(\mathfrak{H}_n)$ satisfying $\pi_\tau(A_\Psi)\Omega_\tau = \Psi$.

Let ω be a state of $\mathscr{B}(\mathfrak{H}_n)$. Since $\hat{\omega}(\Psi) \equiv \omega(A_\Psi)$ is a linear functional on $\Psi \in \mathfrak{H}_\tau$, there exists a vector $\Phi \in \mathfrak{H}_\tau$ such that $\hat{\omega}(\Psi) = (\Phi, \Psi)\,(= \tau(A_\Phi^* A_\Psi))$. Setting $\rho_\omega \equiv \equiv (1/n)(A_\Phi)^*$, we obtain

$$\omega(A) = n\tau(\rho_\omega A) = \mathrm{Tr}(\rho_\omega A) \tag{4.5}$$

for all $A \in \mathscr{B}(\mathfrak{H}_n)$. If there exists another ρ' such that $\omega(A) = \mathrm{Tr}(\rho' A)$, then $\mathrm{Tr}((\rho' - \rho_\omega) A) = 0$ for all A. Setting $A = (\rho' - \rho_\omega)^*$, we obtain $\rho' - \rho_\omega = 0$, which proves the uniqueness of ρ_ω satisfying (4.5).

From the positivity $\omega((A - c)^*(A - c)) \geq 0$ for all complex numbers c, we obtain the hermiticity $\omega(A^*) = \overline{\omega(A)}$. Hence

$$\tau(\rho_\omega A^*) = \frac{1}{n}\omega(A^*) = \overline{\tau(\rho_\omega A)} = \tau((\rho_\omega A)^*) = \tau(A^* \rho_\omega^*) = \tau(\rho_\omega^* A^*), \tag{4.6}$$

which implies the hermiticity $\rho_\omega = \rho_\omega^*$. From the positivity $\omega(A^*A) = \mathrm{Tr}(\rho_\omega A^*A) \geq$ ≥ 0 with a special choice $A = E_{\rho_\omega}(\lambda)$, we see that all eigenvalues of ρ_ω are nonnegative: $\rho_\omega \geq 0$. From the normalization $\omega(1) = 1$, we have $\mathrm{Tr}\, \rho_\omega = 1$.

Conversely,

$$\omega_\rho(A) = \mathrm{Tr}(\rho A) \tag{4.7}$$

is a state of $\mathscr{B}(\mathfrak{H}_n)$ if $\rho \geq 0$ and $\mathrm{Tr}\, \rho = 1$.

Let Ψ_i be a complete orthonormal basis of $\mathfrak{H}_n$ such that $\rho\Psi_i = \lambda_i \Psi_i$, $\lambda_i > 0$ for $1 \leq i \leq m$ and $\lambda_i = 0$ for $i > m$.

The direct sum

$$\mathfrak{H} = \mathfrak{H}^{(1)} \oplus \ldots \oplus \mathfrak{H}^{(m)} \tag{4.8}$$

is a Hilbert space whose element is an m-tuple $\Psi = \Psi^{(1)} \oplus \ldots \oplus \Psi^{(m)}$ of vectors $\Psi^{(j)} \in \mathfrak{H}^{(j)}$ satisfying $(\Psi + \Phi)^{(j)} = \Psi^{(j)} + \Phi^{(j)}$, $(c\Psi)^{(j)} = c\Psi^{(j)}$ and $(\Psi, \Phi) = \sum_j (\Psi^{(j)}, \Phi^{(j)})$.

We consider a special case $\mathfrak{H}^{(1)} = \ldots = \mathfrak{H}^{(m)} = \mathfrak{H}_n$ and define a representation π of $\mathscr{B}(\mathfrak{H}_n)$ by

$$\pi(A)(\Psi^{(1)} \oplus \ldots \oplus \Psi^{(m)}) = A\Psi^{(1)} \oplus \ldots \oplus A\Psi^{(m)}. \tag{4.9}$$

Further define

$$\Omega \equiv \lambda_1^{1/2}\Psi_1 \oplus \ldots \oplus \lambda_m^{1/2}\Psi_m. \tag{4.10}$$

Then we have

$$(\Omega, \pi(A)\Omega) = \sum \lambda_j (\Psi_j, A\Psi_j) = \mathrm{Tr}(\rho A). \tag{4.11}$$

We now check whether Ω is a cyclic vector of $\pi(\mathscr{B}(\mathfrak{H}_n))$ on $\mathfrak{H}$: Let E_j be the one-dimensional projection on Ψ_j: $E_j\Psi = (\Psi_j, \Psi)\Psi_j$. Then by the orthogonality of different Ψ_j, $E_j\Psi_k = 0$ for $j \neq k$. Hence

$$\pi(A)\Omega = A_1\Psi_1 \oplus \ldots \oplus A_n\Psi_n \tag{4.12}$$

for $A = \sum_j \lambda_j^{-1/2} A_j E_j$, $A_j \in \mathscr{B}(\mathfrak{H}_n)$. For any $\Psi^{(j)}$, there exists $A_j \in \mathscr{B}(\mathfrak{H}_n)$ satisfying $A_j\Psi_j = \Psi^{(j)}$. Therefore Ω is cyclic.

From the above analysis, we see that the triplet $\mathfrak{H}, \pi, \Omega$ is canonically associated with ω_ρ (i.e. unitarily equivalent to $\mathfrak{H}_{\omega_\rho}, \pi_{\omega_\rho}, \Omega_{\omega_\rho}$). We now introduce another way of explicitly constructing this triplet.

Consider two Hilbert spaces $\mathfrak{H}^a$ and $\mathfrak{H}^b$ with complete orthonormal bases $\{e_i^a\}$ and $\{e_j^b\}$. The tensor product $\mathfrak{H}^a \otimes \mathfrak{H}^b$ is the Hilbert space with a complete orthonormal basis consisting of symbols $e_i^a \otimes e_j^b$. For $\Psi = \sum c_i e_i^a$ and $\Phi = \sum d_j e_j^b$, we define $\Psi \otimes \Phi$ to be $\sum c_i d_j (e_i^a \otimes e_j^b)$. It has an inner product

$$(\Psi_1 \otimes \Phi_1, \Psi_2 \otimes \Phi_2) = (\Psi_1, \Psi_2)(\Phi_1, \Phi_2). \tag{4.13}$$

(One could start from (4.13) and give a basis-independent definition of $\mathfrak{H}^a \otimes \mathfrak{H}^b$.)

If $A_1 \in \mathscr{B}(\mathfrak{H}^a)$ and $A_2 \in \mathscr{B}(\mathfrak{H}^b)$, $A_1 \otimes A_2$ is defined by

$$(A_1 \otimes A_2)(e_i^a \otimes e_j^b) = (A_1 e_i^a) \otimes (A_2 e_j^b). \tag{4.14}$$

It belongs to $\mathscr{B}(\mathfrak{H}^a \otimes \mathfrak{H}^b)$ and $\|A_1 \otimes A_2\| = \|A_1\| \|A_2\|$. In terms of matrix elements, we have

$$(A_1 \otimes A_2)_{(ij),(kl)} = (A_1)_{ik}(A_2)_{jl} \tag{4.15}$$

where the suffix (ij) corresponds to $e_i^a \otimes e_j^b$.

Let $\mathfrak{H}^a = \mathfrak{H}_n$, $\mathfrak{H}^b = \mathfrak{H}_m$, $\pi(A) = A \otimes 1$ for $A \in \mathscr{B}(\mathfrak{H}_n)$ and $\Omega = \sum_{j=1}^m \lambda_j^{1/2} \Psi_j \otimes e_j$, where $\{e_j\}$ is some orthonormal basis of $\mathfrak{H}_m$. Then (4.11) holds. Further, the cyclicity of Ω can be shown, as before, by considering $\pi(A)\Omega = \sum_j A_j \Psi_j \otimes e_j$ for $A = \sum \lambda_j^{-1/2} A_j E_j$. Thus the triplet

$$\mathfrak{H} = \mathfrak{H}^a \otimes \mathfrak{H}^b, \qquad \pi(A) = A \otimes 1, \qquad \Omega = \sum_{j=1}^m \lambda_j^{1/2} \Psi_j \otimes e_j$$

is also canonically associated with the state ω_ρ. It can be brought into the previous form by identifying mutually orthogonal subspace $\mathfrak{H}^a \otimes e_j$ in $\mathfrak{H}^a \otimes \mathfrak{H}^b$ with the previous $\mathfrak{H}^{(j)}$, and the vector $\Psi \otimes e_j$ with $\Psi \in \mathfrak{H}_n = \mathfrak{H}^{(j)}$.

Let $A \in \pi(\mathscr{B}(\mathfrak{H}^a))'$. From $[\pi(e^{ik}), A] = 0$, it follows that $A_{(kj),(il)} = \delta_{ki} A_{(1j)(1l)}$ and hence $A = 1 \otimes A_2$ for some $A_2 \in \mathscr{B}(\mathfrak{H}^b)$. Namely

$$(\mathscr{B}(\mathfrak{H}^a) \otimes 1)' = 1 \otimes \mathscr{B}(\mathfrak{H}^b). \tag{4.16}$$

Another method of constructing the triplet $\mathfrak{H}$, π, Ω is to set $\mathfrak{H} \subset \mathfrak{H}_\tau$, $\pi = \pi_\tau$, $\Omega = \pi_\tau((n\rho)^{1/2})\Omega_\tau$ and $\mathfrak{H} = \pi_\tau(\mathscr{B}(\mathfrak{H}_n))\Omega$.

We now restrict our attention to the case where a state ω is faithful, i.e., $\omega(A^*A) = 0$ implies $A = 0$. If $m < n$, then $\omega(E_n^*E_n) = 0$ where m is the number of non-zero eigenvalues of ρ_ω (including their multiplicities). If $m = n$, then $\omega(A^*A) = 0$ implies $\mathrm{Tr}((A\rho_\omega^{1/2})^*(A\rho_\omega^{1/2})) = 0$, which implies $A\rho_\omega^{1/2} = 0$ and hence $A = 0$ because $\rho_\omega^{-1/2}$ exists for this case. Therefore ω is faithful if and only if ρ_ω is invertible (i.e. $m = n$).

In this case $h_\omega \equiv -\log\rho_\omega$ exists as a selfadjoint operator in $\mathscr{B}(\mathfrak{H}_n)$ and $\omega(A) = \mathrm{Tr}(e^{-h_\omega}A)$.

We now introduce operators Δ_Ω and J_Ω for $\Omega = \sum \lambda_i^{1/2} \Psi_i \otimes \Psi_i$ by

$$\Delta_\Omega \sum c_{ij} \Psi_i \otimes \Psi_j = \sum (\lambda_i/\lambda_j) c_{ij} \Psi_i \otimes \Psi_j \tag{4.17}$$

$$J_\Omega \sum c_{ij} \Psi_i \otimes \Psi_j = \sum \bar{c}_{ij} \Psi_j \otimes \Psi_i. \tag{4.18}$$

Δ_Ω can be written as

$$\Delta_\Omega = \rho_\omega \otimes \rho_\omega^{-1} = \exp\{-h_\omega \otimes 1 + 1 \otimes h_\omega\}, \tag{4.19}$$

and has the following properties:

$$\Delta_\Omega \Omega = \Omega, \tag{4.20}$$

$$\Delta_\Omega^{-it} \pi(A) \Delta_\Omega^{+it} = \pi(e^{ith_\omega} A e^{-ith_\omega}). \tag{4.21}$$

J_Ω is an antiunitary involution:

$$J_\Omega^2 = 1, \tag{4.22}$$

$$(J_\Omega \Psi, J_\Omega \Phi) = \overline{(\Psi, \Phi)}, \tag{4.23}$$

and has the following properties:

$$J_\Omega \Omega = \Omega, \qquad J_\Omega \Delta J_\Omega = \Delta^{-1} \tag{4.24}$$

$$J_\Omega \pi(\mathscr{B}(\mathfrak{H}_n)) J_\Omega = \pi(\mathscr{B}(\mathfrak{H}_n))' . \tag{4.25}$$

If we define

$$S_\Omega = J_\Omega \Delta_\Omega^{1/2} = \Delta_\Omega^{-1/2} J_\Omega , \tag{4.26}$$

$$F_\Omega = \Delta_\Omega^{1/2} J_\Omega = J_\Omega \Delta_\Omega^{-1/2} , \tag{4.27}$$

then they satisfy

$$S_\Omega A_1 \Omega = A_1^* \Omega, \qquad A_1 \in \pi(\mathscr{B}(\mathfrak{H}_n)), \tag{4.28}$$

$$F_\Omega A_2 \Omega = A_2^* \Omega, \qquad A_2 \in \pi(\mathscr{B}(\mathfrak{H}_n))'. \tag{4.29}$$

In the general case, one considers a von Neumann algebra R on $\mathfrak{H}$ and a vector $\Omega \in \mathfrak{H}$ which is cyclic (i.e. $R\Omega$ is dense in $\mathfrak{H}$) and separating (i.e. $A\Omega = 0$ implies $A = 0$ for any $A \in R$). Then (4.28) and (4.29) define two densely defined antilinear operators, whose closures are denoted by S_Ω and F_Ω. (The closure is the same as S_Ω and F_Ω in the present case.) It then follows that $F_\Omega = S_\Omega^*$.

The modular operator Δ_Ω is then defined by

$$\Delta_\Omega = F_\Omega S_\Omega (= S_\Omega^* S_\Omega = F_\Omega F_\Omega^* = (S_\Omega F_\Omega)^{-1}) \tag{4.30}$$

which is selfadjoint. J_Ω is defined by

$$J_\Omega = \Delta_\Omega^{1/2} S_\Omega (= S_\Omega \Delta_\Omega^{-1/2} = F_\Omega \Delta_\Omega^{1/2} = \Delta_\Omega^{-1/2} F_\Omega) . \tag{4.31}$$

The properties (4.20), (4.22), (4.23), (4.24) and (4.25) hold. (4.21) is replaced by

$$\Delta_\Omega^{-it} R \Delta_\Omega^{+it} = R . \tag{4.32}$$

These results were obtained by Tomita.

5. KMS STATES

A one-to-one map α of a C^*-algebra $\mathfrak{A}$ onto itself is called an automorphism of $\mathfrak{A}$ if it preserves the structure of the $*$-algebra, i.e., if $\alpha(A_1 + A_2) = \alpha(A_1) + \alpha(A_2)$, $\alpha(cA) = c\alpha(A)$, $\alpha(A_1 A_2) = \alpha(A_1)\alpha(A_2)$ and $\alpha(A^*) = \alpha(A)^*$. It is called inner if there exists a unitary $u \in \mathfrak{A}$ such that $\alpha(A) = uAu^*$.

Let $\alpha(x)$, $-\infty < x < \infty$ be a one-parameter group of automorphisms of $\mathfrak{A}$: $\alpha(x_1 + x_2) = \alpha(x_1)\alpha(x_2)$. Assume that for each $A \in \mathfrak{A}$, $\alpha(x)A$ is continuous in x.

A state ω of $\mathfrak{A}$ called $\alpha(x)$-invariant if $\omega(\alpha(x)A) = \omega(A)$ for all $A \in \mathfrak{A}$. In this case there exist unique unitary operators $U_\omega(x)$ such that

$$U_\omega(x) \pi_\omega(A) \Omega_\omega = \pi_\omega(\alpha(x)A) \Omega_\omega . \tag{5.1}$$

It is a continuous one-parameter group of unitary operators, satisfying

$$U_\omega(x)\, \pi_\omega(A)\, U_\omega(x)^* = \pi_\omega(\alpha(x)\, A)\,, \tag{5.2}$$

$$U_\omega(x)\, \Omega_\omega = \Omega_\omega\,. \tag{5.3}$$

A state ω of $\mathfrak{A}$ is called an $(\alpha(x)$-$)$ KMS state of $\mathfrak{A}$ if there exists a function $F_{A,B}(z)$ for each $A, B \in \mathfrak{A}$ such that it is continuous for $0 \leq \operatorname{Im} z \leq \beta$, analytic for $0 < \operatorname{Im} z < \beta$ and satisfies

$$\omega(A \cdot \alpha(x)\, B) = F_{A,B}(x), \tag{5.4}$$

$$\omega((\alpha(x)\, B)\, A) = F_{A,B}(x + i\beta). \tag{5.5}$$

$F_{A,B}$ is necessarily bounded uniformly on the strip $0 \leq \operatorname{Im} z \leq \beta$.

The modular operator Δ_ω of a KMS state of $\mathfrak{A}$ (considering it as a vector state of $\pi_\omega(\mathfrak{A})''$ by the vector Ω_ω) is

$$\Delta_\omega = e^{-\beta H} \quad \text{where} \quad U_\omega(x) = e^{ixH_\omega}, \tag{5.6}$$

while the antiunitary involution J_ω satisfies

$$J_\omega \pi_\omega(A)\, \Omega_\omega = \Delta_\omega^{1/2} \pi_\omega(A^*)\, \Omega_\omega\,. \tag{5.7}$$

This structure was found by Haag, Hugenholtz, and Winnink.

Conversely, for a cyclic and separating unit vector Ω of a W^*-algebra R on $\mathfrak{H}$,

$$\alpha_\Omega(x)\, A = \Delta_{\omega_\Omega}^{-it} A \Delta_{\omega_\Omega}^{+it}, \qquad A \in R \tag{5.8}$$

defines a continuous one-parameter group of automorphisms and ω is the KMS state of R relative to $\alpha_\Omega(x)$ with $\beta = 1$. $\alpha_\Omega(x)$ is called the modular automorphism of R relative to Ω. This was obtained by Takesaki.

We now use $\mathscr{B}(\mathfrak{H}_n)$ to illustrate these results. Let us consider an inner automorphism $\alpha(x)$ of $\mathscr{B}(\mathfrak{H}_n)$ defined by

$$\alpha(x)\, A = e^{ixH} A e^{-ixH}, \qquad H \in \mathscr{B}(\mathfrak{H}_n)\,. \tag{5.9}$$

Let ω be an $\alpha(x)$-KMS state of $\mathscr{B}(\mathfrak{H}_n)$. In the present case, $\alpha(x)A$ is an entire function of x and hence the KMS boundary conditions (5.4) and (5.5) reduce to

$$\omega(A\tau(x + i\beta)\, B) = \omega((\alpha(x)\, B)\, A)\,. \tag{5.10}$$

Let

$$\omega'(A) \equiv \omega(e^{\beta H/2} A e^{\beta H/2})/\omega(e^{\beta H})\,. \tag{5.11}$$

It is a state of $\mathscr{B}(\mathfrak{H}_n)$. Further it satisfies

$$\omega'(AB) = \omega(\{e^{\beta H/2} A e^{\beta H/2}\}\, \alpha(i\beta)\, \{e^{\beta H/2} B e^{-\beta H/2}\})/\omega(e^{\beta H})$$

$$= \omega(\{e^{\beta H/2} B e^{-\beta H/2}\}\, \{e^{\beta H/2} A e^{\beta H/2}\})/\omega(e^{\beta H})$$

$$= \omega'(BA)\,. \tag{5.12}$$

Hence $\omega' = \tau$ by the uniqueness of the central state of $\mathscr{B}(\mathfrak{H}_n)$. Therefore,

$$\omega(e^{\beta H}) = \omega'(e^{-\beta H})^{-1} = \left(\frac{1}{n}\operatorname{Tr} e^{-\beta H}\right)^{-1} \tag{5.13}$$

and

$$\omega(A) = \omega(e^{\beta H})\,\omega'(e^{-\beta H/2} A e^{-\beta H/2})$$

$$= \operatorname{Tr}(e^{-\beta H} A)/\operatorname{Tr} e^{-\beta H}. \tag{5.14}$$

Conversely, it is easily seen by going backwards that (5.14) satisfies the $\alpha(x)$-KMS boundary condition.

In this case, the modular operator Δ_ω is given by (4.19) where $h_\omega = \beta H + \log(\operatorname{Tr} e^{-\beta H})$ and hence

$$H_\omega = H \otimes 1 - 1 \otimes H. \tag{5.15}$$

6. PURE STATES AND MIXTURES

A state ω of a C^*-algebra $\mathfrak{A}$ is called a mixture of states ω_1 and ω_2 of $\mathfrak{A}$ with weights λ and $1 - \lambda$, where $0 \leqq \lambda \leqq 1$, if

$$\omega(A) = \lambda\omega_1(A) + (1 - \lambda)\,\omega_2(A) \tag{6.1}$$

holds for all $A \in \mathfrak{A}$. In this case we write

$$\omega = \lambda\omega_1 + (1 - \lambda)\,\omega_2. \tag{6.2}$$

A state ω is called a pure state if (6.2) never holds except in the trivial cases $\lambda = 0$ or $\lambda = 1$ or $\omega_1 = \omega_2 = \omega$.

Let us consider the simplest example of $\mathfrak{A} = \mathscr{B}(\mathfrak{H}_n)$. Let $\{\Psi_i\}$ be an orthonormal basis of $\mathfrak{H}_n$ consisting of eigenvectors of ρ_ω with eigenvalues λ_i. Then $\omega = \sum \lambda_i\omega_i$ where $\omega_i(A) = (\Psi_i, A\Psi_i)$. Hence ω cannot be pure unless $m = 1$, i.e., ρ_ω is a one-dimensional projection on a vector Ψ_1. On the other hand, a vector state $\omega_\Psi(A) = (\Psi, A\Psi)$ by a vector Ψ in $\mathfrak{H}_n$ is pure: Suppose $\omega_\Psi = \lambda\omega_1 + (1 - \lambda)\,\omega_2$ with $0 < \lambda < 1$. Let E be the one-dimensional projection on Ψ. Then $\omega_\Psi(1 - E) = 0$. Since $(1 - E)^2 = (1 - E)$, $\omega_1(1 - E) \geq 0$, $\omega_2(1 - E) \geqq 0$ and hence $\omega_1(1 - E) = 0$. This implies, by the Schwarz inequality, that $|\omega_1(A(1 - E))|^2 \leqq \omega_1(AA^*) \cdot \omega_1(1 - E) = 0$ and hence $\omega_1(A(1 - E)) = \omega_1((1 - E)A) = 0$. Hence $\omega_1(A) = \omega_1(AE) = \omega_1(EAE) = \omega_1(\omega_\Psi(A)E) = \omega_\Psi(A)\omega_1(E) = \omega_\Psi(A)\omega_1(1) = \omega_\Psi(A)$. Namely, $\omega_1 = \omega_\Psi$ and hence $\omega_2 = \omega_\Psi$ and ω_Ψ is pure. Thus ω is pure if and only if ρ_ω is a one-dimensional projection.

Since $\pi_\omega(\mathscr{B}(\mathfrak{H}_n))' = 1 \otimes \mathscr{B}(\mathfrak{H}_m)$, $\pi_\omega(\mathscr{B}(\mathfrak{H}_n))'' = \mathscr{B}(\mathfrak{H}_\omega)$ if and only if $m = 1$, i.e., ω is pure. If $\pi_\omega(\mathfrak{A})'' = \mathscr{B}(\mathfrak{H}_\omega)$, π_ω is called irreducible. It is generally true that ω is pure if and only if π_ω is irreducible.

7. DECOMPOSITION OF A STATE

It is often important to consider a decomposition of a state within some subset K of states of $\mathfrak{A}$. For example, in statistical mechanics one expects that a KMS state relative to the time translation automorphism represents an equilibrium state and any such KMS state can be decomposed into a mixture of pure phases, where a pure phase is understood as a KMS state which cannot be decomposed into other KMS states any further.

K is called convex if the mixture $\lambda\omega_1 + (1 - \lambda)\omega_2, 0 < \lambda < 1$ of every two points ω_1 and ω_2 in K is again in K. ω in K is called extremal if $\omega = \lambda\omega_1 + (1 - \lambda)\omega_2$ holds for $\omega_1, \omega_2 \in K$ only when $\lambda = 0$ or $\lambda = 1$ or $\omega_1 = \omega_2 = \omega$. The set of all invariant states is convex and its extremal point is called an ergodic state. The set of all KMS states is convex and its extremal point is an extremal KMS state.

Let us consider a decomposition $\omega = \sum \lambda_i \omega_i, \sum \lambda_i = 1, \lambda_i \geq 0$. This can be described conveniently in terms of a measure μ in the space of all states of $\mathfrak{A}$:

$$\mu = \sum \lambda_i \delta_{\omega_i}, \tag{7.1}$$

where δ_{ω_i} is the point measure (the δ function) at ω_i. For a function $F(\phi)$ of state ϕ, its integral with respect to μ is

$$\int F(\phi)\,\mu(\mathrm{d}\phi) = \sum F(\omega_i)\,\lambda_i. \tag{7.2}$$

In particular, the original decomposition of ω is described as

$$\omega = \int \phi\mu(\mathrm{d}\phi). \tag{7.3}$$

If we decompose each ω_i again into other states as $\omega_i = \sum \lambda_{i\alpha}\omega'_\alpha$, then we have a new decomposition $\omega = \sum \lambda'_\alpha \omega'_\alpha$ with $\lambda'_\alpha = \sum \lambda_i \lambda_{i\alpha}$. This corresponds to a measure

$$\nu = \sum \lambda'_\alpha \delta_{\omega'_\alpha}. \tag{7.4}$$

In this case we say that the second decomposition is finer than the first decomposition and write as $\mu \propto \nu$. If $\mu \propto \nu$ and $\nu \propto \mu$ at the same time, then it can be shown by an elementary reasoning that $\mu = \nu$. Since $\omega = \omega$ is a decomposition with the measure δ_ω, we may write $\delta_\omega \propto \mu$ if $\omega = \int \phi\mu(\mathrm{d}\phi)$.

For a general probability measure μ on K, one defines the relation $\mu \propto \nu$ if $\int F(\phi)\,\mu(\mathrm{d}\phi) \leq \int F(\phi)\,\nu(\mathrm{d}\phi)$ for all convex functions F, where F is convex if $F(\lambda\phi_1 + (1 - \lambda)\phi_2) \leq \lambda F(\phi_1) + (1 - \lambda)\,F(\phi_2)$. If μ and ν are discrete, this definition coincides with the previous one.

To find the unique decomposition of a given ω into extremal points, one is interested in the existence and the uniqueness of a maximal measure μ satisfying $\mu \propto \delta_\omega$. For compact convex K (equivalently for closed convex K if $1 \in \mathfrak{A}$), the existence of a maximal measure μ satisfying $\delta_\omega \propto \mu$ for a given ω always holds. If $\mathfrak{A}$ is separable any maximal measure is carried by extremal points of K. If the uniqueness of a maximal measure $\mu \propto \delta_\omega$ for any given ω holds, then K is called a simplex.

The set of all states of $\mathscr{B}(\mathfrak{H}_n)$ is not a simplex for $n > 1$, as is easily seen by the following example. Let Ψ_i, $i = 1, 2$, be two unit vectors in $\mathfrak{H}_n$ such that $(\Psi_1, \Psi_2) \neq 0$, $\Psi_1 \neq c\Psi_2$. Let $\omega = \lambda\omega_{\Psi_1} + (1 - \lambda)\omega_{\Psi_2}$ where ω_{Ψ_i} is the vector state of $\mathscr{B}(\mathfrak{H}_n)$ by $\Psi_i \in \mathfrak{H}_n$ and hence is pure. By our earlier results, $\omega(A) = \mathrm{Tr}(\rho_\omega A)$. If $\{\Phi_i\}$ are mutually orthogonal eigenvectors of ρ_ω belonging to eigenvalues $\lambda_i \neq 0$, then $\omega = \sum_i \lambda_i\omega_{\Phi_i}$, which is again a decomposition into pure states. Hence ω has at least two maximal decompositions which are definitely different. (A simpler example is $\tau = (1/n)\sum \omega_{\Psi_i}$ where Ψ_i can be any orthonormal basis.)

Fortunately, the set of KMS states is always a simplex.

If $\alpha(x)$ has the following asymptotic abelian property: $\lim\limits_{x \to \infty} \| [\alpha(x)A, B] \| = 0$, then the set of all $\alpha(x)$-invariant states is a simplex. We shall illustrate the decomposition of KMS states using an example of a *-algebra on $\mathfrak{H}_n$.

8. *-ALGEBRAS ON $\mathfrak{H}_n$

We first consider a commutative von Neumann algebra Z on $\mathfrak{H}_n$, i.e. $[A, B] = 0$ for any $A, B \in Z$.

If $A = A^* \in Z$ and $A = \sum \lambda E_A(\lambda)$, then $E_A(\lambda_0) = \prod_{\lambda \neq \lambda_0}(A - \lambda)/(\lambda_0 - \lambda) \in Z$. We start with $A_1 = A_1^* \in Z$ and a partition of unity by $E_{A_1}(\lambda) \in Z$, i.e. $\sum E_{A_1}(\lambda) = 1$, $E_{A_1}(\lambda_1)E_{A_1}(\lambda_2) = 0$ if $\lambda_1 \neq \lambda_2$. If any of $ZE_{A_1}(\lambda_1)$ contains $A_2 \neq c\,E_{A_1}(\lambda_1)$, replace $E_{A_1}(\lambda_1)$ by $\{E_{A_2}(\lambda_2)\}$ and $E_{A_1}(\lambda_1) - \sum_{\lambda_2} E_{A_2}(\lambda_2)$ and obtain a finer partition of unity. After a finite number of steps we obtain mutually orthogonal projections E_j with the sum $\sum E_j = 1$, such that any selfadjoint operator $A \in Z$ satisfies $AE_j = \lambda_j E_j$, i.e. $A = \sum \lambda_j E_j$. Since any A can be written as $A_{(1)} + iA_{(2)}$ with $A_{(1)} = \frac{1}{2}(A + A^*) = A_{(1)}^*$, $A_{(2)} = (1/2i)(A - A^*) = A_{(2)}^*$, we have $A = \sum c_j E_j$ for any $A \in Z$, i.e.

$$Z = \{\textstyle\sum c_j E_j\}. \tag{8.1}$$

Let $\omega_j(A) = c_j$ if $A = \sum c_j E_j$. It is a vector state of Z by any unit vector in $E_j\mathfrak{H}_n$. It satisfies

$$\omega_j(A_1A_2) = \omega_j(A_1)\,\omega_j(A_2). \tag{8.2}$$

A state satisfying (8.2) is called a character. From (8.2), it follows that $\mathfrak{H}_{\omega_j}$ is one-dimensional and $\pi_{\omega_j}(A) = \omega_j(A)\,1$. In particular, π_{ω_j} is irreducible and ω_j is pure. Conversely, any pure state of a commutative C^*-algebra is a character.

Any state $\omega \in Z$ then has the unique decomposition into pure states:

$$\omega = \sum \lambda_j\omega_j, \qquad \lambda_j = \omega(E_j). \tag{8.3}$$

(8.1) shows that Z is *-isomorphic to the algebra of functions on the set of all characters, where $A \in Z$ corresponds to the function $c_j = \omega_j(A)$ of ω_j. This isomorphism holds for a general commutative C*-algebra (where one considers the algebra of continuous functions over characters relative to the weak topology of characters) and is called the Gelfand isomorphism.

Next consider an arbitrary von Neumann algebra R on $\mathfrak{H}_n$. The set of elements of R which commute with every element of R is called the center of R and can be written as $R \cap R'$. Since it is abelian,

$$R \cap R' = \{\textstyle\sum c_j E_j\} \tag{8.4}$$

for some partition $\{E_j\}$ of unity. We then have the decomposition

$$\mathfrak{H} = \bigoplus_j \mathfrak{H}_j, \qquad \mathfrak{H}_j = E_j\mathfrak{H}, \tag{8.5}$$

$$R = \bigoplus_j R_j, \qquad R_j = (E_j R)\big|\mathfrak{H}_j, \tag{8.6}$$

where R_j is a von Neumann algebra on $\mathfrak{H}_j$ and any $A \in R$ is written as $A = \sum A_j$, $A_j = AE_j \in R_j$.

Suppose $A \in RE_j$ is in the center of RE_j. Since $A = AE_j$, we have $[A, B] = [AE_j, B] = [A, E_j B] = 0$ for any $B \in R$. Therefore $A \in R \cap R'$ and $A = AE_j = (\sum c_j E_j) E_j = c_j E_j$.

A von Neumann algebra whose center consists of complex multiples of the identity operator is called a factor. R_j is a factor on $\mathfrak{H}_j$. (A state ω such that $\pi_\omega(\mathfrak{A})''$ is a factor is called a primary or factor state.)

Now let R be a factor on $\mathfrak{H}_n$. Since $R \cap R' = \{c1\}$, $(R \cap R')' = (R' \cup R)'' = \mathcal{B}(\mathfrak{H})$, and hence any vector $\Psi \neq 0$ is cyclic with respect to R and R' together. As in the commutative case, one can find a projection E such that $EAE = c_A E$ for some complex number c_A for any $A \in R$. If $F \in R$ is a subprojection of E (i.e. $FE = F$), then $EFE = EF = F = cE$ and F must be either E or 0. A projection E of a von Neumann algebra R without any subprojection in R except E and 0 is called a minimal projection of R. A factor with a minimal projection is called type I.

Let $\Psi \in E\mathfrak{H}_n$, $\|\Psi\| = 1$. Let $\mathfrak{H} = R\Psi$. If $A \in R$ satisfies $A\mathfrak{H} = 0$, then $0 = R'A\mathfrak{H} = AR'\mathfrak{H}$. Since $R'\mathfrak{H} = R'R\Psi$ is total due to $(R \cup R')'' = \mathcal{B}(\mathfrak{H})$, we have $A = 0$. Therefore R is *-isomorphic to its restriction $R_\mathfrak{H}$ on $\mathfrak{H}$. If $\Phi \in \mathfrak{H}$ satisfies $E\Phi = \Phi$, then $\Phi = B\Psi$ for some $B \in R$ and hence $\Phi = E\Phi = EBE\Psi = c_B E\Psi = c_B \Psi$. Therefore $E\mathfrak{H}$ is one-dimensional. Let $A \in (R_\mathfrak{H})'$. Then $A\Psi = AE\Psi = EA\Psi = c\Psi$. Therefore, $A(B\Psi) = BA\Psi = cB\Psi$ for $B \in R$, which implies $A = c1$. Therefore $R_\mathfrak{H} = \mathcal{B}(\mathfrak{H})$. Namely, any factor on a finite dimensional space is *-isomorphic to $\mathcal{B}(\mathfrak{H})$ for some $\mathfrak{H}$. This is true for any type-I factor.

Let R be a *-algebra on $\mathfrak{H}_n$ and ω be a state of R. Let $R \cap R' = \{\sum c_i E_i\}$ and $\omega_i(A) = \omega(E_i A)/\omega(E_i)$ whenever $\omega(E_i) \neq 0$. Then ω_i is a state of R such that $\pi_{\omega_i}(E_i) = 1$

and $\pi_{\omega_i}(R) = \pi_{\omega_i}(RE_i)$ is *-isomorphic to RE_i. Hence ω_i is a primary state and

$$\omega = \sum \lambda_i \omega_i, \qquad \lambda_i = \omega(E_i). \tag{8.7}$$

The decomposition into primary states by diagonalizing the center is called a central decomposition. [In the general case, the central decomposition is the supremum (in the partial ordering $\propto$) of all decompositions $\omega = \sum \lambda_i \omega_i$, $\lambda_i = \omega(B_i)$, $\omega_i(A) = = \omega(AB_i)/\lambda_i$ where B_i is any positive element of $R \cap R'$ with the sum 1, and the sum over i in the decomposition runs over all i with $\omega(B_i) \neq 0$.]

Let us now consider an automorphism

$$\alpha(x)\, A = e^{iHx} A e^{-iHx}, \tag{8.8}$$

where $H = \sum H_j$, $H_j = HE_j$. Let ω be an $\alpha(x)$-KMS state. Since $\alpha(x)\, E_j = E_j$, $\omega(E_j A)/\omega(E_j) \equiv \omega_j(A)$ is again an $\alpha(x)$-KMS state, where we consider j such that $\omega(E_j) \neq 0$. Since ω_j is a KMS state of RE_j, it is uniquely given by

$$\omega_j(A) = \frac{\tau_j(e^{-H_j}(AE_j))}{\tau_j(e^{-H_j})}, \tag{8.9}$$

where τ_j is the central state of RE_j. Thus

$$\omega = \sum \lambda_j \omega_j, \qquad \lambda_j = \omega(E_j). \tag{8.10}$$

Conversely, a state of the form (8.10) with ω_j defined by (8.9) and with arbitrary λ_j satisfying $\lambda_j \geqq 0$, $\sum \lambda_j = 1$ is a KMS state. Therefore, KMS states are not unique, and each one of them has a unique decomposition into extremal KMS states ω_j which happen to be primary states.

For a general KMS state ω, the center of $\pi_\omega(\mathfrak{A})''$ is invariant under $A \to U_\omega(x)\, A \cdot \cdot\, U_\omega(x)^*$ and the decomposition into extremal KMS states coincides with the central decomposition.

Suppose that the C^*-algebra has the following local structure: For each open subset $\mathcal{O}$ in R^3, there corresponds a C^*-subalgebra $\mathfrak{A}(\mathcal{O})$ of $\mathfrak{A}$ such that $\cup_{\mathcal{O}} \mathfrak{A}(\mathcal{O})$ is dense in $\mathfrak{A}$, and $\mathfrak{A}(\mathcal{O}_1)$ commute with $\mathfrak{A}(\mathcal{O}_2)$ if $\mathcal{O}_1 \cap \mathcal{O}_2$ is empty. Then a KMS state ω relative to some time-translation automorphism is an extremal KMS state if and only if it has the following uniform clustering property. For $A \in \mathfrak{A}$ and $\varepsilon > 0$, there exists a (sufficiently large) $\mathcal{O}_\varepsilon$ such that for any $\mathcal{O}$ outside of $\mathcal{O}_\varepsilon$ and for any $Q \in \mathfrak{A}(\mathcal{O})$ with $\|Q\| \leqq 1$,

$$|\omega(AQ) - \omega(A)\,\omega(Q)| < \varepsilon. \tag{8.11}$$

This characterization of an extremal KMS state supports the view that it represents a pure phase.

Two extremal KMS states may correspond physically to "the same" pure phase if they are related by some physical automorphism α of $\mathfrak{A}$ such as spatial rotation or gauge transformation by $\omega_1(A) = \omega_2(\alpha A)$. In this case we speak of a broken symmetry. Otherwise the existence of two different KMS states would imply coexistence of two phases, i.e. a phase transition.

9. THE FREE BOSE GAS

We shall illustrate the general theories explained in the preceding sections by a physical example: the free Bose gas. To make some of the calculations simpler, we use a *-algebra (of unbounded operators) instead of a C*-algebra.

Observables for a system of Bose particles can be described conveniently in terms of creation and annihilation operators: $a^\dagger(x)$ and $a(x)$. Let $\mathscr{S}$ denote the topological vector space of complex-valued C^∞-functions of fast decrease in R^3, and let K denote $\mathscr{S} \oplus \mathscr{S}$: the topological vector space of pairs of functions of the class $\mathscr{S}$. Instead of saying "$a^\dagger$ or a", we use the following notation:

$$B(f \oplus g) = \int a^\dagger(x) f(x)\, dx + \int a(x) g(x)\, dx, \tag{9.1}$$

namely, $B(h)$ denotes a creation operator $\int a^\dagger(x) f(x)\, dx$ if $h = f \oplus 0$, an annihilation operator $\int a(x) g(x)\, dx$ if $h = 0 \oplus g$, and their sum if h is of the general form $f \oplus g$.

We shall introduce in K an indefinite metric,

$$\gamma(h_1, h_2) = \int \overline{f_1(x)} f_2(x)\, dx - \int \overline{g_1(x)} g_2(x)\, dx, \tag{9.2}$$

for $h_j = f_j \oplus g_j$, $j = 1, 2$. We define an operator Γ acting on K by

$$\Gamma(f \oplus g) = \bar{g} \oplus \bar{f}, \tag{9.3}$$

where $\bar{f}$ and $\bar{g}$ are the complex conjugates of f and g. It satisfies $\Gamma^2 = 1$, $\Gamma(h_1 + h_2) = \Gamma h_1 + \Gamma h_2$, $\Gamma(ch) = \bar{c}\Gamma h$ and $\gamma(\Gamma h_1, \Gamma h_2) = -\gamma(h_2, h_1)$.

In terms of these notations, the basic properties of the operators $B(h)$ are

$$B(c_1 h_1 + c_2 h_2) = c_1 B(h_1) + c_2 B(h_2), \tag{9.4}$$

$$B(h)^* = B(\Gamma h), \tag{9.5}$$

$$[B(h_1)^*, B(h_2)] = \gamma(h_1, h_2). \tag{9.6}$$

Let $\mathfrak{A}$ denote the *-algebra consisting of polynomials

$$\sum_{i=1}^{N} c_i B(h_{i1}) \dots B(h_{in_i}), \tag{9.7}$$

where $n_i \geq 0$, c_i are complex numbers, and $B(h)$ satisfies the relations (9.4)–(9.6).

To remember the original distinction between $a^\dagger$ and a, we introduce a projection operator P_0 by

$$P_0(f \oplus g) = f \oplus 0. \tag{9.8}$$

It satisfies

$$P_0^2 = P_0, \tag{9.9}$$

$$\gamma(P_0 h_1, h_2) = \gamma(h_1, P_0 h_2), \qquad \gamma(P_0 h, h) > 0 \quad \text{for} \quad P_0 h \neq 0, \tag{9.10}$$

$$\Gamma P_0 \Gamma = 1 - P_0. \tag{9.11}$$

Let $H_0 = \boldsymbol{P}^2/(2m)$ be the free Hamiltonian for a particle of mass m. Then

$$[\exp(iH_0 t)f](\boldsymbol{x}) = \int \exp(i\boldsymbol{p}\boldsymbol{x}) \exp[i\boldsymbol{p}^2 t/(2m)]\tilde{f}(\boldsymbol{p})\, \mathrm{d}\boldsymbol{p} \in \mathscr{S} \tag{9.12}$$

if

$$f(\boldsymbol{x}) = \int \exp(i\boldsymbol{p}\boldsymbol{x})\tilde{f}(\boldsymbol{p})\, \mathrm{d}\boldsymbol{p} \in \mathscr{S}. \tag{9.13}$$

The time translation *-automorphism $\alpha^T(t)$ of $\mathfrak{A}$ for the free motion is given by

$$\alpha^T(t) \sum c_i \mathbf{B}(h_{i1}) \dots \mathbf{B}(h_{in_i}) = \sum c_i \mathbf{B}(\mathbf{U}^T(t) h_{i1}) \dots \mathbf{B}(\mathbf{U}^T(t) h_{in_i}), \tag{9.14}$$

where

$$\mathbf{U}^T(t)(f \oplus g) = \exp(iH_0 t)f \oplus \exp(-iH_0 t)g. \tag{9.15}$$

The transformation $\mathbf{U}^T(t)$ satisfies

$$\mathbf{U}^T(t)\, \mathbf{U}^T(s) = \mathbf{U}^T(t+s), \tag{9.16}$$

$$[\mathbf{U}^T(t), \Gamma] = 0, \tag{9.17}$$

$$\gamma(\mathbf{U}^T(t) h_1, \mathbf{U}^T(t) h_2) = \gamma(h_1, h_2), \tag{9.18}$$

$$[\mathbf{U}^T(t), P_0] = 0. \tag{9.19}$$

We need another *-automorphism of $\mathfrak{A}$: the gauge transformation $\alpha^G(\lambda)$. It is defined by

$$\alpha^G(\lambda) \sum c_i \mathbf{B}(h_{i1}) \dots \mathbf{B}(h_{in_i}) = \sum c_i \mathbf{B}(\mathbf{U}^G(\lambda) h_{i1}) \dots \mathbf{B}(\mathbf{U}^G(\lambda) h_{in_i}), \tag{9.20}$$

where

$$\mathbf{U}^G(\lambda)(f \oplus g) = e^{i\lambda} f \oplus e^{-i\lambda} g. \tag{9.21}$$

The transformation $\mathbf{U}^G(\lambda)$ satisfies

$$\mathbf{U}^G(\lambda)\, \mathbf{U}^G(\mu) = \mathbf{U}^G(\lambda + \mu), \tag{9.22}$$

$$[\mathbf{U}^G(\lambda), \Gamma] = 0, \tag{9.23}$$

$$\gamma(\mathbf{U}^G(\lambda) h_1, \mathbf{U}^G(\lambda) h_2) = \gamma(h_1, h_2), \tag{9.24}$$

$$[\mathbf{U}^G(\lambda), P_0] = 0, \tag{9.25}$$

$$[\mathbf{U}^G(\lambda), \mathbf{U}^T(t)] = 0. \tag{9.26}$$

Due to (9.26), the two *-automorphisms $\alpha^T(t)$ and $\alpha^G(\lambda)$ commute.

A state ϕ of $\mathfrak{A}$ is a complex-valued linear functional on $\mathfrak{A}$ which is positive ($\phi(A^*A) \geq 0$ for all $A \in \mathfrak{A}$) and normalized ($\phi(1) = 1$). For each state ϕ, there exist a Hilbert space $\mathfrak{H}_\phi$, a dense subset D_ϕ of $\mathfrak{H}_\phi$, a unit vector Ω_ϕ in D_ϕ and a *-representation $\pi_\phi(A)$, $A \in \mathfrak{A}$ such that $\pi_\phi(A)$ is defined on D_ϕ and $D_\phi = \pi_\phi(\mathfrak{A})\Omega_\phi$. (The difference between this and a C^*-algebra is that we cannot assert the boundedness of each operator $\pi_\phi(A)$, and hence its domain of definition is not the entire $\mathfrak{H}_\phi$. $\pi_\phi(A)^*$ coincides with $\pi_\phi(A^*)$ on D_ϕ but has in general a larger domain of definition.)

We shall define an $\alpha(t)$-KMS state ϕ by the following three conditions:

(a) (Invariance): $\phi(\alpha(t)A) = \phi(A)$ for all $A \in \mathfrak{A}$ and $-\infty < t < \infty$.

(b) (Continuity): $\phi(A_1 B(h) A_2)$ is continuous in h for all $A_1, A_2 \in \mathfrak{A}$.

(c) (KMS-condition): Let $\mathfrak{A}_0$ be the sub-*-algebra of $\mathfrak{A}$ generated by $B(f \oplus g)$, such that f and g are in $\tilde{\mathcal{D}}$, the space of Fourier transforms of C^∞-functions with compact supports. (Namely consider f and g such that $\tilde{f}$ and $\tilde{g}$ vanish outside a bounded region.) For any $A_1, A_2 \in \mathfrak{A}_0$, there exists a function $F(z)$ which is continuous for $0 \leq \operatorname{Im} z \leq \beta$, holomorphic for $0 < \operatorname{Im} z < \beta$ and satisfies

$$F(t) = (A_1 \alpha(t) A_2), \tag{9.27}$$

$$F(t + i\beta) = (\{\alpha(t) A_2\} A_1) \tag{9.28}$$

for $-\infty < t < \infty$.

In particular, such ϕ for $\alpha(t) = \alpha^T(t)\alpha^G(-\mu t)$ is called a free KMS state with an inverse temperature β and a chemical potential μ. If $\mu < 0$, it can be uniquely determined as follows.

[Actually, (a) follows from (b) and (c) for this $\alpha(t)$: From the condition (b) and the definition of $\alpha^T(t)$ and $\alpha^G(\lambda)$, it follows that $F(t)$ is polynomially bounded in $\operatorname{Re} t$, and

$$F(t) = \int e^{itq} \tilde{F}(q)\, dq$$

for a tempered distribution $\tilde{F}$ with compact support. If we consider the special case $A_1 = 1$, then (c) implies $F(t) = F(t + i\beta)$, and hence $(1 - e^{-\beta q})\tilde{F}(q) = 0$. Multiplying a function $q(1 - e^{-\beta q})^{-1}$ (defined to be β^{-1} at $q = 0$), we obtain $q\tilde{F}(q) = 0$. Hence $\tilde{F}(q) = c\delta(q)$ and $F(t)$ must be independent of t.]

The definition (9.12) can be extended to an arbitrary complex t if $\tilde{f}$ vanishes outside a bounded region. Therefore, $\alpha^T(t)A$ can be extended to complex t, if $A \in \mathfrak{A}_0$. Obviously, $\alpha^G(\lambda)A$ can be extended to complex λ for any $A \in \mathfrak{A}$. Further, due to (b) and the nuclear theorem (or the Hartogs' theorem), $\phi(A_1\{\alpha(t)A\}A_2)$ is holomorphic in t for $A \in \mathfrak{A}_0$ and $A_1, A_2 \in \mathfrak{A}$. Hence,

$$F(t) = \phi(A_1 \alpha(t) A_2) \tag{9.29}$$

for any t and $A_1, A_2 \in \mathfrak{A}_0$.

We now define an operator J as follows. On $\pi_\phi(\mathfrak{A}_0)\Omega_\phi \equiv D_\phi^0$, it is defined by

$$J\pi_\phi(A)\Omega_\phi = \pi_\phi(\alpha(i\beta/2)A^*)\Omega_\phi, \qquad A \in \mathfrak{A}_0. \tag{9.30}$$

From $[\alpha(t) A]^* = \alpha(\bar{t}) A^*$, we obtain

$$(\pi_\phi[\alpha(i\beta/2) A_1{}^*] \Omega_\phi, \pi_\phi[\alpha(i\beta/2) A_2{}^*] \Omega_\phi) =$$

$$= \phi(\{\alpha(-i\beta/2) A_1\} \{\alpha(i\beta/2) A_2{}^*\}) =$$

$$= \phi(\{\alpha(-i\beta/2) A_1\} \alpha(i\beta) \{\alpha(-i\beta/2) A_2{}^*\}). \qquad (9.31)$$

From the KMS condition (c) and (9.29), this is equal to

$$\phi(\{\alpha(-i\beta/2) A_2{}^*\} \{\alpha(-i\beta/2) A_1\}) = \phi(\alpha(-i\beta/2) \{A_2{}^* A_1\}). \qquad (9.32)$$

Since $\phi(\alpha(t) A)$, $A \in \mathfrak{A}_0$ is holomorphic in t, and is independent of t for $\mathrm{Im}\, t = 0$ due to (a), we obtain $\phi(\alpha(t) A) = \phi(A)$ for any complex t. Hence (9.32) is equal to

$$\phi(A_2{}^* A_1) = (\pi_\phi(A_2) \Omega_\phi, \pi_\phi(A_1) \Omega_\phi). \qquad (9.33)$$

The equality of the first line of (9.31) and the right-hand side of (9.33) prove first that the right-hand side of (9.30) vanishes whenever $\pi_\phi(A) \Omega_\phi = 0$, and secondly that $(J\Psi_1, J\Psi_2) = (\Psi_2, \Psi_1)$ for $\Psi_1, \Psi_2 \in D_\phi^0$. From the continuity (b) and the nuclear theorem, it follows that D_ϕ^0 is dense in D_ϕ and hence in $\mathfrak{H}_\phi$. Therefore J defined by (9.30) on D_ϕ^0 is an anti-isometric mapping from a dense subset D_ϕ^0 of $\mathfrak{H}_\phi$ onto itself. Hence its closure, which shall be denoted again by J, is an antiunitary operator on $\mathfrak{H}_\phi$.

From (9.30) it follows that $J^2 = 1$ on D_ϕ^0 and hence on $\mathfrak{H}_\phi$. By definition, $J\Omega_\phi = \Omega_\phi$.

Let us define operators $\pi_\phi'(A)$ for $A \in \mathfrak{A}_0$ on a vector $\Psi \in D_\phi^0$ by

$$\pi_\phi'(A) \Psi = J\pi_\phi(\alpha_\Gamma A) J\Psi, \qquad (9.34)$$

where

$$\alpha_\Gamma \sum c_j \mathrm{B}(h_{j1}) \ldots \mathrm{B}(h_{jn_j}) = \sum \bar{c}_j \mathrm{B}(\Gamma h_{j1}) \ldots \mathrm{B}(\Gamma h_{jn_j}). \qquad (9.35)$$

The following properties of $\pi_\phi'(A)$ follow immediately from the corresponding properties of π_ϕ and the properties of Γ and J:

$$\pi_\phi'(c_1 A_1 + c_2 A_2) = c_1 \pi_\phi'(A_1) + c_2 \pi_\phi'(A_2); \qquad (9.36)$$

$$\pi_\phi'(A)^* \Psi = \pi_\phi'(A^*) \Psi \quad \text{for} \quad \Psi \in D_\phi^0; \qquad (9.37)$$

$$[\pi_\phi'(\mathrm{B}(h_1))^*, \pi_\phi'(\mathrm{B}(h_2))] = -\gamma(h_1, h_2). \qquad (9.38)$$

Further, $\pi_\phi'(A_1)$ commutes with $\pi_\phi(A_2)$ as can be seen as follows, where A_1 and A_2 are in $\mathfrak{A}_0$: Let $h_1 \in \tilde{\mathscr{D}} \oplus \tilde{\mathscr{D}} \equiv K_0 (\subset K)$ and $A \in \mathfrak{A}_0$. We then have

$$\pi_\phi'(\mathrm{B}(h_1)) \pi_\phi(A) \Omega_\phi = J\pi_\phi(\mathrm{B}(\Gamma h_1)) \pi_\phi(\alpha(i\beta/2) A^*) \Omega_\phi =$$

$$= \pi_\phi(\alpha(i\beta/2) \{\mathrm{B}(h_1)^* \alpha(i\beta/2) A^*\}^*) \Omega_\phi = \pi_\phi(A) \pi_\phi(\alpha(i\beta/2) \mathrm{B}(h_1)) \Omega_\phi. \qquad (9.39)$$

Using (9.39) for $A = \mathrm{B}(h_2) A_2$ and $A = A_2$, we obtain for $\Psi = \pi_\phi(A_2) \Omega_\phi$, $A_2 \in \mathfrak{A}_0$,

$$\pi'_\phi(\mathrm{B}(h_1))\,\pi_\phi(\mathrm{B}(h_2))\,\Psi = \pi'_\phi(\mathrm{B}(h_1))\,\pi_\phi(\mathrm{B}(h_2)\,A_2)\,\Omega_\phi =$$

$$= \pi_\phi(\mathrm{B}(h_2)\,A_2)\,\pi_\phi(\alpha(i\beta/2)\,\mathrm{B}(h_1))\,\Omega_\phi =$$

$$= \pi_\phi(\mathrm{B}(h_2))\,\{\pi_\phi(A_2)\,\pi_\phi(\alpha(i\beta/2)\,\mathrm{B}(h_1))\,\Omega_\phi\} =$$

$$= \pi_\phi(\mathrm{B}(h_2))\,\pi'_\phi(\mathrm{B}(h_1))\,\Psi\;;$$

namely, $\pi'_\phi(\mathrm{B}(h_1))$ and $\pi_\phi(\mathrm{B}(h_2))$ commute on D^0_ϕ and hence

$$[\pi'_\phi(A_1),\pi_\phi(A_2)]\,\Psi = 0 \tag{9.40}$$

for A_1, $A_2 \in \mathfrak{A}_0$ and $\Psi \in D^0_\phi$. (Actually, due to the continuity (b) and the nuclear theorem, D_ϕ is in the domain of the closure of $\pi'_\phi(A_1)$, $A_1 \in \mathfrak{A}_0$ and the latter commutes with $\pi_\phi(A_2)$, $A_2 \in \mathfrak{A}$ on D_ϕ.)

These properties of $\pi'_\phi(A)$ motivate the introduction of a new space $\hat{K}_0 = K_0 \oplus K_0$ where $K_0 \equiv \tilde{\mathcal{D}} \oplus \tilde{\mathcal{D}}$, a new indefinite metric

$$\hat{\gamma}(h_1 \oplus h'_1, h_2 \oplus h'_2) = \gamma(h_1,h_2) - \gamma(h'_1,h'_2), \tag{9.41}$$

a new operator

$$\hat{\Gamma}(h \oplus h') = \Gamma h \oplus \Gamma h' \tag{9.42}$$

and a new *-algebra $\hat{\mathfrak{A}}_0$ which consists of polynomials (9.7). The h_{ik} are now in $\hat{K}_0$, and $\mathrm{B}(h)$, $h \in \hat{K}_0$ satisfies relations (9.4)–(9.6), in which γ and Γ are replaced by $\hat{\gamma}$ and $\hat{\Gamma}$.

From the properties of π_ϕ, (9.30)–(9.38) and (9.39), it follows that

$$\hat{\pi}(\mathrm{B}(h \oplus h')) \equiv \pi_\phi(\mathrm{B}(h)) + \pi'_\phi(\mathrm{B}(h')) \tag{9.43}$$

generates a *-representation of $\hat{\mathfrak{A}}_0$ on $\mathfrak{H}_\phi$ where h, $h' \in K_0$.

The vector Ω_ϕ is in the domain of $\hat{\pi}(A)$, $A \in \hat{\mathfrak{A}}_0$ and is cyclic for $\hat{\pi}(\hat{\mathfrak{A}}_0)$. Due to (9.39) with $A = 1$, Ω_ϕ satisfies

$$\hat{\pi}(\mathrm{B}(k))\,\Omega_\phi = 0 \tag{9.44}$$

whenever

$$k = \{-\,\mathrm{U}^G(-i\beta\mu/2)\,\mathrm{U}^T(i\beta/2)\,h\} \oplus h\,. \tag{9.45}$$

Let $P^{\beta,\mu}$ be an operator on $\hat{K}_0$ defined by

$$P^{\beta,\mu}(h \oplus h') = \{(1 - U^2_{\beta,\mu})^{-1}\,(h + U_{\beta,\mu}h')\} \oplus \{-(1 - U^2_{\beta,\mu})^{-1}\,U_{\beta,\mu}(h + U_{\beta,\mu}h')\},$$

$$\tag{9.46}$$

where $U_{\beta,\mu} \equiv \mathrm{U}^G(-i\beta\mu/2)\,\mathrm{U}^T(i\beta/2)$ maps K_0 onto K_0 by

$$U_{\beta,\mu}(f \oplus g) = \{e^{-\beta(H_0-\mu)/2}f\} \oplus \{e^{\beta(H_0-\mu)/2}g\}\,. \tag{9.47}$$

For $\mu > 0$, $(1 - U^2_{\beta,\mu})^{-1}$ maps K_0 onto K_0 and $P^{\beta,\mu}$ is defined on $\hat{K}_0$. $P^{\beta,\mu}$ satisfies

$$(P^{\beta,\mu})^2 = P^{\beta,\mu}, \tag{9.48}$$

$$\hat{\gamma}(P^{\beta,\mu}k_1, k_2) = \hat{\gamma}(k_1, P^{\beta,\mu}k_2), \tag{9.49}$$

$$\hat{\gamma}(P^{\beta,\mu}k, k) > 0 \quad \text{for} \quad P^{\beta,\mu}k \neq 0, \tag{9.50}$$

$$\hat{\Gamma} P^{\beta,\mu} \hat{\Gamma} = 1 - P^{\beta,\mu}. \tag{9.51}$$

The importance of this operator is due to the consequence that any $B(h \oplus h')$ splits up as a sum

$$B(h \oplus h') = B(P^{\beta,\mu}(h \oplus h')) + B(P^{\beta,\mu}\Gamma(h \oplus h'))^* \tag{9.52}$$

and, due to (9.44),

$$\hat{\pi}_\phi(B(k)^*)\Omega_\phi = 0 \tag{9.53}$$

whenever $k \in P^{\beta,\mu}\hat{K}_0$. Namely $\hat{\pi}(B(k))$ and $\hat{\pi}_\phi(B(k))^*$, $k \in P^{\beta,\mu}\hat{K}_0$ can be regarded as new creation and annihilation operators in a Fock representation. In particular,

$$\hat{\phi}(\hat{A}) \equiv (\Omega_\phi, \hat{\pi}(\hat{A})\Omega_\phi), \qquad \hat{A} \in \hat{\mathfrak{A}}_0 \tag{9.54}$$

is uniquely determined. For example,

$$(\Omega_\phi, \hat{\pi}(B(k_1))^* \, \hat{\pi}(B(k_2))\Omega_\phi) = (\Omega_\phi, [\hat{\pi}(B(P^{\beta,\mu}k_1)^*), \hat{\pi}(B(P^{\beta,\mu}k_2))]\Omega_\phi) =$$

$$= \hat{\gamma}(P^{\beta,\mu}k_1, P^{\beta,\mu}k_2) \tag{9.55}$$

for $k_1, k_2 \in \hat{K}_0$. If $k_j = h_j \oplus 0$, $j = 1, 2$, then we obtain

$$\phi(B(h_1)^* B(h_2)) = \hat{\gamma}(P^{\beta,\mu}(h_1 \oplus 0), P^{\beta,\mu}(h_2 \oplus 0)) =$$

$$= \gamma(h_1, (1 - U_{\beta,\mu}^2)^{-1}h_2). \tag{9.56}$$

Similarly, all values of ϕ on $\mathfrak{A}_0$ are uniquely determined. By the continuity (b), ϕ on $\mathfrak{A}$ is unique.

To see the existence we introduce new creation and annihilation operators $b_j^\dagger$ and $b_j, j = 1, 2$:

$$\hat{\pi}(B(P^{\beta,\mu}\{h \oplus 0\})) = b_1^\dagger(f) + b_2^\dagger(g), \tag{9.57}$$

$$\hat{\pi}(B((1 - P^{\beta,\mu})\{h' \oplus 0\})) = b_1(f) + b_2(g), \tag{9.58}$$

where $h = (\rho_{\beta,\mu} + 1)^{-1/2}f \oplus \rho_{\beta,\mu}^{-1/2}g, h' = \rho_{\beta,\mu}^{-1/2}g \oplus (\rho_{\beta,\mu} + 1)^{-1/2}f$ and

$$\rho_{\beta,\mu} = (e^{\beta(H_0 - \mu)} - 1)^{-1}. \tag{9.59}$$

From the properties of $\hat{\pi}$, the following properties of b's follow immediately:

(α) (Linearity): $b_j^\dagger(f)$ and $b_j(f), j = 1, 2$, are linear in f.

(β) (Adjointness): $\{b_j^\dagger(f)\}^* \Psi = b_j(\bar{f})\Psi \quad$ for $\quad \Psi \in D_\phi^0, \quad j = 1, 2$.

(γ) (CCR): $[b_i(f), b_j(g)] = [b_i^\dagger(f), b_j^\dagger(g)] = 0$ and

$$[b_i(f), b_j^\dagger(g)] = \delta_{ij}\int f(x)\, g(x)\, \mathrm{d}x. \tag{9.60}$$

(δ) (Cyclicity): The algebra generated by $b_j^\dagger(f)$ and $b_j(g)$, $j = 1, 2$, $f, g \in \tilde{\mathscr{D}}$ is $\hat{\pi}(\hat{\mathfrak{A}}_0)$, for which Ω_ϕ is cyclic.

(ε) $b_j(f)\Omega_\phi = 0$ for $j = 1, 2$, $f \in \tilde{\mathscr{D}}$.

Due to the property (ε), $b_j^\dagger$ and b_j are in the Fock representation and can be extended to $f \in L_2(R^3)$ by continuity in f. We shall also extend the domains of $b_j(f)$ and $b_j(f)$, which originally were D_ϕ^0, by closure.

The relations between the original a^+ and a and the new $b_j^\dagger$ and b_j can be computed from

$$\pi_\phi(a^+(f)) = \hat{\pi}(\mathbf{B}(\{f \oplus 0\} \oplus 0)),$$

$$\pi_\phi(a(f)) = \hat{\pi}(\mathbf{B}(\{0 \oplus f\} \oplus 0)),$$

for $f \in \tilde{\mathscr{D}}$, and by continuity for a general f in $\mathscr{S}$. They are given by

$$\pi_\phi(a^+(f))\,\Psi = b_1^\dagger((\rho_{\beta,\mu} + 1)^{1/2}f)\Psi + b_2(\rho_{\beta,\mu}^{1/2}f)\,\Psi, \tag{9.61}$$

$$\pi_\phi(a(f))\,\Psi = b_2^\dagger(\rho_{\beta,\mu}^{1/2}f)\,\Psi + b_1((\rho_{\beta,\mu} + 1)^{1/2}f)\,\Psi, \tag{9.62}$$

for $\Psi \in D_\phi$.

For the existence proof, we start from the creation and annihilation operators $b_j^\dagger$ and b_j in the Fock representation and define a *-representation π_ϕ of a^+ and a by (9.61) and (9.62). It is then straightforward to verify the conditions (a), (b) and (c) for the vector state by the Fock vacuum Ω_ϕ of $b_j^\dagger$ and b_j, the verification being left to the reader.

The infinite volume limit of the grand canonical ensemble average has been computed by Araki and Woods. The comparison of their formulas (4.18) and (4.17) with our (9.61) and (9.62) shows that the limit is given by the KMS state determined above. (There is some discrepancy of notations in connection with $(2\pi)^3$ in the Fourier transform.)

We now turn our attention to the case $\mu = 0$. In this case, $\rho_{\beta,0}$ defined by (9.59) is an unbounded selfadjoint operator on $L_2(R^3)$ and $\mathscr{S}$ is in the domain of $\rho_{\beta,0}^{1/2}$ and $(\rho_{\beta,0} + 1)^{1/2}$. It is then straightforward to verify that the vector state by the Fock vacuum of $b_j^\dagger$ and b_j, $j = 1, 2$ for the *-representation given by (9.61) and (9.62) is a KMS state. This shows the existence. The uniqueness, however, is broken in this case.

To find all KMS states for $\mu = 0$, we follow the above analysis up to (9.40). We shall extend the domains of $\pi_\phi(f)$ and $\hat{\pi}(h)$ by closure for our convenience in calculation.

Instead of $\hat{K}_0$, we consider now a subspace $\hat{K}_1$ of $\hat{K}_0$, which consists of

$$k = (h - U_{\beta,0}h') \oplus (h' - U_{\beta,0}h), \tag{9.63}$$

where $h, h' \in K_0$. It is $\hat{\Gamma}$-invariant. The sub-*-algebra of $\hat{\mathfrak{A}}_0$, generated by $\mathbf{B}(k)$, $k \in \hat{K}_1$, is denoted by $\hat{\mathfrak{A}}_1$.

Since (9.63) vanishes only when $h = h' = 0$, we define a projection operator $P^{\beta,0}$ acting on $\hat{K}_1$ by

$$P^{\beta,0}k = h \oplus - U_{\beta,0}h, \tag{9.64}$$

where k is given by (9.63). It has the same properties as $P^{\beta,\mu}$, $\mu > 0$, on $\hat{K}_1$. Hence we define $b_j^\dagger$ and b_j, $j = 1, 2$ by (9.57) and (9.58), where f and g run over $H_0^{1/2}\tilde{\mathcal{D}}$. The last restriction is due to the fact that $h \oplus 0 \in \hat{K}_1$ and $h \in K_0$ if and only if $h \in \{1 - (U_{\beta,0})^2\} K_0 = H_0\tilde{\mathcal{D}} \oplus H_0\tilde{\mathcal{D}}$.

On the Fock representation, we can define $b_j^\dagger(f)$ and $b_j(f)$ for all $f \in L_2(R^3)$ by continuity in f. Using the extended b's (which are defined on a dense subset of the closure of $\pi_\phi(\hat{\mathfrak{A}}_1)\Omega_\phi$ for the moment), we define $\pi_{\beta,0}(a^+(f))$ and $\pi_{\beta,0}(a(f))$ for all $f \in \mathscr{S}$ by (9.61) and (9.62), where Ψ is any vector for which the right-hand side is meaningful. We define

$$c^+(f) = \pi_\phi(a^+(f)) - \pi_{\beta,0}(a^+(f)), \tag{9.65}$$

$$c(f) = \pi_\phi(a(f)) - \pi_{\beta,0}(a(f)). \tag{9.66}$$

Since the domain of $\hat{\pi}(\mathrm{B}(k))$ has been extended by closure, it contains $\pi_{\beta,0}(a(f))\Omega_\phi$ and

$$\hat{\pi}(\mathrm{B}(k))(c(f_1)\Omega_\phi) = c(f_1)\hat{\pi}(\mathrm{B}(k))\Omega_\phi = 0 \tag{9.67}$$

for $k \in (1 - P^{\beta,0})\hat{K}_1$ and $f_1 \in \mathscr{S}$. By the same argument as before, $b_j^\dagger(f)$ and $b_j(f)$, $j = 1, 2, f \in \hat{K}_1$ are in the Fock representation on the closure of $\hat{\pi}(\hat{\mathfrak{A}}_1)c(f_1)\Omega_\phi$ with $c(f_1)\Omega_\phi$ as their Fock vacuum. Proceeding successively, we reach the same conclusion for any vector $\mathscr{P}\Omega_\phi$ where $\mathscr{P}$ is any polynomial of 1, $c(f_1)\ldots c(f_n)$, $c^+(f_1)\ldots c^+(f_n)$ and $f_1, \ldots, f_n \in \mathscr{S}$.

Therefore, we have a tensor product decomposition

$$\mathfrak{H}_\phi = \mathfrak{H}_F \otimes \mathfrak{H}_c, \tag{9.68}$$

$$b_j^\dagger(f) = b_{jF}^\dagger(f) \otimes 1, \qquad j = 1, 2 \tag{9.69}$$

$$b_j(f) = b_{jF}(f) \otimes 1, \qquad j = 1, 2 \tag{9.70}$$

$$c^+(f) = 1 \otimes c_0^\dagger(f), \tag{9.71}$$

$$c(f) = 1 \otimes c_0(f), \tag{9.72}$$

$$\Omega_\phi = \Omega_F \otimes \Omega_c, \tag{9.73}$$

where $b_{jF}^\dagger$ and b_{jF} are in the Fock representation on $\mathfrak{H}_F$ with Ω_F as their Fock vacuum. The domain of each $c_0(f)$ contains D_c, which is a dense linear subset of $\mathfrak{H}_c$, and consists of $\mathscr{P}\Omega_c$, $\mathscr{P}$ being a polynomial of 1, $c_0(f_1), \ldots, c_0(f_n)$ $(f_1, \ldots, f_n \in \mathscr{S})$. $c_0^\dagger(f)$ and $c_0(f)$ further have the following properties:

(i) $(\Psi, c_0(f)\Phi)$ is linear and continuous in $f \in \mathscr{S}$ for each $\Psi, \Phi \in D_c$,

(ii) $c_0(f)^* \supset c_0^\dagger(f)$, $\qquad c_0^\dagger(f)^* \supset c_0(f)$,

(iii) $[c_0(f), c_0(g)] \Psi = [c_0^\dagger(f), c_0(g)] \Psi = [c_0^\dagger(f), c_0^\dagger(g)] \Psi = 0$ for $\Psi \in D_c$,

(iv) $c_0(f) = 0$ for $f \in H_0 \tilde{\mathscr{D}}$.

Conversely, for any such c_0, the vector state by Ω_ϕ for the representation

$$\pi_\phi(a^\dagger(f)) = \pi_{\beta,0}(a^\dagger(f)) + c^\dagger(f),\tag{9.74}$$

$$\pi_\phi(a(f)) = \pi_{\beta,0}(a(f)) + c(f)\tag{9.75}$$

gives a KMS state for $\mu = 0$.

To investigate the structure of $c(f)$ in more detail, we shall restrict our attention to the case where closures of $c_0^\dagger(f)$ and $c_0(f)$ for all f have a simultaneous spectral decomposition. Such KMS states may be decomposed into an integral over their extremal states and the latter are characterized by $\dim \mathfrak{H}_c = 1$, i.e. $c_0^\dagger(f)$ and $c_0(f)$ are complex-valued functionals of f. From (i) and (iv) we see that c_0 must be a tempered distribution solution of

$$\Delta c_0(x) = 0.\tag{9.76}$$

It follows that $c_0(x)$ is a polynomial.

To select states of physical interest, we consider the average particle density, given by

$$\langle \rho \rangle = \langle \rho \rangle_0 + \lim_{V \to 0} V^{-1} \int_V d^3x |c_0(x)|^2,\tag{9.77}$$

where $\langle \rho \rangle_0$ is the value for $c_0(x) = 0$ and is given by

$$\langle \rho \rangle_0 = (2\pi)^{-3} \int d^3p \, [\exp(p^2/2m) - 1]^{-1}.\tag{9.78}$$

The solution of (9.76) with a finite $\langle \rho \rangle$ is a constant. Thus, for each given value $\langle \rho \rangle \geqq \langle \rho \rangle_0$, there exists a one-parameter family of KMS states with $\mu = 0$ for which $c_0(x) = (\langle \rho \rangle - \langle \rho \rangle_0) e^{i\theta}$, $0 \leq \theta < 2\pi$. We denote the corresponding KMS states by $\phi_{\beta,\langle\rho\rangle,\theta}$. They are distinct for distinct θ if $\langle \rho \rangle > \langle \rho \rangle_0$ and transform transitively among themselves under the gauge transformation. Therefore there exists a unique gauge-invariant mixture of these states, given by

$$\phi_{\beta,\langle\rho\rangle}(A) = (2\pi)^{-1} \int_0^{2\pi} \phi_{\beta,\langle\rho\rangle,\theta}(A)\, d\theta.\tag{9.79}$$

The limit of $\phi_{\beta,\langle\rho\rangle}$ as $\beta \to \infty$ (i.e. the zero-temperature limit) coincides with the infinite volume limit of the ground state of the free Bose gas computed by Araki and Woods.

$\phi_{\beta,\langle\rho\rangle,\theta}$ is a typical case of a broken symmetry and (9.79) is a decomposition of Gibbs states into extremal KMS states of broken symmetry. The appearance of the gauge-noninvariant phase $\phi_{\beta,\langle\rho\rangle,\theta}$ is known as the Bose condensation of the free Bose gas.

REFERENCES

Text Books

[1] J. Dixmier, "Les algèbres d'operateurs dans l'espace hilbertien", 2nd ed. Gauthier–Villars, Paris (1969).

[2] J. Dixmier, "Les C^*-algèbres et leurs représentations", 2nd ed. Gauthier–Villars, Paris (1969).

[3] I. Kaplansky, "Rings of Operators", W. A. Benjamin, New York (1968).

[4] M. Naimark, "Normed Rings" (English translation). P. Noordhoff, Groningen (1964).

[5] C. E. Rickart, "General Theory of Banach Algebras". Van Nostrand, Princeton (1960).

[6] S. Sakai, "C^*-Algebras and W^*-Algebras". Springer, Berlin (1971).

[7] J. T. Schwarz, "W^*-Algebras". Gordon and Breach, New York (1967).
Another textbook is being prepared by M. Takesaki.
A brief summary of the properties of a C^*-algebra is given in the appendix of the following basic paper in this field:

[8] R. Haag and D. Kastler, *J. Math. Phys.*, **5**, 848–861 (1964).

KMS States

[9] R. Haag, N. M. Hugenholtz, and M. Winnink, *Commun. Math. Phys.,* **5**, 215–236 (1967).

[10] N. M. Hugenholtz and J. D. Wieringa, *Commun. Math. Phys.,* **11**, 183–197 (1969).

[11] M. Takesaki, "Tomita's Theory of Modular Hilbert Algebras and Its Application". Springer-Verlag, Berlin (1970).

[12] M. Winnink, in: "Cargèse Lectures in Physics", Vol. 4, ed. D. Kastler. Gordon and Breach, New York (1970).

Decomposition of States

[13] D. Ruelle, a lecture in: "Cargèse Lectures in Physics", Vol. 4, ed. D. Kastler. Gordon and Breach, New York (1970).

Free Bose Gas

[14] H. Araki and E. J. Woods, *J. Math. Phys.*, **4**, 637–662 (1963).

Classification of Factors

HUZIHIRO ARAKI

Research Institute for Mathematical Sciences, Kyoto (Japan)

Abstract

A classification of von Neumann algebra factors in terms of their asymptotic ratio set is described.

In this talk, I would like to describe a classification of von Neumann algebras in terms of their asymptotic ratio sets. For proofs of statements, see references [1], [2] and [3].

1. INFINITE TENSOR PRODUCT OF VON NEUMANN ALGEBRA

Consider a one-dimensional spin lattice, where the total spin $J_n (\in Z_+/2)$ at the lattice site $n = 0, \pm 1, \pm 2, \ldots$ is not necessarily constant. The algebra $\mathfrak{A}_n$ of observables at each lattice site n is taken to be the algebra of all (bounded) linear operators on a Hilbert space of dimension $2J_n + 1 (\in Z_+ + 1)$. For our purpose, it is enough to consider the mixture state

$$\phi_\Lambda = \sum_m \lambda_m \phi_m^n \tag{1.1}$$

of the pure state ϕ_m^n, which corresponds to the value m of the Z-component of the spin at the lattice site n, where Λ denotes the set $\{\lambda_m\}$ (called the eigenvalue list), $m = -J_n, -J_n + 1, \ldots, J_n, \lambda_m \geqq 0, \sum \lambda_m = 1$.

A product state $\phi = \underset{n}{\otimes} \phi_{\Lambda^n}$ of the algebraic tensor product $\mathfrak{A}$ of $\mathfrak{A}_n$ (i.e. the algebra generated by $\cup \mathfrak{A}_n$ where $A_n \in \mathfrak{A}_n$ and $A_m \in \mathfrak{A}_m$ are assumed to commute for $m \neq n$) is defined by

$$\phi \left(\sum_{j=1}^{N} c_j A_{jn_1} \ldots A_{jn_k} \right) = \sum_{j=1}^{N} c_j \prod_{l=1}^{k} \phi_{\Lambda^{n_l}}(A_{jn_l}) \tag{1.2}$$

for $A_{j n_l} \in \mathfrak{A}_{n_l}$, where $\Lambda^n = \{\lambda_l^n\}$. Then there exists a Hilbert space $\mathfrak{H}_\phi$, a *-representation π_ϕ of $\mathfrak{A}$ (by bounded linear operators on $\mathfrak{H}_\phi$) and a cyclic unit vector $\Omega_\phi \in \mathfrak{H}_\phi$ such that $(\Omega_\phi, \pi_\phi(A)\Omega_\phi) = \phi(A)$ for all $A \in \mathfrak{A}$. Let $R(\{\Lambda^n\})$ denote the von Neumann algebra generated by $\pi(A)$, $A \in \mathfrak{A}$. It is a factor. In the above definitions, we may allow the case $J_n = \infty$.

A special case of $R(\{\Lambda^n\})$ with $J_n = 1/2$ and $\Lambda^n = \{\lambda, 1 - \lambda\}$ (the translationally invariant case) has been considered by von Neumann [9] and has been shown to be mutually non-*-isomorphic for $0 \leq \lambda \leq 1/2$ by Powers [10]. Let us denote this special one-parameter family of factors (von Neumann algebras with a trivial center) by R_x, where x is related to λ by $x = \lambda/(1 - \lambda)$, $0 \leq \lambda < 1$. Since the interchange of the name "up" and "down" of the spin changes nothing essential, it is obvious that $R_x = R_{x^{-1}}$ for $x \neq 0$.

2. ASYMPTOTIC RATIO SET

For an arbitrary von Neumann algebra R, the asymptotic ratio set $r_\infty(R)$ is defined to be the set of x such that R is *-isomorphic to the tensor product $R \otimes R_x$. This set is obviously an invariant under *-isomorphisms of R and gives a label to distinguish non-*-isomorphic von Neumann algebras.

From the result to be described in Section 3, it immediately follows that nonzero elements of $r_\infty(R)$ are a subgroup of the multiplicative group of positive real numbers. From the result to be described in Section 4, it also follows that $r_\infty(R)$ is closed if R is a von Neumann algebra on a separable space. Therefore, $r_\infty(R)$ for such R is limited to one of the following sets:

$$
\begin{aligned}
S_\phi &= \text{empty} \\
S_0 &= \{0\} \\
S_{01} &= \{0, 1\} \\
S_1 &= \{1\} \\
S_x &= S_0 \cup \{x^n; n = 0, \pm 1, \pm 2, \ldots\}, \qquad 0 < x \neq 1 \\
S_\infty &= [0, \infty).
\end{aligned}
$$

All of them appear as $r_\infty(R)$ for some R.

If we restrict our attention to $R(\{\Lambda^n\})$, then $r_\infty(R_1) = r_\infty(R_2) \neq S_{01}$ implies that R_1 is *-isomorphic to R_2. Namely the asymptotic ratio set gives a complete classification of *-isomorphism classes of $R(\{\Lambda^n\})$ except for the class S_{01}. The class S_{01} among $R(\{\Lambda^n\})$ contains $R_0 \otimes R_1$ and infinitely many (type III) non-*-isomorphic factors. A partial classification can be given by the set $\rho(R)$ of all x such that $R \otimes R_x$ is *-isomorphic to R_x.

The factor R_x itself has an asymptotic ratio set $r_\infty(R) = S_x$ for $0 \leq x < \infty$. A factor $R = R(\{\Lambda^n\})$ with $r_\infty(R) = S_\infty$, which is unique up to a *-isomorphism, is denoted by R_∞. An example of Λ^n such that $r_\infty(R(\Lambda^n)) = S_\infty$ is given in Section 3.

The origin of the name "asymptotic ratio set" is in its original definition for $R(\Lambda^n)$, which we shall briefly describe:

For any finite subset I of lattice points and given $\{\Lambda^n\}$, let $\Lambda(I)$ denote the product of Λ^n, $n \in I$, namely the set of numbers $\prod_{n \in I} \lambda^n_{l_n}$, where $\Lambda^n = \{\lambda^n_l\}_{l = -J_n, -J_n+1, \ldots, J_n}$ and $\{l_n\}_{n \in I}$ runs over all possibilities. Then a nonnegative real number x is in the asymptotic ratio set $r_\infty(R(\{\Lambda^n\}))$ if and only if there exist a sequence I_k, $k = 1, 2, \ldots$ of mutually disjoint finite subsets of lattice points, two disjoint subsets K_{k1} and K_{k2} of $\Lambda(I_k)$ and a one-to-one correspondence α_k between K_{k1} and K_{k2} such that

$$\lim_k \left\{ \sup_{\lambda \in K_{k1}} \left| \frac{\alpha_k \lambda}{\lambda} - x \right| \right\} = 0 \,,$$

$$\sum_k \left\{ \sum_{\lambda \in K_{k1}} \lambda \right\} = \infty \,.$$

3. TENSOR PRODUCT FORMULA

Let (a, b) denote the greatest common divisor of a and b if a and b are rationally related (i.e. the largest number c such that both a and b are integer multiples of c) and $(a, b) = \infty$ otherwise. We shall define $(a, 0) = a$ and $(a, \infty) = \infty$ for any a. We then have for nonzero x and y the following formula:

$$R_x \otimes R_y \sim R_z \quad \text{for} \quad \log z = (\log x, \log y) \,, \tag{3.1}$$

where $\sim$ denotes a *-isomorphism. We also have

$$R_x \otimes R_0 \sim R_x \quad \text{if} \quad x \neq 1 \,. \tag{3.2}$$

To see some physical significance of $\log x$ for R_x, consider the translationally invariant case for spin 1/2. If the energy difference of spin up and down states at each lattice site is Δ, then the Gibbs state with an inverse temperature β is the product state $\phi = \otimes \phi_{\Lambda^n}$ with

$$\Lambda^n = ((1 + e^{-\Delta\beta})^{-1}, e^{-\Delta\beta}(1 + e^{-\Delta\beta})^{-1}), \tag{3.3}$$

which gives rise to R_x with $\log x = \pm \Delta\beta$; namely, $\log x$ is proportional to β.

If we consider the case where the energy difference of spin up and down states is 1 at even lattice sites and Δ at odd lattice sites, then we obtain $R_x \otimes R_y$ where $\log x = \beta$ and $\log y = \Delta\beta$. The physical significance of (3.1) for such a system is not known.

It is interesting to note that another physical quantity—entropy—gives a complete classification of the triplet of the vector Ω_ϕ, the commutative sub-von-Neumann algebra of R_x generated by z-components of spin operators, and the lattice shift automorphism (by one lattice unit), a recent result of D. Ornstein. This result has been extended by S. Ito, H. Murata, and H. Totoki to a state satisfying a cluster property.

4. PROPERTY L'_λ

In order to streamline the nonisomorphism proof of R_x, Powers has introduced the following notion. A von Neumann algebra R has the property L_λ if for any $\varepsilon > 0$ and for any normal state ω, there exists $N \in R$ such that $N^2 = 0, N^*N + + NN^* = 1$ and

$$|(1 - \lambda)\,\omega(QN) - \lambda\omega(NQ)| \leqq \varepsilon\,\|Q\| \tag{4.1}$$

for all $Q \in R$. The asymptotic ratio set is characterized by modifying the above notion slightly: We say that R has the property L'_λ if for any $\varepsilon > 0$ and for a finite number of normal states $\omega_j, j = 1, \ldots, n$ there exists $N \in R$ such that $N^2 = 0, N^*N + + NN^* = 1$ and (4.1) holds for $\omega = \omega_j, j = 1, \ldots, n$. For a von Neumann algebra R on a separable Hilbert space, $\lambda(1 - \lambda)^{-1} \in r_\infty(R)$ if and only if R has the property L'_λ.

Property L_0 and property L'_0 are equivalent while the properties $L_{1/2}$ and $L'_{1/2}$ are not equivalent, because the property $L_{1/2}$ is satisfied by all finite continuous von Neumann algebras and there is a finite continuous von Neumann algebra not satisfying the property $L'_{1/2}$. The properties L_λ and L'_λ have been shown to be different for $\lambda > 0$ by Connes.

REFERENCES

[1] H. Araki and E. J. Woods, *Publ. Res. Inst. Math. Sci.,* **4**, 51–130 (1968).

[2] H. Araki, *Publ. Res. Inst. Math. Sci.,* **4**, 585–593 (1968).

[3] H. Araki, *Publ. Res. Inst. Math. Sci.,* **6**, 443–460 (1970).

[4] H. Araki, "Some Topics in the Theory of Operator Algebras", in: *Proc. Intern. Congr. of Mathematicians,* Nice, 1970, Vol. 2, 379–382 (1971).

[5] W. Krieger, *J. Functional Analysis,* **6**, 1–13 (1970).

[6] W. Krieger, "Lecture Notes in Math.", Vol. 160, pp. 158–177. Springer, Berlin (1970).

[7] W. Krieger, "On Hyperfinite Factors and Non-Singular Transformations of a Measure Space".

[8] O. A. Nielsen, "The Asymptotic Ratio Set and Direct Integral Decompositions of a von Neumann Algebra", *Canadian J. Math.,* **23**, 598–607 (1971).

[9] J. von Neumann, *Comp. Math.,* **6**, 1–77 (1938).

[10] R. T. Powers, *Annals of Math.,* **86**, 138–171 (1967).

[11] J. J. Williams, "Non-Isomorphic Tensor Products of von Neumann Algebras". (to appear in *Canadian J. Math.*).

Algebraic Aspects of Wightman Field Theory

H. J. BORCHERS

Institut für Theoretische Physik, Universität Göttingen (Germany)

Abstract

Wightman field theory is reformulated as a theory of positive linear functionals over a test-function algebra. Some properties of this algebra are derived. Examples are given showing that this setting of quantum field theory is suitable for investigating the nonlinear aspects of field theory.

FOREWORD

Axiomatic quantum field theory has been in existence for 20 years by now [10], [6]. Although many interesting results have been obtained, the general theory is still in an unsatisfactory state. This, in my opinion, is due to the fact that the linear program leads to mathematical problems which are abstractly solvable, but which cannot be handled in the concrete case. Therefore, any reformulation of Wightman quantum field theory which puts the emphasis on points different from the linear program might be of help in reactivating this theory.

The C^*-algebra approach to field theory, invented by Araki, Haag and Kastler [1, 4], replaced Wightman field theory for two reasons: firstly the nonlinear aspects were built into the theory in an essential way, and secondly a sufficiently sophisticated mathematical theory was available. The price which one had to pay for these advantages was the loss of concreteness. The underlying algebraic structure is only defined abstractly by a set of axioms and not in concrete terms.

In 1963 I showed that an algebraic formulation of the Wightman axioms is possible [2]. This setting has the advantage that the algebra is given in concrete terms, but it has the disadvantage that it leads to a mathematical object which, to my knowledge, has not been investigated systematically by mathematicians. This was the main reason for my not pursuing these ideas. Two years ago W. Wyss and his Boulder lecture notes [11] convinced me that there are enough mathematical techniques around which, when put together, could lead to a manageable theory. Since last spring I have been collecting elementary facts which could serve as a basis for a theory. During the summer school I will try to give an introduction to this field and the conference talk will be concerned with first results.

I. WIGHTMAN AXIOMS AND FIELD ALGEBRA

I.1. Wightman's Axioms

It is not my intention to provide an introduction to Wightman quantum field theory. This can be found in textbooks $[\![2]\!]$, $[\![6]\!]$. For this reason we will start immediately with the axioms. We also restrict ourselves to the simplest possible case, which means that we treat one neutral field defined on the Schwartz space $\mathscr{S}$. This special choice might serve as a prototype for a more general scheme which includes several fields, as well as different test function spaces as long as they are nuclear spaces.

We now start with the axioms of Wightman [10]. The form given here differs somewhat from Wightman's original version but it presents the most convenient and most often used form of these axioms.

Axiom 1:

Let $\mathscr{H}$ be a Hilbert space with elements $\{\psi, \phi, \chi, \ldots\}$ and $D \subset \mathscr{H}$ a linear dense subspace of $\mathscr{H}$. Let $\mathscr{S} = \mathscr{S}(\mathbb{R}^4)$ be the Schwartz space of strongly decreasing C^∞-functions with elements $\{f, g, h, \ldots\}$.

For every $f \in \mathscr{S}$ an operator $A(f)$ shall exist with the properties:

a) if $D_{A(f)}$ denotes the domain of definition of $A(f)$, then

$$D \subset D_{A(f)};$$

b) for all $\psi \in D$ we have

$$A(f)\psi \in D;$$

c) for all pairs $\psi, \phi \in D$ we have the relation:

$$(\psi, A(f)\phi) = (A(\bar{f})\psi, \phi)$$

($\bar{f}$ denotes the complex conjugate function of f, and $(\cdot, \cdot)$ the scalar product in $\mathscr{H}$);

d) for all $\psi, \phi \in D$ the map

$$f \to (\psi, A(f)\phi)$$

is a continuous linear functional on $\mathscr{S}$.

This implies the relation:

$$A(f + \lambda g)\psi = A(f)\psi + \lambda A(g)\psi; \quad f, g \in \mathscr{S}, \psi \in D, \lambda \in \mathbb{C}.$$

Axiom 2:

Let G denote the vector group of $\mathbb{R}^4$ in the usual topology and let us denote by f_a the function $f_a(x) = f(x - a)$. There exists a strongly continuous unitary representation $\mathcal{U}(a)$, $a \in G$ acting on the Hilbert space $\mathcal{H}$ such that

a) $\mathcal{U}(a)\, D \subset D$ for all $a \in G$;

b) $\mathcal{U}(a)\, A(f)\, \mathcal{U}^{-1}(a)\, \psi = A(f_a)\, \psi$, $a \in G$, $f \in \mathcal{S}$, $\psi \in D$.

Axiom 3:

There exists a vector $\Omega \in D$ with the properties

a) $\mathcal{U}(a)\, \Omega = \Omega$ for all $a \in G$;

b) The set of vectors

$$\{\Omega,\, A(f)\,\Omega,\, A(f_1)\, A(f_2)\Omega,\, \dots A(f_1)\, A(f_2)\dots A(f_i)\Omega,\, \dots; i = 0, 1, 2 \dots, f_k \in \mathcal{S}\}$$

is total in $\mathcal{H}$.

Axiom 4:

Let us furnish $\mathbb{R}^4$ with the Minkowski metric, i.e. for $x, y \in \mathbb{R}^4$ define $(x, y) =$

$$= x^0 y^0 - \sum_{i=1}^{3} x^i y^i$$ and denote by V^+ the forward light cone

$$V^+ = \{x \in \mathbb{R}^4; (x)^2 > 0, x^0 > 0\}$$

and by $\overline{V^+}$, the closure of V^+. Let

$$\mathcal{U}(a) = \int_{\mathbb{R}^4} e^{i(pa)}\, E(dp)$$

be the Stone representation of $\mathcal{U}(a)$; then we require

$$\text{support } E(dp) \subset \overline{V^+}.$$

Axiom 5:

We call two functions $f, g \in \mathcal{S}$ spacelike separated if their supports are spacelike separated, i.e. the inequality $(x - y)^2 < 0$ holds for all pairs $x \in \text{supp } f$ and $y \in \text{supp } g$. We require

$$A(f)\, A(g)\, \psi = A(g)\, A(f)\, \psi \quad \text{for all} \quad \psi \in D$$

whenever $f, g \in \mathcal{S}$ are spacelike separated.

Remarks:

1) Axiom 2 is called the *axiom of translational invariance*. Axiom 3 requires the existence of a cyclic vacuum state. Axiom 4 is called the *spectrum condition* and axiom 5 is known by the name *locality axiom*.

2) Besides the translational invariance, one usually considers also the invariance under the homogeneous Lorentz group. Since, however, the most important consequences of this symmetry, namely the invariance of spectrum condition and locality, are incorporated into the axioms, we will make no further use of this symmetry.

I.2. The Field Algebra

We want to reformulate the axioms in such a way that their algebraic content is placed in the foreground. In this section we proceed to introduce the algebra in question.

I.2.1. *Definition*:

Let us denote by $\mathscr{S}_0$ the space of complex numbers and by $\mathscr{S}_n$ the space $\mathscr{S}(\mathbb{R}^{4n})$ of strongly decreasing C^∞ function on $\mathbb{R}^{4n}$ [5].

We denote by $\mathscr{S}$ the space of sequences

$$f = \{f_0, f_1, \ldots f_n \ldots\}$$

with the properties

a) $f_i \in \mathscr{S}_i$;

b) only a finite number of the f_i's are different from zero.

If we furnish $\mathscr{S}_i$ with the usual topology, then $\mathscr{S}$ becomes a topological vector space by identifying it with the topological direct sum of the $\mathscr{S}_i$:

$$\mathscr{S} = \sum \oplus \mathscr{S}_i$$

(see, e.g. N. Bourbaki [1]).

Next we want to furnish $\mathscr{S}$ with an algebraic structure.

I.2.2. *Definition*:

Let f, g be two elements of $\mathscr{S}$ with

$$f = \{f_0, f_1, \ldots\}$$
$$g = \{g_0, g_1, \ldots\};$$

then we define:

a) the sum

$$f + g = \{f_0 + g_0, f_1 + g_1, \ldots\};$$

b) the product with a complex number

$$\lambda f = \{\lambda f_0, \lambda f_1, \ldots\};$$

c) the product of two elements

$$f \times g = \{f_0 g_0, f_0 g_1 + f_1 g_0, \ldots, \sum_{i+h=l} f_i \times g_h, \ldots\},$$

where $f_i \times g_h$ is a function of $4(i + h)$ variables defined as follows:

$$(f_i \times g_h)(x_1, \ldots x_{i+h}) = f_i(x_1, \ldots x_i) g(x_{i+1}, \ldots x_{i+h}); \qquad x_j \in \mathbb{R}^4;$$

d) involution

$$f^* = \{f_0^*, f_1^*, \ldots\},$$

where $f_0^* = \bar{f}_0$ and f_i^* is a function of $4i$ variables defined by the relation

$$(f_i^*)(x_1, \ldots x_i) = \overline{f_i(x_i, x_{i-1}, \ldots x_1)}; \qquad x_j \in \mathbb{R}^4.$$

We find immediately

I.2.3. *Lemma*:

The space $\mathscr{S}$ is a $*$-algebra under the operations defined in I.2.2.

Proof:

a) The linearity is trivial.

b) The distributive law:

$$f \times (g + \lambda h) = \left\{ \ldots \sum_{i+h=l} f_i \times (g_h + \lambda h_h) \ldots \right\}$$

$$= \left\{ \ldots \sum_{i+h=l} (f_i \times g_h + \lambda f_i \times h_h) \ldots \right\}$$

$$= \left\{ \ldots \sum_{i+h=l} f_i \times g_h \ldots \right\} + \lambda \left\{ \ldots \sum_{i+h=l} f_i \times h_h \right\}$$

$$= f \times g + \lambda f \times h.$$

Here we have used the relation

$$(f_i \times (g_h + \lambda h_h))(x_1 \ldots x_{i+h}) = f_i(x_1 \ldots x_i)(g_h(x_{i+1} \ldots x_{i+h}) + \lambda h_h(x_{i+1} \ldots x_{i+h})),$$

which implies the distributive law for the components.

c) The associative law:

Since the distributive law holds, we need to show the associative law only for the components:

$$f_i \times (g_j \times h_l)(x_1 \ldots x_i \ x_{i+1} \ldots x_{i+j+l}) = f_i(x_1 \ldots x_i)(g_j \times h_l)(x_{i+1} \ldots x_{i+j+l}) =$$

$$= f_i(x_1 \ldots x_i) g_j(x_{i+1} \ldots x_{i+j}) h_l(x_{i+j+1} \ldots x_{i+j+l}) =$$

$$= (f_i \times g_j) \times h_l (x_1 \ldots x_{i+j+l}).$$

d) Double involution:

$$(f^*)^* = \{\ldots f_i \ldots\}^{**} = \{\ldots f_i^* \ldots\}^* = \{\ldots f_i^{**} \ldots\} = \{\ldots f_i \ldots\} = f.$$

e) Involution and sum:

$$(f + \lambda h)^* = \{\ldots (f_i + \lambda h_i)^* \ldots\} = \{\ldots f_i^* + \bar{\lambda} h_i^* \ldots\} = f^* + \bar{\lambda} h^*.$$

f) Product and involution:

The relation

$$(f_i \times g_j)^*(x_1 \ldots x_{i+j}) = \overline{(f_i \times g_j)(x_{i+j} \ldots x_1)} = \overline{f_i(x_{i+j} \ldots x_{j+1})}\, \overline{g_j(x_j \ldots x_1)} =$$

$$= g_j^*(x_1 \ldots x_j)\, f_i^*(x_{j+1} \ldots x_{j+i}) = (g_j^* \times f_i^*)(x_1 \ldots x_{j+i})$$

implies $(f \times g)^* = g^* \times f^*$ by linearity.

Next we want to list some elementary algebraic properties of the algebra $\mathscr{S}$.

I.2.4. *Lemma*:

 a) $\mathscr{S}$ contains an identity.

 b) It is free of divisors of zero, i.e. $f \neq 0$ and

$$g \times f = 0 \quad \text{resp.} \quad f \times g = 0 \quad \text{implies} \quad g = 0.$$

 c) The center of $\mathscr{S}$ coincides with the scalar multiples of the identity.

 d) The identity is the only idempotent different from zero.

 e) The only invertible elements of $\mathscr{S}$ are scalar multiples of the identity.

 f) The radical of $\mathscr{S}$ is trivial.

Proof:

 a) It is clear from the definition of the product that $1 = \{1, 0, 0, \ldots\}$ is an identity in $\mathscr{S}$.

 b) Since the second case can be reduced to the first case by involution, we have to deal only with the latter. Since $f \neq 0$, there exists a lowest index i such that $f_i \neq 0$ but $f_j = 0$ for $j < i$. From this it follows that $(g \times f)_i = g_0 f_i = 0$, hence $g_0 = 0$.

 Assume $g_j = 0$ for $j = 0, 1 \ldots h - 1$; then we get

$$0 = (g \times f)_{h+i}\,(x_1 \ldots x_{h+i}) = (g_h \times f_i)(x_1 \ldots x_{h+i}) = g_h(x_1 \ldots x_h) f_i(x_{h+1} \ldots x_{h+i})$$

and hence $g_h = 0$. By induction, this implies that $g = 0$.

 c) Let g be from the center of $\mathscr{S}$; then we have

$$g \times f = f \times g \quad \text{for all} \quad f \in \mathscr{S}.$$

Assume that g has a highest component $g_i \neq 0$ with $i > 0$; then we get for every $f \in \mathscr{S}_1$ the relation

$$g_i(x_1 \ldots x_i) f(x_{i+1}) = f(x_1) g_i(x_2 \ldots x_{i+1}).$$

Since the supports of all $f \in \mathscr{S}_i$ have empty intersection, it follows that $g_i = 0$. This contradicts the assumption. Hence

$$g = \{g_0, 0, 0 \ldots\} = g_0 \{1, 0, 0, ..\}^* = g_0 1.$$

 d) Let f be different from zero and $f \times f = f$; it then follows that $(f - 1) \times f = 0$ and hence by b)

$$f - 1 = 0.$$

 e) follows from the relation:

Index of the highest component of $g \times f =$ Index of the highest component of $g +$ Index of the highest component of f.

Hence $f \times g = 1$ implies

$$f = f_0 \cdot 1 \quad \text{and} \quad g = g_0 \cdot 1 \quad \text{with} \quad f_0 g_0 = 1.$$

f) The radical is defined as the set of all elements $g \in \mathscr{S}$ with the property that $(1 + f \times g)$ has an inverse for all $f \in \mathscr{S}$. Hence by e) the radical is defined as those elements g, for which

$$f \times g = \lambda 1 \quad \text{for all} \quad f \in \mathscr{S} \quad \text{and hence again by e)} \ g = 0.$$

Having introduced the field algebra $\mathscr{S}$, we proceed giving further definitions and notations.

I.2.5. *Definitions*:
  a) We denote by $\mathscr{S}_h$ the set of all hermitian elements in $\mathscr{S}$, i.e.,

$$\mathscr{S}_h = \{ f \in \mathscr{S}; f^* = f \}.$$

  b) We call an element $g \in \mathscr{S}$ positive if it can be represented in the form

$$g = \sum_i f_i^* \times f_i; \qquad f_i \in \mathscr{S},$$

where $\sum_i$ denotes either a finite or a convergent sum. We denote by $\mathscr{S}^+$ the set of all positive elements.

  c) We introduce in $\mathscr{S}_h$ a semi-ordering by the relation

$$f > g \Leftrightarrow f - g \in \mathscr{S}^+.$$

Before stating some simple consequences of the last definition, we give a preparatory lemma.

I.2.6. *Lemma*:
  Let $f \in \mathscr{S}^+$; then:
  a) The lowest index i_l and the greatest index i_g for which f_{i_l} resp. f_{i_g} is not zero are even numbers.
  b) If $i_l = 0$ then $f_0 > 0$.
Let i_l resp. $i_g = 2n$, $n \neq 0$; then we can write

$$(x_1 \ldots x_{2n}) = (X, Y) \quad \text{with} \quad X = (x_1 \ldots x_n), \qquad Y = (x_{n+1} \ldots x_{2n})$$

and define $X' = (x_n \ldots x_1)$. With these notations we get for f_{i_l} resp. f_{i_g} the following inequalities:

$$\sum_{i,j} f_{2n}(X_i' X_j) \bar{c}_i c_j \geq 0, \qquad c_i \in \mathbb{C},$$

which implies as a special case

$$| f_{2n}(X, Y) |^2 \leq f_{2n}(X, X') \cdot f_{2n}(Y', Y).$$

Proof:

Let $f \in \mathscr{S}$; then from the definition of the product it follows that the lowest and the highest index of $f^* \times f$ are even. If $2n$ is the lowest or highest index, then we have:

$$(f \times f)_{2n}(X', X) = \overline{f_n(X)} \cdot f_n(X) \geq 0.$$

Let now $g = \sum_{i=1}^{N} f_i^* \times f_i$ and l resp. g the lowest resp. highest index in the set $\{f_i\}$; then the positivity along the "diagonal" implies

$$i_l = 2l \quad \text{and} \quad i_g = 2g.$$

If now $g = \sum_{i=1}^{N} f_i \times f_i$ then $g^N = \sum_{i=1}^{\infty} f_i \times f_i$ is a bounded sequence and hence the set of indices i for which $g_i \neq 0$ is bounded independently of N. Since a bounded set of even numbers has a lowest resp. highest element which is also even, it follows together with the positivity along the "diagonal" that i_l and i_g are even. This proves a).

Let now $2n$ be the lowest resp. highest index of g. Then n is the lowest resp. highest index of the family f_i. From this we get

$$g_{2n} = \sum_{i} (f_i)_n^* \times (f_i)_n$$

and hence

$$\sum_{h,j} g_{2n}(X'_j X_h)\, \overline{c}_j c_h = \sum_{h,j} \sum_{i} \overline{(f_i)_n(X_j)}\, (f_i)_n(X_h) \cdot \overline{c}_j c_h =$$

$$= \sum_{i} \left\{ \sum_{j} \overline{c}_j \overline{(f_i)_n(X_j)} \right\} \cdot \left\{ \sum_{h} c_h (f_i)_n(X_h) \right\} \geq 0.$$

This proves the first inequality. Putting $i, h = 1, 2$ we get

$$|c_1|^2\, g_{2n}(X'_1 X_1) + 2\,\mathrm{Re}\, \{\overline{c}_1 c_2 g_{2n}(X'_1 X_2)\} + |c_2|^2\, g_{2n}(X'_2 X_2) \geq 0,$$

which gives the Cauchy inequality.

The properties of $\mathscr{S}_h$ and $\mathscr{S}^+$ are listed in the following

I.2.7. *Proposition*:
   a) $\mathscr{S}_h$ is a real vector space and we have

$$\mathscr{S} = \mathscr{S}_h + i\mathscr{S}_h.$$

   b) $\mathscr{S}^+$ is a convex cone with the property

$$\mathscr{S}^+ \cap -\mathscr{S}^+ = \{0\}.$$

   c) $\mathscr{S}_h = \mathscr{S}^+ - \mathscr{S}^+.$

Proof:
   a) From the equation $(f + \lambda g)^* = f^* + \overline{\lambda} g^*$ it follows immediately that $\mathscr{S}_h$ is a real vector space.

b) Let $g = \sum f_i^* \times f_i$ and $h = \sum k_j^* \times k_j$ be two positive elements. Then we have for $r_1, r_2 > 0$

$$r_1 g + r_2 h = \sum \left(\sqrt{r_1} f_i\right)^* \times \sqrt{r_1} f_i + \sum \left(\sqrt{r_2} k_j\right)^* \times \sqrt{r_2} k_j \in \mathscr{S}^+.$$

Hence $\mathscr{S}^+$ is a convex cone.

Next we have to show that $\mathscr{S}^+$ is a proper cone. Assume $g \in \mathscr{S}^+ \cap - \mathscr{S}^+$. Let g_{2n} be the lowest component of g. Then from Lemma I.2.6 we have $g_{2n}(X', X) \geqq 0$ since $g \in \mathscr{S}^+$, and also $g_{2n}(X', X) \leqq 0$ since $- g \in \mathscr{S}^+$. Hence we get $g_{2n}(X', X) = 0$ and from this, $g_{2n}(X, Y) = 0$ on the basis of Lemma I.2.6, b. This implies that g has no lowest nonvanishing component, and hence $g = 0$.

c) Let $f = f^* \in \mathscr{S}_h$. Then it follows that

$$f = \tfrac{1}{4}(1 + f) \times (1 + f) - \tfrac{1}{4}(1 - f) \times (1 - f) \in \mathscr{S}^+ - \mathscr{S}^+.$$

We will conclude this section with some further notations and some remarks.

I.2.8. *Definition*:
1) We denote by Aut $\mathscr{S}$ the set of all mappings

$$\alpha: \mathscr{S} \to \mathscr{S}$$

with the properties:
   a) α is a continuous linear mapping of the vector space $\mathscr{S}$ into $\mathscr{S}$.
   b) α is compatible with the algebraic structure of $\mathscr{S}$, i.e.,

$$\alpha(f_1 \times f_2) = (\alpha f_1) \times (\alpha f_2); \qquad f_1, f_2 \in \mathscr{S}.$$

   c) α commutes with the involution, i.e.

$$(\alpha f)^* = \alpha f^*$$

   d) α is a mapping of $\mathscr{S}$ onto $\mathscr{S}$ and its inverse is also continuous.
2) We denote by Hom $\mathscr{S}$ the set of mappings

$$\alpha: \mathscr{S} \to \mathscr{S}$$

fulfilling only the conditions 1 a), b) and c).

I.2.9. *Remarks and Examples*:
a) Let $\alpha \in$ Aut $\mathscr{S}$; then also $\alpha^{-1} \in$ Aut $\mathscr{S}$. Hence Aut $\mathscr{S}$ is a group. Hom $\mathscr{S}$ contains Aut $\mathscr{S}$ as a subset and it is only a semi-group.
b) Let $a \in \mathbb{R}^4$; then the map

$$\alpha_a f_i(x_1 \ldots x_i) = f_i(x_1 - a, \ldots x_i - a)$$

defines a continuous isomorphism from $\mathscr{S}_i$ onto $\mathscr{S}_i$ $[\![5]\!]$.

Hence α_a defines an isomorphism from $\mathscr{S}$ onto $\mathscr{S}$ if we put the restriction of α_a to $\mathscr{S}_0$ equal to the identity mapping.

It is easy to check that α_a preserves the algebraic structure and commutes with the involution, hence:

$$\alpha_a \in \operatorname{Aut} \underline{\mathscr{S}}.$$

c) Let $\mathscr{F}_i : \mathscr{S}_i \to \mathscr{S}_i$ be the Fourier operator. $\mathscr{F}_i$ is a continuous isomorphism [5]. Define

$$\mathscr{F}\{f_0, f_1, f_2 \ldots\} = \{f_0, \mathscr{F}_1 f_1, \mathscr{F}_2 f_2 \ldots\};$$

then $\mathscr{F}$ is a continuous linear isomorphism from $\underline{\mathscr{S}}$ onto $\underline{\mathscr{S}}$. Furthermore, $\mathscr{F}$ preserves the algebraic structure, i.e.

$$\mathscr{F}(f \times g) = (\mathscr{F}f) \times (\mathscr{F}g).$$

But because of the relation $\exp(+i(hx)) = \exp(-i(hx))$, it follows that $\mathscr{F}$ does not commute with the involution. We get instead the relation

$$(\mathscr{F}f)^* = \mathscr{F}^{-1} f^*$$

which implies that $\mathscr{F}$ is not a $*$-automorphism.

I.3. The Topology on the Field Algebra

It is the aim of this section to collect the main properties of $\underline{\mathscr{S}}$ as topological vector space. Furthermore, we want to study the continuity properties of the product and the involution. We suppose that the reader is familiar with the theory of topological vector spaces.

We start with the Schwartz spaces $\mathscr{S}_n$ [5]. Let $f \in \mathscr{S}_r$ and k be a multi-index

$$k = \{k_{i,v}; i = 1 \ldots r, v = 0, 1, 2, 3\};$$

then we define

$$D^k = \prod_{i,v} \frac{\partial^{k_{i,v}}}{(\partial x_i^v)^{k_{i,v}}},$$

$$|k| = \sum_{i,v} k_{i,v} \quad \text{and} \quad |x|^2 = \sum_{i,v} (x_i^v)^2.$$

The usual topology on $\mathscr{S}_r$ is given by the family of norms

$$q_{n,m}(f(x_1 \ldots x_r)) = \sup_x \max_{|k| \leq m} (1 + |x|)^n |D^k f(x_1 \ldots x_r)|.$$

These norms are not adapted to the algebraic structure of $\underline{\mathscr{S}}$. For this reason we prefer to work with an equivalent family of norms.

I.3.1. *Definition:*

On $\mathscr{S}_r$ we define a family of norms as follows:

$$P_{n,m}(f(x_1, \ldots, x_r)) = \sup_x \max_{|k_i| \leq m} \left| \left\{ \prod_{i=1}^r (1 + x_i)^n D_{x_i}^{k_i} \right\} f(x_1 \ldots x_r) \right|$$

$$|x_i|^2 = \sum_{v=0}^{3} (x_i^v)^2, \qquad k_i = \{k_i^v\}_{v=0}^{3}.$$

We get immediately:

I.3.2. *Lemma*:

The two families of norms $\{q_{n,m}\}$ and $\{P_{n,m}\}$ define the same topology on $\mathscr{S}_r$.

Proof:

From $\prod_i (1 + |x_i|)^n \leqq (1 + |x|)^{r \cdot n}$ it follows that

$$P_{n,m}(f) \leqq q_{rn,rm}(f).$$

From $\prod_i (1 + |x_i|)^n \geqq (1 + \sum_i |x_i|)^n \geqq (1 + |x|)^n$ it follows that

$$P_{n,m}(f) \geqq q_{n,m}(f).$$

This implies the coincidence of the two topologies.

Next we want to define the topology on $\mathscr{S}$.

I.3.3. *Definition*:

Let $\{n_i\}, \{m_i\}, \{c_i\}$ be three sequences of numbers subject to the conditions:

a) $n_0 = 0, n_i \in \mathbb{Z}^+, n_i \leqq n_{i+1}$.

b) $m_0 = 0, m_i \in \mathbb{Z}^+, m_i \leqq m_{i+1}$.

c) $c_i \in \mathbb{R}^+, c_0 = 1, c_i \leqq c_{i+1}, c_j \xrightarrow[j \to \infty]{} \infty$.

Let $f = \{f_0, f_1 \ldots\} \in \mathscr{S}$. Then we define a family of norms by

$$Q_{\{n_i\}, \{m_i\}, \{c_i\}}(f) = \sup_i c_i P_{n_i, m_i}(f_i).$$

The topology on $\mathscr{S}$ is the topology defined by this family of norms.

I.3.4. *Theorem*:

The vector space $\mathscr{S}$ furnished with the above defined topology has the following properties:

a) $\mathscr{S}$ is an L–F space.

b) $\mathscr{S}$ is nuclear and dual-nuclear, which implies
   i. $\mathscr{S}$ is complete;
   ii. $\mathscr{S}$ is an M-space;
   iii. $\mathscr{S}$ is reflexive;
   iv. $\mathscr{S}$ is barrelled.

c) $\mathscr{S}$ is a bornological space.

d) $\mathscr{S}$ is separable (but its topology is not separable).

Proof:

The properties b), c) and d) are well known for $\mathscr{S}_n$. Furthermore, $\mathscr{S}_n$ is an F-space.

Since $\mathscr{S}$ is the topological direct sum of a countable family of such spaces, the properties of Theorem I.3.4 follow (see, e.g., $[\![1]\!]$ and $[\![3]\!]$).

Now we can give the continuity properties of the algebraic relations.

I.3.5. *Theorem*:

a) The product $\mathscr{S} \times \mathscr{S} \to \mathscr{S}$ is separately continuous.

b) The product from $\left(\sum_{i=0}^{N} \oplus \mathscr{S}_i\right) \times \left(\sum_{i=0}^{M} \oplus \mathscr{S}_i\right)$ to $\sum_{i=0}^{M+N} \oplus \mathscr{S}_i$ is jointly continuous.

This implies in particular that the product is jointly continuous on bounded sets.

c) The involution $* : \mathscr{S} \to \mathscr{S}$ is continuous.

Proof:

Statement c) follows trivially from the definition of the norms $Q_{\{n_i\},\{m_i\},\{c_i\}}$. Hence it is sufficient to prove a) for fixed first factor. Since the linear operations are continuous, we need to prove a) at zero. Let $f \in \mathscr{S}$ be a fixed element and N be its highest nonvanishing component. Let $Q_{\{n_i\},\{m_i\},\{c_i\}}$ be a given norm.

Define

$$d_k = \sum_i P_{n_k,m_k}(f_i)$$

$$n'_k = n_{k+N}, \qquad m'_k = m_{k+N}, \qquad c'_k = d_k \cdot c_{k+N}.$$

We get:

$$d_k c_{k+N} P_{n_k+N,m_k+N}(g_k) \leqq Q_{\{n_i\},\{m_i\},\{c_i\}}(g) \equiv Q'(g).$$

This implies the relation:

$$Q_{\{n_i\},\{m_i\},\{c_i\}}(f \times g) = \max_h c_h \sum_{i+j=h} P_{n_h,m_h}(f_i) P_{n_h,m_h}(g_j) \leqq$$

$$\leqq \max_h c_h \sum_{i+j=h} P_{n_h,m_h}(f_i) \frac{1}{d_h c_h} \cdot d_h c_{j+N} P_{n_j+N,m_j+N}(g_j) \leqq$$

$$\leqq \max_h \frac{1}{d_h} \sum_{i \leqq h} P_{n_h,m_h}(f_i) Q'(g) \leqq Q'(g).$$

This implies statement a).

For proving b), note that there is a number d with $d^h \geqq c_h$ for $h = 1 \ldots N + M$. Hence we get

$$Q_{\{h_i\},\{m_i\},\{c_i\}}(f \times g) = \max_h c_h \sum_{i+j=h} P_{n_h,m_h}(f_i) P_{n_h,m_h}(g_j) \leqq$$

$$\leqq \max_h \sum_{i+j=h} d^i P_{n_{N+M},m_{N+M}}(f_i) P_{n_{N+M},m_{N+M}}(g_j) \leqq$$

$$\leqq \max_h Q'(f) Q'(g)(h+1) \leqq (N+M+1) Q'(f) Q'(g),$$

where $Q' = Q_{\{n_i\},\{m_i\},\{c_i\}}$ with $n'_i = n_{N+M}, m'_i = n_{N+M}, c'_i = d^i$.

This implies that the product is jointly continuous in $\left(\sum_{i=0}^{N} \oplus \mathscr{S}_i\right) \times \left(\sum_{i=0}^{M} \oplus \mathscr{S}_i\right)$.
Since every bounded set is contained in $\sum_{i=0}^{N} \oplus \mathscr{S}_i$ for some finite N, it follows that the product is jointly continuous on bounded sets.

From this theorem we get the following

I.3.6. *Corollary*:

a) The closure of any left or right or two-sided ideal is again a left or right or two-sided ideal.

b) The closure of any subalgebra is again a subalgebra.

c) If a subalgebra or two-sided ideal is invariant under involution, then this remains true for its closure.

Proof:

Let $\mathscr{L}$ be a left-ideal, and $\bar{\mathscr{L}}$ its closure. Let l be an element of $\bar{\mathscr{L}}$ and $f \in \mathscr{S}$. We have to show that $f \times l \in \bar{\mathscr{L}}$. Let $\mathscr{U}$ be a neighborhood of $f \times l$. Then there exists a neighborhood $\mathscr{U}_1$ of l such that $f \times \mathscr{U}_1 \subset \mathscr{U}$. Now $\mathscr{U}_1 \cap \mathscr{L} \neq \varnothing$, hence $\mathscr{U} \cap \mathscr{L} \neq \varnothing$ which implies $f \times l \in \bar{\mathscr{L}}$. Since the same arguments are true for a fixed second factor, we get statement a). Let now $\mathscr{A}$ be a subalgebra of $\mathscr{S}$. Then we get with the same arguments first $\mathscr{A} \times \bar{\mathscr{A}} \subset \bar{\mathscr{A}}$ and repeating the arguments again we find $\bar{\mathscr{A}} \times \bar{\mathscr{A}} \subset \bar{\mathscr{A}}$. Hence statement b) is proved. Statement c) is trivial, since the involution is continuous.

I.4. Gelfand–Segal Construction

It is the aim of this section to link the algebra $\mathscr{S}$ to fields fulfilling Axiom 1 of Section I.1. We start again with some definitions.

I.4.1. *Definitions*:

a) We denote by $\mathscr{S}'$ the space of continuous linear functionals, i.e. the set of linear mappings

$$T : \mathscr{S} \to \mathbb{C},$$

which are continuous.

b) We denote by $\mathscr{S}'_h$ the set of real elements in $\mathscr{S}'$, i.e.,

$$\mathscr{S}'_h = \{T \in \mathscr{S}'; (T, f) = \overline{(T, f^*)}\}.$$

c) We denote by $\mathscr{S}'^+$ the set of positive elements of $\mathscr{S}$, i.e.,

$$\mathscr{S}'^+ = \{T \in \mathscr{S}'; (T, p) \geqq 0 \,\forall\, p \in \mathscr{S}^+\}.$$

d) We denote by $E(\mathscr{S})$ the set of states defined by

$$E(\mathscr{S}) = \{T \in \mathscr{S}'^{+}\,;\ (T, 1) = 1\}.$$

I.4.2. *Lemma*:

a) $\mathscr{S}'_h$ is a real linear vector space and

$$\mathscr{S}' = \mathscr{S}'_h + i\mathscr{S}'_h\,.$$

b) i. $\mathscr{S}'^{+} \subset \mathscr{S}'_h\,.$

ii. Let $T \in \mathscr{S}'^{+}$ and $f, g \in \mathscr{S}$. Then the Cauchy–Schwartz inequality $|(T, f \times g)|^2 \le (T, f \times f)(T, g \times g)$ holds.

iii. $(T, 1) = 0 \Rightarrow T = 0$.

c) $\mathscr{S}'^{+}$ is a cone with $E(\mathscr{S})$ as a convex base.

Remark:

Let C be a cone. A subset $B \subset \mathbb{C}$ is called base of C if for every $x \in \mathbb{C}$ with $x \ne 0$ the ray $\{\rho x;\ \rho \in \mathbb{R}^{+}\}$ has a nonempty intersection with B containing only one point.

Proof:

a) For every $T \in \mathscr{S}'$ one defines an element $T^{*} \in \mathscr{S}'$ by the relation

$$(T^{*}, f) = \overline{(T, f^{*})}.$$

Since $f \to f^{*}$ and $\lambda \to \bar{\lambda}$ are antilinear, it follows that T^{*} is linear. The continuity of the involution, Theorem I.3.5, implies the continuity of T^{*}. From the definition of T^{*} it follows immediately that $T^{**} = T$.

Hence we get $T = \frac{1}{2}(T + T^{*}) + i(T - T^{*})/2i$. This implies a).

b) i. From

$$0 \le (T, (1 + f)^{*} \times (1 + f)) = (T, 1) + (T, f^{*}) + (T, f) + (T, f^{*} \times f)$$

it follows that $(T, f^{*}) + (T, f)$ is real.

Hence we get $(T, f^{*}) = (T, f)$, which implies $T \in \mathscr{S}'_h\,.$

ii. The Cauchy inequality follows from the fact that

$$0 \le (T, (\lambda f + \mu g)^{*} \times (\lambda f + \mu g)) = |\lambda|^2 (T, f^{*} \times f) + \bar{\lambda}\mu (T, f^{*} \times g) +$$

$$+ \lambda\bar{\mu} (T, g^{*} \times f) + |\mu|^2 T(g^{*} \times g),$$

which means that

$$\begin{pmatrix} (T, f^{*} \times f), & (T, f^{*} \times g) \\ (T, g^{*} \times f), & (T, g^{*} \times g) \end{pmatrix}$$

is a nonnegative hermitian matrix, which must therefore have a nonnegative determinant. This gives b) ii.

iii. follows from ii because $(T, 1) = 0$ implies

$$|(T,f)|^2 = |(T, 1 \times f)|^2 \leq (T, 1)(T,f^* \times f) = 0$$

and hence $T = 0$.

c) Let $T_1, T_2 \in \mathscr{S}'^+$; $\rho_1, \rho_2 \geq 0$. Then

$$(\rho_1 T_1 + \rho_2 T_2, f^* \times f) = \rho_1 (T_1, f^* \times f) + \rho_2 (T_2, f^* \times f) \geq 0.$$

Hence $\mathscr{S}'^+$ is a convex cone. Let $T \in \mathscr{S}'^+$ and $T \neq 0$; then $(T, 1) > 0$ and hence $(T, 1)^{-1} T \in E(\mathscr{S})$, which means that $E(\mathscr{S})$ is a base. The convexity of $E(\mathscr{S})$ is trivial.

I.4.3. *Definition*:

Let T be a state, i.e. $T \in E(\mathscr{S})$. We define the left-kernel $L(T)$ of T as follows:

$$L(T) = \{f \in \mathscr{S}; \ (T, f^* \times f) = 0\}.$$

I.4.4. *Lemma*:

Let T be a state. Then its left-kernel $L(T)$ is a closed left-ideal.

Proof:

The proof is based on the Cauchy–Schwarz inequality, Lemma 1.4.2.

This implies for $f \in L(T)$ and g arbitrary

$$(T, g \times f) = (T, f^* \times g) = 0.$$

Hence we get for $f, g \in L(T)$

$$(T, (f + \lambda g)^* \times (f + \lambda g)) =$$
$$= (T, f^* \times f) + \bar{\lambda}(T, g^* \times f) + \lambda(T, f^* \times g) + |\lambda|^2 (T, g^* \times g) = 0,$$

which implies that $L(T)$ is a vector space.

Let now $f \in L(T)$ and $g \in \mathscr{S}$. Then

$$(T, (g \times f)^* \times (g \times f)) = (T, f^* \times g^* \times g \times f) = (T, \{f^* \times g^* \times g\} \times f) = 0,$$

and hence $L(T)$ is a left-ideal. Let now g_α be a Cauchy net in $L(T)$ with limit g.

Then we get by Theorem I.3.5

$$(T, g^* \times g) = \lim_\alpha (T, g^* \times g_\alpha) = \lim_\alpha 0 = 0,$$

which means that $L(T)$ is closed.

After these preparations we are able to give the connection between fields and the algebra $\mathscr{S}$.

I.4.5. *Theorem*:

a) Every state $T \in E(\mathscr{S})$ defines in a canonical manner a representation A_T of $\mathscr{S}$ in a Hilbert space $\mathscr{H}_T$ such that the restriction of A_T to $\mathscr{S}_1$ satisfies Axiom 1 of Section I.1 (Gelfand–Segal construction).

b) Let $A(f), f \in \mathscr{S}_1$ be a field defined on a dense domain D in a Hilbert space $\mathscr{H}$

satisfying Axiom 1. Then every vector $\psi \in D$ with $\|\psi\| = 1$ defines a state $T_\psi \in E(\mathscr{S})$ by continuous linear extension of the equation

$$(T_n, f_1 \times f_2 \times \ldots \times f_n) = (\psi, A(f_1) A(f_2) \ldots A(f_n) \psi), \qquad f_i \in \mathscr{S}_1.$$

c) Let $A(f), f \in \mathscr{S}_1$ be a field fulfilling Axiom 1 and $\psi \in D$ with $\|\psi\| = 1$. Define $D_\psi = $ linear span of the vectors $\{\psi, A(f)\psi, A(f_1) A(f_2) \psi, \ldots\}$, $\mathscr{H}_\psi = $ closure of D_ψ and $A_\psi(f) = $ restriction of $A(f)$ to D_ψ. Then $\{A_\psi(f), D_\psi, \mathscr{H}_\psi\}$ is unitarily equivalent to the field $A_{T_\psi}(f)$ constructed by the state T_ψ.

Proof:

a) Let $T \in E(\mathscr{S})$ and $L(T)$ its left-kernel. Assume $f, g \in \mathscr{S}$ and $l_1, l_2 \in L(T)$. Then it follows that

$$(T, (f + l_1)^* \times (g + l_2)) = (T, f^* \times g) + (T, l_1^* \times g) + (T, (f + l_1)^* \times l_2) = (T, f^* \times g).$$

Furthermore, $(T, f \times f) = 0$ implies $f \in L(T)$.

Both equations together imply that T defines a nondegenerate positive bilinear form on $\mathscr{S}/L(T)$ by the relation:

$$[f], [g] \in \mathscr{S}/L(T)$$

$$([f], [g]) = (T, f^* \times g).$$

We define $\mathscr{H}_T$ to be the completion of the pre-Hilbert space $\mathscr{S}/L(T)$. On $\mathscr{S}/L(T)$ we define a representation A_T of $\mathscr{S}$ by the equation

$$A(f)[g] = [f \times g].$$

Since $L(T)$ is a left-ideal, it follows that for $g_1 - g_2 \in L(T)$ also $f \times g_1 - f \times g_2 \in L(T)$. This means the definition of $A(f)$ does not depend on the choice of the representative in the class $[g]$. This representation has the following properties:

 i. $\mathscr{S}/L(T)$ is dense in $\mathscr{H}_T$;
 ii. $A(f)$ is linear in f;
 iii. $A(f) \mathscr{S}/L(T) \subset \mathscr{S}/L(T)$;
 iv. for $[g], [h] \in \mathscr{S}/L(T)$ we get the relation

$$([g]), A(f)[h]) = ([g], [f \times h]) = (T, g^* \times f \times h) =$$

$$= (T, (f^* \times g)^* \times h) = ([f^* \times g], [h]) = (A(f^*)[g], [h]);$$

 v. $A(f) A(g)[h] = A(f \times g)[h]$;
 vi. the equation $([g], A(f)[h]) = (T, g^* \times f \times h)$ implies together with Theorem I.3.5 that the matrix elements $([g], A(f)[h])$ for fixed g and h are continuous linear functionals on $\mathscr{S}$. Hence statement a) is proved.

b) Let now $\{A(f), \mathscr{H}, D\}$ be a field fulfilling Axiom 1 and $\psi \in D$ with $\|\psi\| = 1$. Let $f_1, \ldots f_n \in \mathscr{S}_1$. Then, by assumption, the expression

$$(\psi, A(f_1)\, A(f_2)\dots A(f_n)\,\psi)$$

is a separately continuous linear n-form on

$$\underbrace{\mathscr{S}_1 \times \mathscr{S}_1 \times \dots \times \mathscr{S}_1}_{n \text{ factors}}.$$

By the nuclear theorem [9] we have the joint continuity and hence there exists a unique distribution $T_n \in \mathscr{S}'_n$ with

$$(T_n, f_1 \times f_2 \times \dots \times f_n) = (\psi, A(f_1)\, A(f_2)\dots A(f_n)\,\psi).$$

Define T to be the sequence

$$T_\psi = \{1, T_1, T_2, \dots, T_n \dots\}$$

which is a continuous linear functional on $\mathscr{S}$. It remains to show that T_ψ is a state. Since $(T_\psi, 1) = 1$, we need to show that T_ψ is positive.

Let $f = \{f_0, f_1, f_2, \dots f_n, 0, 0, 0, \dots\} \in \mathscr{S}$.

Since $\mathscr{S}_i$ has a separable topology, there exists a sequence

$$f_i^\beta = \sum_\alpha g_1^{\beta,\alpha} \times g_2^{\beta,\alpha} \times \dots g_i^{\beta,\alpha} \qquad g_i^{\beta,\alpha} \in \mathscr{S}_1$$

such that $\lim\limits_{\beta \to \infty} f_i^\beta = f_i$ holds.

Let now $f_\beta = \{f_0, f_1, f_2^\beta, \dots f_n^\beta, 0, 0, \dots\}$. Then it follows from Theorem I.3.5 that

$$\lim_{\beta \to \infty} f_\beta^* = f^* \quad \text{and} \quad \lim_{\beta \to \infty} f_\beta^* \times f_\beta = f^* \times f$$

also holds.

By the construction of T_ψ it follows that

$$(T_\psi, f_\beta^* \times f_\beta) \geqq 0$$

and hence

$$(T_\psi, f^* \times f) = \lim_{\beta \to \infty} (T_\psi, f_\beta^* \times f_\beta) \geqq 0.$$

This proves statement b).

For proving the last statement we establish a mapping V by the equation

$$V \sum_i A(f_1^i)\dots A(f_{n_i}^i)\,\psi = \left[\sum_i f_1^i \times f_2^i \dots f_{n_i}^i\right]$$

and we get

$$\left\| \left[\sum_i f_1^i \times \dots \times f_{n_i}^i\right] \right\|^2 = \left(\left[\sum_i f_1^i \times \dots \times f_{n_i}^i\right], \left[\sum_i f_1^i \times \dots \times f_{n_i}^i\right]\right)$$

$$= \left(T_\psi, \left\{\sum_i f_1^i \times \dots \times f_{n_i}^i\right\}^* \times \left\{\sum_i f_1^i \times \dots \times f_{n_i}^i\right\}\right)$$

$$= \sum_{i,j} (\psi, A(f_{n_i}^i)^* \dots A(f_1^i)^* A(f_1^j)\dots A(f_{n_j}^j)\dots A(f_{n_j}^j)\,\psi)$$

$$= \left(\sum_i A(f_1^i)\dots A(f_{n_i}^i)\,\psi, \sum_j A(f_1^j)\dots A(f_{n_j}^j)\,\psi\right)$$

$$= \left\| \sum_j A(f_1^j)\dots A(f_{n_j}^j)\,\psi \right\|^2.$$

This implies that V is an isometric mapping from D onto a dense subset of $\mathcal{S}/L(T_\psi)$ and has therefore an extension to a unitary operator. Furthermore we get:

$$VA(f)V^{-1}[f_1 \times f_2 \times \ldots \times f_n] = VA(f)A(f_1)\ldots A(f_n)\psi =$$

$$= [f \times f_1 \times \ldots \times f_n] = A_{T_\psi}(f)[f_1 \times f_2 \times \ldots f_n].$$

This means that V is an intertwining operator. This proves statement c).

I.4.6. *Remark*:

Since, in the general case, the operators $A(f)$ are unbounded, it follows that Axiom 1 is more general than the concept of states on $\mathcal{S}$. However, this difference is not important for physics as long as one is only interested in asymptotic fields and the S-matrix.

I.5. Wightman Functionals

The aim of this section is to show that Axioms 2–5 can also be translated into the language of positive functionals on $\mathcal{S}$.

I.5.1. *Definition*:

Let $a \in \mathbb{R}^4$ and $\alpha_a \in \text{Aut}\,\mathcal{S}$ be the translation introduced in example I.2.9.b. A state $T \in E(\mathcal{S})$ is called translation-invariant if the relation

$$(T, \alpha_a f) = (T, f)$$

holds for all $a \in \mathbb{R}^4$ and all $f \in \mathcal{S}$.

I.5.2. *Theorem*:

a) Let $T \in E(\mathcal{S})$ be a translation-invariant state. Then the canonical representation $A_T(f)$ satisfies Axioms 1–3.

b) Assume on the other hand that the system $\{A(f), D, \Omega \in D\}$ satisfies the Axioms 1–3. Then the state T_Ω defined by

$$(T_\Omega, f) = (\Omega, A(f)\Omega)$$

is a translation-invariant state.

Proof:

a) We define on $\mathcal{S}/L(T)$ a mapping

$$\mathcal{U}(a)[f] = [\alpha_a f].$$

Since T is translation-invariant, it follows that

$$(\mathcal{U}(a)[f], \mathcal{U}(a)[g]) = (T, (\alpha_a f)^* \times \alpha_a g) = (T, \alpha_a f^* \times g) = (T, f \times g) = ([f], [g]).$$

This implies that $\mathscr{U}(a)$ is an isometric mapping. Since in particular $\alpha_a L(T) = L(T)$, it follows that $\mathscr{U}(a)$ is a linear operator. Since α_a is an automorphism, it follows that $\mathscr{U}(a)\mathscr{S}/L(T) = \mathscr{S}/L(T)$, and hence $\mathscr{U}(a)$ can be extended to a unitary operator on all of $\mathscr{H}_T$.

From

$$\mathscr{U}(a)\,\mathscr{U}(b)\,[f] = \mathscr{U}(a)\,[\alpha_b f] = [\alpha_a \alpha_b f] = [\alpha_{a+b} f] = \mathscr{U}(a+b)\,[f]$$

it follows that $\mathscr{U}(a)$ is a representation of the translation group.

Since α_a and T are both continuous, it follows that

$$([g], \mathscr{U}(a)\,[f]) = (T, g \times \alpha_a f) \underset{a \to 0}{\longrightarrow} (T, g^* \times f) = ([g],[f]),$$

which implies that $\mathscr{U}(a)$ is a continuous representation of the translation group.

The equation

$$\mathscr{U}(a)\,A_T(f)\,\mathscr{U}(-a)\,[g] = \mathscr{U}(a)\,A_T(f)\,[\alpha_{-a} g] = \mathscr{U}(a)\,[f \times \alpha_{-a} g] =$$

$$= [\alpha_a(f \times \alpha_{-a} g)] = [(\alpha_a f) \times g] = A_T(\alpha_a f)\,[g]$$

shows that $\{A_T(f), \mathscr{U}(a), \mathscr{S}/L(T)\}$ fulfills Axiom 2.

If we identify Ω with $[1]$, we get

$$\mathscr{U}(a)\,\Omega = \mathscr{U}(a)\,[1] = [\alpha_a 1] = [1] = \Omega.$$

Hence the cyclic vector is invariant and thus Axiom 3 is also fulfilled.

b) Assume now $\{A(f), \mathscr{U}(a), D, \Omega \in D\}$ fulfills Axioms 1–3. Define

$$(T_\Omega, f) = (\Omega, A(f)\,\Omega).$$

Then we get:

$$(T_\Omega, \alpha_a f) = (\Omega, A(\alpha_a f)\,\Omega) = (\Omega, \mathscr{U}(a)\,A(f)\,\mathscr{U}(-a)\,\Omega) = (\Omega, A(f)\,\Omega) = (T_\Omega, f).$$

This implies that T_Ω is a translation-invariant state.

Next we want to consider the spectral condition. We start with a remark. Let $f \in \mathscr{S}_1$; then we define

$$(f(a), \alpha_a g) \underset{\text{def}}{=} \int d^4 a\, f(a)\,\alpha_a g \qquad g \in \mathscr{S}.$$

The element $(f(a), \alpha_a g)$ again belongs to $\mathscr{S}$.

I.5.3. *Definition*:

Let $V^+ \subset \mathbb{R}^4$ be the forward light cone and denote

$$\mathscr{S}_1(CV^+) = \{f(x) \in \mathscr{S}_1; f(x) = 0 \quad \text{for} \quad x \in V^+\}.$$

We denote by $\mathscr{F}$ the Fourier transformation.

Define Sp to be the smallest closed left-ideal containing the set of elements

$$\{(F(a), \alpha_a f); \quad f \in \mathcal{S} \quad \text{and} \quad f_0 = 0, \quad F(a) \in \mathcal{F}\mathcal{S}_1(CV^+)\}.$$

With these notations we are able to formulate the spectral condition in terms of functionals.

I.5.4. *Theorem*:

a) Let $T \in E(\mathcal{S})$ be a translation-invariant state annihilating the left-ideal Sp; i.e., $(T, Sp) = 0$. Then the representation $\{A_T(f), \mathcal{S}/L(T), \mathcal{U}(a), \Omega\}$ defined by T fulfills Axioms 1–4.

b) Conversely, let $\{A(f), D, \mathcal{U}(a), \Omega\}$ be a field satisfying Axioms 1–4. Then the state T_Ω is translation-invariant and annihilates the left-ideal Sp.

Proof:

a) Since T is a translation-invariant state, there exists, by Theorem 2 of this section, a continuous unitary representation of the translation group defined by

$$\mathcal{U}(a)[g] = [\alpha_a g].$$

Let now $f \in \mathcal{F}\mathcal{S}_1(CV^+)$. Then we get

$$\int d^4 a f(a)\, \mathcal{U}(a)[g] = \int d^4 a f(a)[\alpha_a g] = [(f(a), \alpha_a g)] = 0,$$

since $(f(a), \alpha_a g) \in Sp \subset L(T)$. Since g was arbitrary, it follows that

$$\int d^4 a f(a)\, \mathcal{U}(a) = 0$$

and hence support $\mathcal{F}^{-1}\mathcal{U}(a) \subset V^+$, which is equivalent to the spectral condition.

b) Assume now that $\{A(f), \mathcal{U}(a), D, \Omega\}$ satisfies Axioms 1–4. Then we have for every $f \in \mathcal{F}\mathcal{S}_1(CV^+)$,

$$\int f(a)\, \mathcal{U}(a)\, d^4 a = 0.$$

This implies for every $g \in \mathcal{S}$,

$$0 = \int d^4 a f(a)\, \mathcal{U}(a)\, A(g)\, \Omega = \int d^4 a f(a)\, A(\alpha_a g)\, \Omega = A((f(a), \alpha_a g))\, \Omega = 0.$$

This means $(f(a), \alpha_a g) \in L(T_\Omega)$, and since $L(T_\Omega)$ is a closed left-ideal, it follows that $Sp \subset L(T_\Omega)$.

For the translation of the locality axiom into the language of functionals we have to introduce some notations.

I.5.5. *Definition*:

a) Let $f_n \in \mathscr{S}_n$ and $g_m \in \mathscr{S}_m$. We say that f_n and g_m have spacelike separated supports if

$$f(x_1 \ldots x_n)\, g(y_1 \ldots y_m) = 0$$

for

$$(x_i - y_j)^2 \geqq 0, \qquad i = 1 \ldots n, \quad j = 1 \ldots m, \qquad x^2 = (x^0)^2 - (\boldsymbol{x})^2.$$

b) Let $f, g \in \mathscr{S}$. We say that f and g have spacelike separated supports if $f_0 = g_0 = 0$ and if f_i, g_j have spacelike separated supports.

c) We denote by I_c the smallest closed two-sided ideal in $\mathscr{S}$ containing all elements of the form $f \times g - g \times f$; f and g have spacelike separated supports.

I.5.6. *Theorem*:

a) Let $T \in E(\mathscr{S})$ be a state annihilating I_c. Then the representation

$$\{A_T(f), \mathscr{S}/L(T)\}$$

satisfies the locality Axiom 5.

b) Conversely, let $\{A(f), D\}$ be a field satisfying Axiom 5. Then every vector state $T_\psi, \psi \in D$ annihilates the ideal I_c.

Proof:

a) Let $f, g \in \mathscr{S}_1$ with spacelike separated supports. Then we get

$$\| [A_T(f), A_T(g)] [h] \|^2 = (T, h \times (g \times f - f \times g) \times (g \times f - f \times g) \times h) = 0,$$

since $g \times f - f \times g \in I_c$. Hence $A_T(f)$ fulfills Axiom 5.

b) On the other hand, let now $\{A(f), D\}$ be a field fulfilling Axioms 1 and 5. Let $\psi, \phi \in D$ and $f, g \in \mathscr{S}$ with spacelike separated supports. Then we have $(\psi, A(f) A(g) \phi) = (\psi, A(g) A(f) \phi)$. This implies in particular

$$(T_\psi, h_1 \times (f \times g - g \times f) \times h_2) = (A(h_1^*)\psi, (A(f) A(g) - A(g) A(f)) A(h_2)\psi) = 0.$$

Thus we get $(T_\psi, I_c) = 0$.

These results together imply that we can formulate Wightman's axioms in terms of functionals over the algebra $\mathscr{S}$. This leads us to the following notation:

I.5.7. *Definition*:

a) A functional W over the algebra $\mathscr{S}$ is called a Wightman state if it fulfills the following conditions:

    i. $W \in E(\mathscr{S})$;

    ii. W is a translation-invariant state;

    iii. W annihilates the left-ideal Sp;

    iv. W annihilates the two-sided ideal I_c.

b) We denote by $\mathscr{W}$ the set of all Wightman states.

I.5.8. *Theorem*:

$\mathscr{W}$ is a convex closed subset of $E(\mathscr{S})$.

Proof:

The set of states fulfilling one of the conditions ii, iii, iv is obviously convex and closed. Hence $\mathscr{W}$ is convex and closed.

II. EXAMPLES AND SUPPLEMENTS

In this chapter we want to acquire some familiarity with the algebra $\mathscr{S}$. This will be done by studying some examples and developing some more notations.

II.1. Extension of the Algebraic Structure to the Dual Space

II.1.1. *Definition*:

a) Denote by $\mathscr{S}'_0 = \mathbb{C}$ the space of complex numbers and by $\mathscr{S}'_n$ the dual space of $\mathscr{S}_n$, i.e. the space of tempered distributions over $\mathbb{R}^{4n}$. We denote by $\underline{\mathscr{S}}'$ the space of sequences

$$F = \{F_0, F_1, \ldots F_n \ldots\} \qquad F_i \in \mathscr{S}'_i.$$

If we furnish $\underline{\mathscr{S}}'$ with the product topology $\underline{\mathscr{S}}' = \prod_{n=0}^{\infty} \mathscr{S}'_n$, then $\underline{\mathscr{S}}'$ can be identified as the dual space of $\underline{\mathscr{S}}$ by means of the relation

$$(F,f) = \sum_i (F_i, f_i), \qquad F \in \underline{\mathscr{S}}', \qquad f \in \underline{\mathscr{S}}.$$

b) Let $F \in \mathscr{S}'(\mathbb{R}^n)$, $G \in \mathscr{S}'(\mathbb{R}^m)$. Then the direct product $F \times G$ is well-defined as a distribution $F \times G \in \mathscr{S}'(\mathbb{R}^{n+m})$. This means that we can define a product in $\underline{\mathscr{S}}'$ by analogy with Definition I.2.2.

Defining an involution in $\underline{\mathscr{S}}'$ by the equation $(F^*, f) = (F, f^*)$, we get an extension of the algebraic relations from $\underline{\mathscr{S}}$ to $\underline{\mathscr{S}}'$. Since we have used only the algebraic relations in the proof of Lemma I.2.3., we can use these results unchanged for $\underline{\mathscr{S}}'$ and get:

II.1.2. *Lemma*:

$\underline{\mathscr{S}}'$ is a $*$-algebra. If we identify $\underline{\mathscr{S}}$ in the natural manner with a subset of $\underline{\mathscr{S}}'$ then $\underline{\mathscr{S}}$ is sub-$*$-algebra of $\underline{\mathscr{S}}'$.

The main difference between $\underline{\mathscr{S}}$ and $\underline{\mathscr{S}}'$ is that in $\underline{\mathscr{S}}$ we can have only polynomials, while in $\underline{\mathscr{S}}'$ also power series can occur.

II.1.3. *Lemma*:

Let $p(x) = \sum a_n x^n$ be a formal power series and $F \in \underline{\mathscr{S}}'$ with $F_0 = 0$. Then

$$P(F)_\times = \sum a_n (F^n)_\times$$

exists, with

$$(F^n)_\times = \underbrace{F \times F \times \ldots \times F}_{n \text{ times}}.$$

Proof:

Since the topology in $\mathscr{S}'$ is the topology of componentwise convergence, it suffices to show that the components converge. Since by assumption $F_0 = 0$ it follows that

$$((F^n)_\times)_i = 0 \quad \text{for} \quad i = 0, 1, \ldots, \qquad n - 1.$$

Hence we get

$$\left(\sum_{n=0}^{\infty} a_n (F^n)_\times \right)_i = \left(\sum_{n=0}^{i} a_n (F^n)_\times \right)_i.$$

This means that the n-th component of the sum is only a finite sum and thus convergent.

If, on the contrary, $F_0 \neq 0$ then not every power series is convergent. We get instead the following result:

II.1.4. *Theorem*:

Let $\sum a_n x^n$ be a power series. In order that the series $\sum a_n (F^n)_\times$ be convergent for all $F \in \mathscr{S}'$ with $|F_0| \leq a$, it is necessary and sufficient that $\sum a_n x^n$ have a radius of convergence R with $R > a$.

Proof:

Put $F = F_0 + F'$ where F' has a vanishing zeroth component. Note that F_0 is in the center of $\mathscr{S}'$. Thus we get (suppressing the cross of the product)

$$\sum_n a_n (F^n)_\times = \sum_n a_n (F_0 + F')^n =$$

$$= \sum_n a_n \sum_k \binom{n}{k} F_0^{n-k} F'^k =$$

$$= \sum_k \left\{ \sum_{n \geq k} \binom{n}{k} a_n F_0^{n-k} \right\} F'^k.$$

According to Lemma II.1.3, the absolute convergence of the sums

$$\sum_{n \geq k} \binom{n}{k} a_n F_0^{n-k}, \qquad k = 0, 1, \ldots$$

is necessary and sufficient to guarantee the statement of Theorem II.1.4.

From this we get

$$\lim_{j \to \infty} \left\{ \frac{(k+j)!}{k!\, j!} |a_{k+j}| \right\}^{-(1/j)} > a \qquad k = 0, 1, \ldots$$

and hence for $k = 0$, the relation $R > a$.

Since f and $(1/k!)(\partial^k/\partial z^k)f(z)$ have the same radius of convergence, it follows that

$$\lim_{j \to \infty} \left\{ \frac{(k+j)!}{k!\,j!} |a_{k+j}| \right\}^{-1/j} = R, \quad \text{independent of } k.$$

Next we want to introduce the space of multiplicative operators on $\mathscr{S}$. Let O_n^M be the space of slowly increasing C^∞ functions over $\mathbb{R}^{4n}$. For $f \in \mathscr{S}_n$ and $F \in O_n^M$ the product $F \cdot f$ is defined by the equation

$$F \cdot f = F(x_1, x_2, \ldots x_n) \cdot f(x_1, x_2, \ldots x_n).$$

This product defines again an element of $\mathscr{S}_n$ [5].

II.1.5. *Definition*:

We denote by O^M the space of sequences

$$F = \{F_0, F_1, \ldots F_n, \ldots\}$$

with $F_0 \in \mathbb{C}$ and $F_i \in O_i^M$.

Furthermore, we define a "product" from $O^M \times \mathscr{S}$ to $\mathscr{S}$ by the relation

$$F \cdot f = \{F_0 f_0, F_1 f_1, \ldots F_n f_n, \ldots\}.$$

It is not necessary to introduce a topology in O^M. The easiest way to do this would be to identify O^M with the topological direct product of the spaces $\{O_n^M\}_{n=0}^\infty$. From the fact that for every fixed F_n the product $f_n \to F_n \cdot f_n$ is a continuous mapping from $\mathscr{S}_n$ to $\mathscr{S}_n$, it follows that for every fixed $F \in O^M$ the product $\{F, \mathscr{S}\} \to \mathscr{S}$ is also continuous in $\mathscr{S}$.

Next we ask for which elements $F \in O^M$ the mapping $f \to F \cdot f$, $f \in \mathscr{S}$ is also compatible with the algebraic structure of $\mathscr{S}$.

II.1.6. *Theorem*:

Let $F \in O^M$, with $F \neq 0$. Then the mapping

$$f \to F \cdot f$$

defines a homomorphism $\alpha_F \in \text{Hom}(\mathscr{S})$ exactly if F is of the form

$$F = \frac{1}{1 - F_1}\Big|_\times = \sum_0^\infty (F_1^n)_\times$$

with $F_1 \in O_1^M$ and $\bar{F}_1 = F_1$.

Proof:

From $\alpha_F 1 \in$ center of $\mathscr{S}$ it follows, with $\alpha_F \neq 0$, that $\alpha_F 1 = r \neq 0$. From $r = \alpha_F 1 = \alpha_F(1 \times 1) = (\alpha_F 1) \times (\alpha_F 1) = r^2$ follows $r = 1$. From $\alpha_F f_1 = F_1 \cdot f_1$ follows

$$(\alpha_F f_1)^* = \alpha_F f_1^* = F \cdot \bar{f}_1 = \overline{F \cdot f_1} = \overline{F} f_1.$$

Since this holds for all $f_1 \in \mathscr{S}_1$, we get $\overline{F}_1 = F_1$. Let now $f_i \in \mathscr{S}_1$. Then we get

$$\alpha_F f_1 \times f_2 \times \ldots \times f_n = F_n \cdot f_1 \times f_2 \times \ldots \times f_n =$$

$$= (\alpha_F f_1) \times (\alpha_F f_2) \times \ldots \times (\alpha_F f_n) = (F_1^n)_\times \cdot f_1 \times f_2 \times \ldots \times f_n.$$

Comparing both expressions, we get

$$F_n = (F_1^n)_\times$$

and consequently

$$F = \sum (F_1^n)_\times = \left. \frac{1}{1 - F_1} \right|_\times .$$

On the other hand, let $F_1 = \overline{F}_1 \in O_1^M$ and

$$F = \left. \frac{1}{1 - F_1} \right|_\times .$$

Then

$$(F \cdot f)^* = \{ \ldots (F_1^n)_\times \cdot \overline{f_n(x_n \ldots x_1)}, \ldots \} = F \cdot f^*.$$

Furthermore we get

$$(F \cdot f) \times (F \cdot g) = \{ \ldots . \sum_{i+h=l} (F_1^i) \times f_i \times (F_1^h)_\times g_h, \ldots \} =$$

$$= \{ \ldots ., (F_1^l)_\times \sum_{i+h=l} f_i \times g_h, \ldots \} = F \cdot f \times g.$$

Hence F defines a homomorphism α_F of $\mathscr{S}$ and the theorem is proved.

To end this section, we want to introduce also the space of convolution operators. We denote by O_n^c the space of strongly decreasing distributions over $\mathbb{R}^{4n}$ and define for $F \in O_n^c$ and $f \in \mathscr{S}_n$ the convolution product as usual $[\![5]\!]$:

$$F * f.$$

This is again an element of $\mathscr{S}_n$.

II.1.7. *Definition*:

We denote by O^c the space of sequences

$$F = \{ F_0, F_1, \ldots F_n, \ldots \}$$

with $F_0 \in \mathbb{C}$ and $F_n \in O_n^c$.

Furthermore, we define for $F \in O^c$ and $f \in \mathscr{S}$ the convolution product

$$F * f = \{ F_0 f_0, F_1 * f_1, \ldots, F_n * f_n, \ldots \}.$$

If one wants to introduce a topology on O^c, it can easily be done by identifying O^c with the topological product of the spaces $\{ O_n^c \}_0^\infty$. From the fact that for every fixed F_n the map $f_n \to F_n * f_n$ is a continuous mapping from $\mathscr{S}_n$ to $\mathscr{S}_n$ it follows that for fixed $F \in \boldsymbol{O}^c$ the convolution product

$$f \to F * f$$

is a continuous linear mapping from $\mathscr{S}$ to $\underline{\mathscr{S}}$.

We now ask again which $F \in \boldsymbol{O}^c$ define a $*$-homomorphism of $\underline{\mathscr{S}}$.

II.1.8. *Theorem*:

Let $F \in \boldsymbol{O}^c$ and $F \neq 0$. Then the mapping

$$f \to F * f$$

defines a $*$-homomorphism of $\mathscr{S}$ exactly if F is of the form

$$F = \frac{1}{1 - F_1}\Big|_\times = \sum (F_1^n)_\times$$

with $F_1 \in O_1^c$ and $F^* = F$.

Proof:

We remark first that

$$(F_n * f_n)(x_1 \ldots x_n) = (F_n(y), \, f_n(x_1 - y_1, x_2 - y_2, \ldots x_n - y_n)).$$

This implies the relation

$$(F_n * f_n)^* = F_n^* * f_n^*.$$

First, we get from

$$(F * 1) = (F * 1) \times (F * 1)$$

the value of F_0, which must be equal to 1. Second, we get from

$$F_1^* * f_1^* = (F_1 * f_1)^* = F * f_1^* = F_1 * f_1^*$$

the relation $F_1^* = F_1$. Furthermore, we get for $f_i \in \mathscr{S}_1$

$$F_n * (f_1 \times f_2 \times \ldots \times f_n) = F * (f_1 \times f_2 \times \ldots \times f_n) =$$

$$= (F * f_1) \times (F * f_2) \times \ldots \times (F * f_i) =$$

$$= (F_1^n)_\times * (f_1 \times f_2 \times \ldots \times f_n),$$

which implies

$$F_n = (F_1^n)_\times .$$

From this follows

$$F = \frac{1}{1 - F_1}\Big|_\times .$$

The converse of this result can be proved by analogy with Theorem II.1.6.

II.2. Abelian Functionals

In this section we want to study a simple class of states, namely such states whose canonical representation gives rise to an abelian representation of $\mathscr{S}$.

II.2.1. *Definition*:

a) An element $T \in E(\mathscr{S})$ is called an abelian state if for all f, g the relation

$$(T, f \times g) = (T, g \times f)$$

holds.

b) An abelian state is called a character if the relation

$$(T, f \times g) = (T, f) \cdot (T, g)$$

holds.

We first want to study the structure of characters.

II.2.2. *Theorem*:

A state $T \in E(\mathscr{S})$ is a character exactly if there exists a $t \in (\mathscr{S}'_1)_n$ with

$$T = \frac{1}{1 - t} \Big|_{\times}$$

Proof:

Assume first $T = \dfrac{1}{1 - t} \Big|_{\times}$ with $t \in \mathscr{S}'_1$ and $t^* = t$. Then we get

$$(T, f \times g) = \sum_n (t^n_{|\times}, \sum_{i+k=n} f_i \times g_k) = \sum_n \sum_{i+k=n} (t^i_{|\times}, f_i)(t^k_{|\times}, g_k)$$

$$\{\sum_i (t^i_{|\times}, f_i)\} \{\sum_j (t^j_{|\times}, g_j)\} = (T, f) \cdot (T, g).$$

Since $t^* = t$, it follows that $(T, f^* \times f) = (T, f^*) \cdot (T, f) = \overline{(T, f)} \cdot (T, f) \geqq 0$. This implies, together with $T_0 = 1$, that T is a character.

Assume, on the other hand, that T is a character. Then it follows that $T_0 = 1$ and furthermore $T_1^* = T_1$. From the relation

$$(T, f_1 \times f_2 \times \ldots \times f_n) = (T, f_1) \cdot (T, f_2) \cdot \ldots (T, f_n)$$

it follows for $f_i \in \mathscr{S}_1$ that

$$T_n = T^n_{1|\times}.$$

Thus $T = \dfrac{1}{1 - T_1} \Big|_{\times}$.

If T is a character then the mapping $T: \mathscr{S} \to \mathbb{C}$ is a continuous homomorphism of the algebra $\mathscr{S}$ into the complex numbers. This means that the dimension of $\mathscr{S}/L(T) =$ $= 1$. One can easily generalize these investigations to the case where $\dim \mathscr{S}/L(T)$ is finite. But it is helpful to study a more general case.

II.2.3. *Theorem*:

Let T be an abelian state such that

$$\text{dimension } \mathcal{S}_1/L(T) \cap \mathcal{S}_1 = N < \infty.$$

Then there exist N distributions $t_i \in (\mathcal{S}'_1)_h$ and a sequence of numbers

$$C(i_1, i_2, \ldots i_k) \qquad i_j = 1, 2, \ldots N, \qquad k = 1, 2, 3, \ldots$$

which is symmetric in the arguments and of positive type (i.e. the sequence of an N-dimensional moment problem) such that

$$T_k = \sum_{i_1, i_2, \ldots i_k = 1}^{N} C(i_1, i_2, \ldots i_k)\, t_{i_1} \times t_{i_2} \times \ldots \times t_{i_k}.$$

Proof:

Since $L(T)$ is invariant by involution, there exist N real elements $g_1 \ldots g_N$ such that the cosets generated by the $g_1 \ldots g_N$ span all of $\mathcal{S}_1/L(T) \cap \mathcal{S}_1$.

Denote by $A_i = A(g_i)$ the representatives of g_i and by ψ_0 the cyclic vector in the Gelfand–Segal representation of T. Define now

$$C(i_1, i_2, \ldots i_k) = (\psi_0, A_{i_1} A_{i_2} \ldots A_{i_k} \psi_0).$$

This sequence is of positive type and it is symmetric in the indices since T is abelian. Since the cosets generated by the g_i span all of $\mathcal{S}_1/L(T) \cap \mathcal{S}_1$, it follows for $f \in \mathcal{S}_1$ that $A(f)$ is a linear combination of A_i:

$$A(f) = \sum_{i=1}^{N} F_i(f)\, A_i.$$

The linearity of $A(f)$ in f and the continuity of the matrix elements of $A(f)$ imply that $F_i(f)$ is linear and continuous in f. Hence there exist $t_i \in \mathcal{S}'_1$ such that

$$(t_i, f) = F_i(f) \quad i = 1 \ldots N.$$

Since $A(f)$ is symmetric for real f, it follows that $t_i^* = t_i$.

By writing out the equation, $f_i \in \mathcal{S}_1$,

$$(T, f_1 \times f_2 \times \ldots \times f_k) = (\psi_0, A(f_1) \ldots A(f_k)\, \psi_0) =$$

$$= \sum_{i, \ldots i_k = 1}^{N} t_{i_1}(f_1) \ldots t_{i_k}(f_k)\, (\psi_0, A_{i_1} \ldots A_{i_k} \psi_0),$$

we get for T_k the equation given in the theorem.

Now we want to show that the set of abelian finite-order states is dense in the set of all states. We start first with a remark.

II.2.4. *Lemma*:

The set of abelian states is a closed subset of $E(\mathcal{S})$.

Proof:

Abelian states are characterized by $(T, f \times g - g \times f) = 0$ for all $f, g \in \mathscr{S}$. Since this relation stays true upon taking limit-points it follows that the set of abelian states is closed.

II.2.5. *Lemma*:

Let us call a state with $\dim \mathscr{S}_1/L(T) \cap \mathscr{S}_1 = N < \infty$ a state of order N. Then we have:

The set of abelian states of finite order is dense in the set of all abelian states.

Proof:

Let T be the given abelian state and let us denote by $A(f)$ the representation of $\mathscr{S}$ induced by T, and denote by Ω the cyclic vector. Choose N arbitrary real elements $f_i \in \mathscr{S}_1$ and denote by $\mathscr{S}_1^N$ the subspace generated by the f_i. Because of the linearity, we get for arbitrary $g \in \mathscr{S}_1^N$

$$A(g) = \sum_{j=1}^{M} F_j(g) \, A(f_j),$$

where $\{A(f_j)\}_{j=1}^{M}$ is a minimal subset of linearly independent elements in $\{A(f_i)\}_{i=1}^{N}$ which spans all of $\{A(f_i)\}_{i=1}^{N}$.

It is clear that $F_j(g)$ are real linear functionals on $\mathscr{S}_1^N$. By the Hahn–Banach theorem there exist real linear functionals $t_j \in \mathscr{S}_1'$ such that

$$(t_j, g) = F_j(g) \qquad \text{for} \qquad g \in \mathscr{S}_1^N.$$

Define now a field theory by the relation

$$A'(g) = \sum_{j=1}^{M} (t_j, g) \, A_j, \qquad g \in \mathscr{S}_1$$

with $A_j = A(f_j)$, $j = 1 \ldots M$, and by $T(f_1, \ldots f_N)$ the vector state T_Ω over the field $A'(g)$. It is clear that $T(f_1 \ldots f_N)$ is a state of order M.

Now it remains to show that the set of states

$$T(f_1, \ldots f_N); \qquad f_i \in \mathscr{S}_1, \qquad N = 1, 2, \ldots$$

defines a Cauchy net converging to T.

We introduce a semi-ordering of the indices by

$$(f_1, \ldots, f_N) < (g_1, \ldots g_M) \Leftrightarrow$$

$\Leftrightarrow$ Linear space generated by $(f_1, \ldots f_N) \subset$ linear space generated by $(g_1 \ldots g_M)$.

Furthermore, by construction, T and $T(f_1, \ldots f_N)$ coincide on $\sum_k \oplus (\mathscr{S}_1^N)^k_{|\times}$. It follows that $T(f_1 \ldots f_N)$ is a Cauchy net converging on $\sum_{k=0}^{\infty} \oplus (\mathscr{S}_1)^k_{|\times}$ to T. But this implies that the limit is separately continuous on $(\mathscr{S}_1)^k_{|\times}$ and hence by the nuclear theorem continuous on $\mathscr{S}_k$. Thus T is the limit of this net.

II.2.6. *Remark*:

At this point one would like to know whether every abelian state could also be approximated by convex combinations of characters. This is not true in general. The reason is that not every n-dimensional moment problem has a solution. We do not want to elaborate on this question. The interested reader might consult the paper by Powers [8].

We want to end this section with the question of the existence of abelian Wightman states. We start with some preparation.

II.2.7. *Lemma*:

Let T be an abelian Wightman state and $\{A(f), \mathscr{H}, \mathscr{U}(a), \Omega\}$ the Gelfand–Segal representation associated with T. Then $\mathscr{U}(a) = 1$ for all $a \in \mathbb{R}^4$.

Proof:

From the equation

$$(\Omega, A(f)\,\mathscr{U}(a)\,A(g)\,\Omega) = (\Omega, A(f)\,A(\alpha_a g)\,\Omega) =$$

$$= (\Omega, A(\alpha_a g)\,A(f)\,\Omega) = (\Omega, A(g)\,\mathscr{U}(-a)\,A(f)\,\Omega)$$

it follows that the spectrum of $\mathscr{U}(a)$ is invariant under the substitution $p \to -p$. Together with the spectral condition this gives

$$\text{spectrum } \mathscr{U}(a) = \{0\}.$$

But this implies $\mathscr{U}(a) = 1$.

II.2.8. *Theorem*:

Let T be an abelian Wightman state. Then there exists a positive-type sequence $\{c_i\}$ with $c_0 = 1$ such that

$$T = \sum_{i=0}^{\infty} c_i \left(\int \right)^i_{|\times},$$

where $\int$ denotes the functional $f \to \int f(x)\,d^4x \qquad f \in \mathscr{S}_1.$

Proof:

Let $A(f)$ be the representative of f for $f \in \mathscr{S}_1$. Then from Lemma II.2.7 it follows that

$$A(\alpha_a f) = A(f).$$

Hence we get for $g \in \mathscr{S}_1$

$$A(g * f) = \int g(a) \, A(\alpha_a f) \, d^4 a = A(f) \int g(a) \, d^4 a =$$

$$= \int f(a) \, A(\alpha_a g) \, da = A(g) \int f(a) \, d^4 a.$$

Taking $\int g(a) \, d^4 a = 1$, we obtain

$$A(f) = 0 \quad \text{if} \quad \int f(a) \, d^4 a = 0.$$

This implies

$$A(f) = A \int f(x) \, d^4 x.$$

Hence T is a state of order one, and the result follows from Theorem II.2.3.

II.3. Matrix States

In this section we want to show that the pathologies which occur in the abelian case are not present in the general situation. In particular we want to show that a certain family of very simple states, which we will call matrix states, is dense in the set of all states. We start with describing the matrix states.

II.3.1. *Definition*:

a) Let $T \in E(\mathscr{S})$ be a state. We denote by $I(T)$ the maximal two-sided ideal which is contained in $L(T)$. $I(T)$ is obviously closed.

b) A state $T \in E(\mathscr{S})$ is called a matrix state if $\mathscr{S}/I(T)$ is a finite-dimensional algebra.

II.3.2. *Lemma*:

Let $T \in E(\mathscr{S})$ be a state. Then $I(T)$ is a symmetric ideal, i.e. $I^*(T) = I(T)$, and consequently $\mathscr{S}/I(T)$ is a $*$-algebra.

Proof:

Let $f \in I(T)$ and $g, h \in \mathscr{S}$. We have to show that $g \times f^* \times h$ belongs to $L(T)$. Since $I(T)$ is a two-sided ideal, it follows that

$$(g \times f^* \times h)^* \times (g \times f^* \times h) = h^* \times f \times g^* \times g \times f^* \times h \in I(T) \subset L(T).$$

Hence $g \times f^* \times h \in L(T)$ which means $f^* \in I(T)$.

II.3.3. *Remark*:

a) Let A be an n-dimensional $*$-algebra with unit. Then A is isomorphic to a sub-$*$-algebra of N by N matrices, where N is a suitable number.

b) In the algebra of N by N matrices there exists a basis e_{ih}, $i, h = 1, \ldots N$, with the properties

$$e_{ik}e_{lm} = \delta_{k,l}e_{im},$$

$$e_{ik}^* = e_{ki},$$

$$\sum_{i=1}^{N} e_{ii} = 1.$$

c) Every state over the algebra of N by N matrices is a vector state. Hence for the characterization of the structure of matrix states it is sufficient to study the homomorphisms of $\mathscr{S}$ into the algebra of N by N matrices.

II.3.4. *Theorem*:

Let β be a continuous $*$-homomorphism of $\mathscr{S}$ into the algebra of N by N matrices such that $\beta 1 = 1$.

Then there exist elements $T^{ih} \in \mathscr{S}'$, $i, h = 1 \ldots N$ with the properties $(T^{ih})^* = T^{hi}$, $(T^{ih})_0 = \delta_{ih}$;

$$(T^{ih})_n = \sum_{l_1 \ldots l_{n-1}} t^{il_1} \times t^{l_1 l_2} \times \ldots \times t^{l_{n-1},h},$$

where $t^{ih} = (T^{ih})_1 \in \mathscr{S}'_1$ such that

$$\beta f = \sum_{i,h=1}^{N} (T^{ih}, f) e_{ih} \quad (e_{ih} \text{ is the basis of the matrix algebra}).$$

Proof:

Let e_{ik} be a basis of the matrix algebra with the properties

$$e_{ik}^* = e_{ki}, \qquad e_{ik}e_{lm} = \delta_{kl}e_{im}, \qquad \sum e_{ii} = 1.$$

Then we get

$$\beta f = \sum_{ik} c_{ik}(f) e_{ik}.$$

Since β is linear and continuous, it follows that $c_{ik}(f)$ is linear and continuous. Hence there exist $T^{ik} \in \mathscr{S}'$ with

$$c_{ik}(f) = (T^{ik}, f).$$

From the relation $(\beta f)^* = \beta f^*$ it follows that

$$\sum_{ik} \overline{(T^{ik}, f)}\, e_{ki} = \sum_{ik} (T^{ki}, f^*)\, e_{ki}$$

and consequently $\overline{(T^{ik}, f)} = (T^{ki}, f^*)$, which means that $T^{ik*} = T^{ki}$.

Since β is also an algebraic homomorphism we must get for $f, g \in \mathscr{S}$ the relation

$$\beta(f \times g) = (\beta f) \cdot (\beta g).$$

Inserting the expression for β, we get

$$\sum_{ik} (T^{ik}, f \times g)\, e_{ik} = \left\{ \sum_{ik} (T^{ik}, f)\, e_{ik} \right\} \left\{ \sum_{lm} (T^{lm}, g)\, e_{lm} \right\} =$$

$$= \sum_{ik} \left\{ \sum_{l} (T^{il}, f)(T^{lk}, g) \right\} e_{ik}.$$

Comparing the coefficient of e_{ik}, we get the relation

$$(T^{ik}, g \times f) = \sum_{l} (T^{il}, g)(T^{lk}, f).$$

Inserting on both sides functions of the form $f^1 \times f^2 \times \ldots \times f^n, f^i \in \mathscr{S}_1$, we find for the n-th component of T^{ik}:

$$(T^{ik})_n = \sum_{l_1 \ldots l_{n-1}} t^{il_1} \times t^{l_1 l_2} \times \ldots \times t^{l_{n-1}k}$$

with $t^{ik} = (T^{ik})_1 \in \mathscr{S}'_1$.

Next we want to study finite-order states which we will introduce by analogy with finite-order states in the abelian case. The finite-order states contain the matrix states as a subset. We will then show that the difficulties of the abelian case do not appear in the general situation, i.e. we will show that every finite-order state is a limit of matrix states.

II.3.5. *Definition*:

Let $T \in E(\mathscr{S})$ be a state and $\{A(f), \mathscr{H}, \psi_0\}$ the representation of $\mathscr{S}$ defined by T. We call T a state of order N if the family of operators $\{A(f); f \in \mathscr{S}_1\}$ contains only N linearly independent elements. This is equivalent to saying $\dim \mathscr{S}_1/I(T) \cap \mathscr{S}_1 = N$.

II.3.6. *Lemma*:

Let T be a state of order N. Then there exist N real distributions $t^i \in \mathscr{S}'_1$ and N symmetric operators A^i defined on a cyclic domain such that for $f \in \mathscr{S}_1$

$$A(f) = \sum_{i=1}^{N} (t^i, f)\, A^i.$$

Proof:

Exactly as the proof of Theorem II.2.3.

II.3.7. *Lemma*:

Every finite-order state is the limit of a sequence of matrix states.

Proof:

Let T be a state of order N and

$$A(f) = \sum_{i=1}^{N} (t^i f)\, A^i$$

for $f \in \mathscr{S}_1$. Then the Hilbert space $\mathscr{H}$ is the Hilbert space generated by the vectors

$$\{\psi_0, A^i\psi_0, A^{i_1}A^{i_2}\psi_0, \ldots, A^{i_1}A^{i_2}\ldots A^{i_r}\psi_0, \ldots; i_j = 1, 2, \ldots N, r = 0, 1, 2, \ldots\}.$$

Let us denote by $\mathscr{H}_n$ the vector space spanned by the vectors

$$\{A^{i_1}\ldots A^{i_r}\psi_0; i_j = 1, 2, \ldots N, r = 0, 1, 2, \ldots, n\}.$$

Since $\dim \mathscr{H}_n \leq \sum_{i=0}^{n} N^i$ is finite, it follows that $\mathscr{H}_n$ is a closed subspace of $\mathscr{H}$, and moreover that A^i restricted to $\mathscr{H}_n$ is bounded. Let us denote by E^n the projection onto $\mathscr{H}_n$. Then

$$f \to \sum_{i=1}^{N} (t^i, f)\, E^n A^i E^n, \qquad f \in \mathscr{S}_1$$

defines a continuous $*$-homomorphism β^n into a finite matrix algebra.

Denote

$$(T^n, f) = (\psi_0, \beta^n f \psi_0);$$

T^n is a matrix state. Let $k \leq n$ and $f^i \in \mathscr{S}_1$. Then

$$(T^n, f^1 \times f^2 \times \ldots \times f^k) = (\psi_0, E^n A^{i_1} E^n \ldots E^n A^{i_k} E^n \psi_0).$$

$$\sum_{i_1\ldots i_k = 1}^{N} (t^{i_1}, f^1)(t^{i_2}, f^2)\ldots(t^{i_k}, f^k) = (\psi_0; A^{i_1}A^{i_2}\ldots A^{i_k}\psi_0).$$

$$\sum_{i_1\ldots i_k = 1}^{N} (t^{i_1}, f^1)\ldots(t^{i_k}, f^k) = (T, f^1 \times f^2 \times \ldots \times f^k).$$

This means T and T^n coincide on $\mathscr{S}_h$ for $h = 0, 1, \ldots, n$. Hence $T = \lim_{n \to \infty} T^n$.

II.3.8. *Theorem*:

The finite-order states and hence also the matrix states are dense in the set of all states.

Proof:

Since we already know that the set of matrix states is dense in the set of finite-order states, we need only prove that the set of finite-order states is dense in the set of all states. But the proof of this statement is exactly the proof of Lemma II.2.5.

II.4. Elementary Operations with States

By operations with states we want to understand formulas which allow us to calculate a new state from one or more given states.

A. Let α be a homomorphism from $\mathscr{S}$ into $\mathscr{S}$, i.e.

$$\alpha \in \mathrm{Hom}\,(\mathscr{S}), \qquad \alpha \neq 0.$$

Then the transposed homomorphism

$$(\alpha' T, f) = (T, \alpha f)$$

defines a mapping from states to states

$$\alpha' : E(\mathscr{S}) \to E(\mathscr{S}).$$

In section II.1, we studied two families of special homomorphisms. We want to study under which additional conditions these homomorphisms map Wightman states into Wightman states. We start with multiplicative homomorphisms.

II.4.1. *Lemma*:

Let $F \in O^M$ and assume F defines a homomorphism of $\mathscr{S}$ into $\mathscr{S}$, i.e. $\alpha_F f = F \cdot f$.

By Theorem II.1.6, F is of the form $F = \dfrac{1}{1 - F_1}\Big|_{\times}$ with $F_1 \in O_1^M$ and $F^* = F$. α'_F

maps Wightman states into Wightman states if $F_1 = c = \text{const}$ with $c \in \mathbb{R}$.

Proof:

Since F acts multiplicatively, F leaves the locality ideal I_c invariant. In order to preserve the spectral condition, F must act as a convolution in momentum space. In order to map all functions vanishing in the forward light cone onto functions with the same property, the support of $\mathscr{F}^{-1}(F_1)$ must be contained in the backward light cone. Since F_1 is real, it follows that support $\mathscr{F}^{-1}(F_1) = \{0\}$. This implies that F_1 is a polynomial. Now it remains to preserve translational invariance. This means

$$\prod_i P(x_i + a) = \prod_i P(x_i)$$

and hence

$$P(x) = c = \text{const with} \quad c \in \mathbb{R}.$$

II.4.2. *Lemma*:

Assume $F_1 = F_1^* \in O_1^c$ and $F = \dfrac{1}{1 - F_1}\Big|_{\times} \in O^c$. Then the transpose of the homomorphism

$$\alpha_F f = F * f$$

maps Wightman states into Wightman states if F_1 is of the form

$$F_1 = \text{Pol}\left(\frac{\partial}{\partial x_i}\right)\delta,$$

where Pol denotes a real polynomial in four variables and δ the Dirac measure at some point.

Proof:

In order that F_1 map all pairs $f, g \in \mathscr{S}_1$ with spacelike separated supports onto functions with spacelike separated supports, the support of F_1 must consist only of one point, i.e.

$$F_1 = \mathrm{Pol}\left(\frac{\partial}{\partial x_i}\right)\delta.$$

From the condition $F_1^* = F_1$ it follows that Pol is a real polynomial. Since F acts multiplicatively in momentum space, it follows that the spectral ideal Sp also stays invariant under F. Since differentiation and translations commute, it follows that α'_F maps translation-invariant states into invariant states.

B. The automorphism described in Lemma II.4.1 corresponds in the language of operators to the mapping

$$A(f) \to cA(f) \qquad f \in \mathscr{S}_1.$$

In the language of states, let

$$T = \{1, T_1, \ldots, T_n, \ldots\}.$$

Then

$$\alpha'_c T = \{1, cT_1, \ldots, c^n T_n, \ldots\}.$$

This transformation can be generalized.

II.4.3. *Lemma*:

Let $\mathscr{C} = \{c_i\}$ be a sequence of positive type with $c_0 = 1$.
Let $T = \{1, T_1, \ldots, T_n, \ldots\} \in E(\mathscr{L})$ be a state. Then

$$\mathscr{C} \cdot T = \{1, c_1 T_1, \ldots c_n T_n, \ldots\}$$

is also a state. If in particular $T \in \mathscr{W}$ (a Wightman state), then this is true also for $\mathscr{C} \cdot T$, i.e. $\mathscr{C} \cdot T \in \mathscr{W}$.

Proof:

Since $\mathscr{C}$ is a sequence of positive type, there exists a Hilbert space $\mathscr{K}$ and a symmetric operator B acting on a dense set of vectors in $\mathscr{K}$ such that c_n is a moment sequence of B, i.e. $c_n = \{\phi_0, B^n \phi_0\}, \phi_0 \in \mathscr{K}$.

Let now $\{A(f), \mathscr{H}, \Omega\}$ be the canonical representation induced by T. Define

$$\mathscr{H}' = \mathscr{H} \otimes \mathscr{K};$$

$$A'(f) = A(f) \otimes B \qquad \text{for} \qquad f \in \mathscr{S}_1,$$

and $D' = D \times D_B^\infty$. Now $\{A'(f), \mathscr{H}', D'\}$ satisfies Axiom 1. By Theorem I.4.5, $T_{\Omega \times \phi_0}$ defines a state. If $A(f)$ is a Wightman field, then $A'(f)$ is also a Wightman field, and hence if T is a Wightman state then this is true also for $T_{\Omega \times \phi_0}$. Computing now for

$f^i \in \mathcal{S}_1$, we get:

$$(T_{\Omega \times \phi_0}, f^1 \times f^2 \times \ldots \times f^i) = (\Omega \times \phi_0, A(f_1) \otimes B \ldots A(f_i) \otimes B\Omega \times \phi_0) =$$

$$= (\Omega, A(f_1) \ldots A(f_i) \Omega) \cdot (\phi_0, B^i \phi_0) =$$

$$= (T, f_1 \times f_2 \times \ldots \times f_i) \cdot c_i.$$

Hence the lemma is proved.

C. Next we want to link two states.

Let T_1, T_2 be two states and $\{A_i(f), \mathcal{H}_i, D_i, \Omega_i\}$ $i = 1, 2$ the canonical representations constructed from these states. We now form the Hilbert space $\mathcal{H}_1 \otimes \mathcal{H}_2$ and for $f \in \mathcal{S}_1$ the operators

$$A_1(f) \otimes 1_2 + 1_1 \otimes A_2(f).$$

These operators are defined on $D_1 \times D_2$ and leave this set invariant. It is easy to check that the triplet

$$\{A_1(f) \otimes 1_2 + 1_1 \otimes A_2(f), \mathcal{H}_1 \otimes \mathcal{H}_2, D_1 \times D_2\}$$

fulfills Axiom 1. This gives us the possibility of constructing new states.

II.4.4. *Definition*:

Let $T_1, T_2 \in E(\mathcal{S})$. We define a new state denoted by $T_1 s T_2$ as the vector state $T_{\Omega_1 \times \Omega_2}$ of the field

$$\{A_1(f) \otimes 1_2 + 1_1 \otimes A_2(f), \mathcal{H}_1 \otimes \mathcal{H}_2, D_1 \times D_2\},$$

where Ω_i denotes the cyclic vector obtained from T_i. According to Theorem I.4.5, $T_1 s T_2$ is again a state. If in addition $A_1(f)$ and $A_2(f)$ are Wightman fields then also $A_1(f) \otimes 1_2 + 1_1 \otimes A_2(f)$ is a Wightman field. This implies $T_1, T_2 \in \mathcal{W} \Rightarrow T_1 s T_2 \in \mathcal{W}$.

II.4.5. *Lemma*:

The mapping $\{T_1, T_2\} \to T_1 s T_2$ obeys the following rules:

a) $T_1 s T_2 = T_2 s T_1$;

b) $(T_1 s T_2) s T_3 = T_1 s (T_2 s T_3)$;

c) Let α be a homomorphism of $\underline{\mathcal{S}}$ into $\mathcal{S}$ such that

$$\alpha \mathcal{S}_i \subset \mathcal{S}_i, \qquad i = 0, 1, 2 \ldots;$$

then we have

$$\alpha'(T_1 s T_2) = (\alpha' T_1) s (\alpha' T_2);$$

d) $T_1 \cdot T_2 \in \mathcal{W}$ implies $T_1 s T_2 \in \mathcal{W}$.

Proof:

a) and b) are nothing else but the symmetry resp. associative properties of the direct product of Hilbert spaces.

68 H.J. BORCHERS

c) follows for $f \in \mathcal{S}_1$ from the equation

$$\alpha'(A_1(f) \otimes 1_2 + 1_1 \otimes A_2(f)) = (A_1(\alpha f) \otimes 1_2 + 1_1 \otimes A_2(\alpha f))$$
$$= (\alpha' A_1(f)) \otimes 1_2 + 1_1 \otimes (\alpha' A_2(f)).$$

d) is already proved in the definition.

Next we want to get computational expressions for the new state $T_1 s T_2$. To this end we have to introduce a new product in the spaces $\mathcal{S}$ and $\mathcal{S}'$. For easy writing we will introduce for $T \in \mathcal{S}'_n$ the notation $T(x_1, x_2, \ldots x_n)$.

II.4.6. *Definition*:

Let $(x_1, x_2, \ldots x_{n+m})$ be an ordered set of $n + m$ numbers. Then we denote by a partition P_{nm} an ordered splitting into two sets

$$P_{n,m}(x_1, \ldots x_{n+m}) = (x_{i_1}, \ldots x_{i_n})(x_{j_1}, \ldots x_{j_m})$$

with the properties

a) $\{i_1, \ldots, i_n\} \cup \{j_1, \ldots j_m\} = \{1, 2, \ldots n + m\};$

$\{i_1, \ldots, i_n\} \cap \{j_1, \ldots j_m\} = \emptyset;$

b) $i_1 < i_2 < \ldots < i_n,$

and $j_1 < j_2 < \ldots < j_m.$

II.4.7. *Definition*:

a) Let $T(x_1 \ldots x_n) \in \mathcal{S}'_n$ and $S(x_1 \ldots x_m) \in \mathcal{S}'_m$.

Then we define $T s S \in \mathcal{S}'_{n+m}$ by the equation

$$T s S(x_1, \ldots x_{n+m}) = \sum_{P_{n,m}} T(x_{i_1}, \ldots x_{i_n}) S(x_{j_1} \ldots x_{j_m}),$$

where the sum runs over all partitions $P_{n,m}$.

b) For $T, S \in \mathcal{S}'$ we define

$$(T s S)_n = \sum_{i+k=n} T_i s S_k.$$

Examples:

$$T(x_1) s S(x_2) = T(x_1) S(x_2) + S(x_1) T(x_2);$$

$$T(x_1, x_2) s S(x_3) = T(x_1, x_2) S(x_3) + T(x_1, x_3) S(x_2) + T(x_2, x_3) S(x_1).$$

Before we go on to study the new product (see also Ruelle $[\![4]\!]$), we want to study its relation to the composition of two states given in Definition II.4.4.

II.4.8. *Lemma*:

The composition of two states given in Definition II.4.4 coincides with the product given in Definition II.4.7.

Proof:

Let

$$B(x) = A_1(x) \otimes 1_2 + 1_1 \otimes A_2(x)$$

and calculate the vector state $T_{\Omega_1 \times \Omega_2}$ of $B(x)$. We get for the *n*-th component:

$$
\begin{aligned}
T_{\Omega_1 \times \Omega_2}(x_1, \ldots x_n) &= (\Omega_1 \times \Omega_2, B(x_1) B(x_2) \ldots B(x_n) \Omega_1 \times \Omega_2) \\
&= (\Omega_1 \times \Omega_2, \{A_1(x_1) \otimes 1_2 + 1_1 \otimes A_2(x_1)\} \ldots \\
&\qquad \{A_1(x_n) \otimes 1_2 + 1_1 \otimes A_2(x_n)\} \, \Omega_1 \times \Omega_2) \\
&= \sum_{l+k=n} \sum_{P_{l,k}} (\Omega_1, A_1(x_{i_1}) \ldots A_1(x_{i_l}) \Omega_1)(\Omega_2, A_2(x_{j_1}) \ldots A_2(x_{j_k}) \Omega_2) \\
&= \sum_{l+k=n} \sum_{P_{l,k}} T_l^{(1)}(x_{i_1}, \ldots x_{i_l}) \, T_k^{(2)}(x_{j_1}, \ldots x_{j_k}) \\
&= \sum_{l+k=n} T_l^{(1)} s T_k^{(2)} = (T^{(1)} s T^{(2)})_n.
\end{aligned}
$$

We now turn back to the *s*-product. Its properties are listed in the following theorem:

II.4.9. *Theorem*:

The *s*-product introduced in Definition II.4.7 has the following properties $(T, S \in \mathscr{S}')$:

a) It is abelian:

$$TsS = SsT;$$

b) $Ts(S + U) = TsS + TsU;$

c) $Ts(SsU) = (TsS)sU;$

d) let $P(x) = \sum a_n x^n$ be a formal power series and $T \in \mathscr{S}'$ with $T_0 = 0$; then

$$P(T)_{|s} = \sum a_n T_{|s}^n$$

converges in $\mathscr{S}'$;

e) let $P(x)$ be a power series. Then

$$P(T)_{|s}$$

converges in $\mathscr{S}'$ for all $T \in \mathscr{S}'$ with $|T_0| \leq a$ if and only if $P(x)$ has a radius of convergence $R > a$.

Proof:

a) Let $T \in \mathscr{S}'_n$ and $S \in \mathscr{S}'_m$. Then we have

$$
\begin{aligned}
TsS(x_1 \ldots x_{n+m}) &= \sum_{P_{n,m}} T(x_{i_1} \ldots x_{i_n}) S(x_{j_1} \ldots x_{j_m}) \\
&= \sum_{P_{n,m}} S(x_{j_1} \ldots x_{j_m}) T(x_{i_1} \ldots x_{i_n}) = SsT(x_1 \ldots x_{n+m}).
\end{aligned}
$$

Hence

$$(T\dot{s}S)_n = \sum_{i+k=n} T_i \dot{s} S_k = \sum_{i+k=n} S_k \dot{s} T_i = (S\dot{s}T)_n.$$

b) This statement is trivial.

c) Let $T \in \mathscr{S}'_n,\, S \in \mathscr{S}'_m,\, U \in \mathscr{S}'_p$; then we get

$$(T\dot{s}S)\dot{s}U = \sum_{P_{n+m,p}} \sum_{P_{n,m}} T(x_{i_1} \ldots x_{i_n})\, S(x_{j_1} \ldots x_{j_m})\, U(x_{k_1} \ldots x_{k_p})$$

$$= \sum_{P_{n,m+p}} \sum_{P_{m,p}} T(x_{i_1} \ldots x_{i_n})\, S(x_{j_1} \ldots x_{j_m})\, U(x_{k_1} \ldots x_{k_p}) = T\dot{s}(S\dot{s}U).$$

This gives together with b)

$$T\dot{s}(S\dot{s}U) = (T\dot{s}S)\dot{s}U.$$

d) We have to prove that the k-th component converges in $\mathscr{S}'_k$. But since $T_0 = 0$, we get

$$(T^i_{|s})_k = 0 \quad \text{for} \quad k = 0, 1, \ldots i - 1.$$

This implies

$$(\sum_{n=0}^{\infty} a_n T^n_{|s})_k = (\sum_{n=0}^{k} a_n T^n_{|s})_k,$$

which means that $(P(T)_{|s})_k$ is only a finite sum.

e) In the general case write

$$T = T_0 + T'$$

where T' has a vanishing zeroth component. Then we get

$$\sum a_n T^n_{|s} = \sum a_n (T_0 + T')^n_{|s} = \sum_n a_n \sum_k \binom{n}{k} T_0^{n-k} T'^k_{|s}$$

$$= \sum_k \left\{ \sum_{n \geq k} \binom{n}{k} a_n T_0^{n-k} \right\} T'^k_{|s}.$$

Hence by d) the sum converges if and only if

$$\sum_{n>h} \binom{n}{h} a_n T_0^{n-h}$$

converges for $h = 0, 1, 2, \ldots$.

These are the same sequences as appear in the proof of Theorem II.1.4. and hence we get the same result.

Remark:

Let $T \in E(\mathscr{S})$. Then we have $T_0 = 1$. Hence $\log T_{|s} = \log((T-1) + 1)_{|s}$ can be developed into a power series with respect to $(T-1)$. This operation is called truncation, i.e.

$$T' = \sum \frac{(-1)^{n+1}}{n} (T-1)^n_{|s} = \log T_{|s}.$$

Since the s-product is abelian, we get

$$(T^{(1)} s T^{(2)})^t = (T^{(1)})^t + (T^{(2)})^t.$$

II.5. Generalized Free Fields

In this section we want to treat generalized free fields (short g.f.f.) as an example of Wightman fields. These fields have been introduced by O. W. Greenberg [3] and Licht and Toll [7]. We will not start from the definition given by Greenberg but from a definition adapted to our previous investigations.

Before we start we need a lemma:

II.5.1. *Lemma*:

A net $T^{(\alpha)} \in E(\mathscr{S})$ converges to an element $T \in E(\mathscr{S})$ if and only if the net of truncated functionals $T^{(\alpha)t}$ is convergent. If both nets converge we get the relation

$$T^t = \lim_\alpha T^{(\alpha)t}.$$

Proof:

If $T^{(\alpha)}$ converges, so does $(T^{(\alpha)} - 1)$. Furthermore, an easy inspection shows that if $T^{(\alpha)}$ and $S^{(\alpha)}$ are two convergent nets then $T^{(\alpha)} s S^{(\alpha)}$ is also a convergent net. This implies that

$$T^{(\alpha)t} = \sum \frac{(-1)^{n+1}}{n} (T^{(\alpha)} - 1)^n_{|s}$$

is a convergent net, since the i-th component is only a polynomial of finite degree.

Assume, conversely, that $(T^{(\alpha)})^t$ converges; then this is true also for $T^{(\alpha)} = e^{T^{(\alpha)t}}_{|s}$, since $T^{(\alpha)t}$ has a vanishing zeroth component and hence only a polynomial of finite degree contributes to the n-th component of $T^{(\alpha)}$. Let now $T = \lim T^{(\alpha)}$; then we get

$$(T^t)_i = \left(\sum_{n=0}^{\infty} \frac{(-1)^{n+1}}{n} (T - 1)^n_{|s} \right)_i = \left(\sum_{n=0}^{i} \frac{(-1)^{n+1}}{n} (T - 1)^n_{|s} \right)_i$$

$$= \left(\sum_{n=0}^{i} \frac{(-1)^{n+1}}{n} \lim_\alpha (T^{(\alpha)} - 1)^n_{|s} \right)_i = \lim_\alpha \left(\sum_{n=0}^{i} \frac{(-1)^{n+1}}{n} (T^{(\alpha)} - 1)^n_{|s} \right)_i$$

$$= \lim_\alpha T^{(\alpha)t}_i.$$

We now take a state $T \in E(\mathscr{S})$ and get a sequence of states by taking the power of T with respect to the s-product:

$$T, T^2_{|s}, \ldots, T^n_{|s}, \ldots$$

Since the truncated functions are of the form

$$T^t, 2T^t, \ldots, nT^t, \ldots$$

the above sequence of states is a divergent sequence. But we get a convergent sequence if we apply to $T_{|s}^n$ the multiplicative automorphism $\alpha_{1/n}$.

In this case we get

$$T, \alpha_{1/2} T_{|s}^2, \ldots, \alpha_{1/n} T_{|s}^n,$$

respectively

$$T^t, 2\alpha_{1/2} T^t, \ldots, n\alpha_{1/n} T^t.$$

Computing now the i-th component, we get

$$(T^t)_i, \frac{1}{2^{i-1}} (T^t)_i, \ldots, \qquad \frac{1}{n^{i-1}} (T^t)_i, \ldots \qquad i = 1, 2, \ldots.$$

This shows the convergence of the sequence and we get

$$\lim_{n \to \infty} n\alpha_{1/n} T^t = \{0, T_1, 0, 0, \ldots\}.$$

Hence

II.5.2. *Lemma*:

The sequence

$$\alpha_{1/n} T_{|s}^n$$

converges to the character

$$e_{|s}^{T_1} = \frac{1}{1 - T_1} \Big|_\times,$$

where T_1 is the first component of T.

If it happens that the first component T_1 of T vanishes then we get the state $e_{|s}^0$. In this case one can choose a different α to get a convergent sequence. We consider now the sequence

$$T, \quad \alpha_{1/\sqrt{2}} T_{|s}^2, \ldots, \qquad \alpha_{1/\sqrt{n}} T_{|s}^n$$

which has the sequence of truncated functions

$$T^t, 2\alpha_{1/\sqrt{2}} T^t, \ldots, \qquad n\alpha_{1/\sqrt{n}} T^t$$

and for the components

$$(T^t)_i, \quad \frac{2}{\sqrt{2^i}} (T^t)_i, \ldots \quad \frac{n}{\sqrt{n^i}} (T^t)_i, \ldots \qquad i = 2, 3, \ldots.$$

This shows that the sequence converges, and we get

$$\lim_{n \to \infty} n\alpha_{1/\sqrt{n}} T^t = \{0, 0, T_2, 0, 0, \ldots\}.$$

II.5.3. *Lemma*:

Let $T \in E(\mathscr{L})$ be a state with vanishing first component, $T_1 = 0$; then the sequence

$$\alpha_{1/\sqrt{n}} T_{|s}^n$$

converges to the element $e_{|s}^{T_2}$, where T_2 is the second component of T.

II.5.4. *Definition*:

A state $T \in E(\underline{\mathscr{S}})$ is called a generalized free field state (g.f.f. state) if T^t is of the form

$$T^t = \{0, 0, T_2, 0, 0, \dots\}.$$

Our next aim is to show that g.f.f. states do exist. We start with some preparation.

II.5.5. *Lemma*:

Denote by $G \subset E(\mathscr{S})$ the set of g.f.f. states and by G^t the set of truncated g.f.f. states. G^t is a closed convex cone.

Proof:

Let $T^{(1)}, T^{(2)} \in G$ and $\rho_1, \rho_2 \in \mathbb{R}^+$. Then we get for the truncated function

$$\{(\alpha_{\sqrt{\rho_1}} T^{(1)}) s (\alpha_{\sqrt{\rho_2}} T^2)\}^t = \alpha_{\sqrt{\rho_1}} T^{(1)t} + \alpha_{\sqrt{\rho_2}} T^{(2)t} =$$

$$= \{0, 0, \rho_1 T^{(1)}_2, 0, 0, \dots\} + \{0, 0, \rho_2 T^{(2)}_2, 0 \dots 0\} =$$

$$= \{0, 0, \rho_1 T^{(1)}_2 + \rho_2 T^{(2)}_2, 0, 0 \dots\} \in G^t.$$

Hence G^t is a convex cone. The closure follows directly from Lemma II.5.1.

II.5.6. *Theorem*:

Let $T_2 \in \mathscr{S}'_2$. Then $e^{T_2}_{|s}$ defines a g.f.f. state exactly if T_2 is positive, i.e. if

$$(T_2, f^* \times f) \geqq 0 \quad \text{for all} \quad f \in \mathscr{S}_1.$$

Proof:

We show first that T_2 can have the form $T_2 = F^* \times F$ with $F \in \mathscr{S}'_1$. To this end consider a state of order two, i.e.

$$A(f) = (F_1, f) A_1 + (F_2, f) A_2 \quad \text{with} \quad F_i = F_i^*,$$

$$A_1 = \begin{pmatrix} 0 & 1 \\ 1 & 0 \end{pmatrix}, A_2 = \begin{pmatrix} 0 & i \\ -i & 0 \end{pmatrix} \quad \text{and} \quad \Omega = (1, 0). \qquad \text{We get}$$

$$T_1 = 0$$

$$T_2 = (\Omega, A_1^2 \Omega) F_1 \times F_1 + (\Omega, A_1 A_2 \Omega) F_1 \times F_2 +$$

$$+ (\Omega, A_2 A_1 \Omega) F_2 \times F_1 + (\Omega, A_2^2 \Omega) F_2 \times F_2$$

$$= F_1 \times F_1 - i F_1 \times F_2 + i F_2 \times F_1 + F_2 \times F_2$$

$$= (F_1 + i F_2) \times (F_1 - i F_2) = (F_1 - i F_2)^* \times (F_1 - i F_2).$$

Since F_1 and F_2 are arbitrary elements of $(\mathscr{S}'_1)_h$, it follows that $(F_1 - i F_2)$ is an arbitrary element in $\mathscr{S}'_1$. Hence by Lemma II.5.3 we get $e^{F^* \times F}_{|s} \in G \; \forall F \in \mathscr{S}'_1$. Let now $(T_2, f^* \times f) \geqq 0$. Then $(f, g) = (T_2, f^* \times g)$ defines a bilinear form on $\mathscr{S}_1$.

Let $N = \{f; (T_2, f^* \times f) = 0\}$. Then $\mathscr{S}_1/N$ is a pre-Hilbert space. Denote its completion by $\mathscr{H}$. Since $\mathscr{S}_1$ is separable, it follows that $\mathscr{H}$ is separable. Let e_n be a basis of $\mathscr{H}$. Then $(e_n, [f])$ is a continuous linear functional, since $\mathscr{S}_1 \to \mathscr{S}_1/N$ is continuous.

Denote by (T_n, f) the matrix element $(e_n, [f])$. Then we get

$$(T_2, f^* \times g) = \sum_n ([f], e_n)(e_n, [g]) = \sum_n \overline{(T_n, f)}(T_n, g) =$$

$$= (\sum_n T_n^* \times T_n, f^* + g).$$

This implies

$$T_2 = \sum_n T_n \times T_n \in G^t;$$

which proves the theorem.

Next we want to show that our definition coincides with Greenberg's definition.

II.5.7. *Theorem*:

Let T be a state with $T_1 = 0$ and let T be a g.f.f. state. Then for $f, g \in \mathscr{S}_1$ and $h, l \in \mathscr{S}$, T satisfies the relation

$$(T, h \times (f \times g - g \times f) \times l) = (T, h \times l) \cdot (T, f \times g - g \times f) \cdot \qquad (*)$$

Proof:

Assume that T is a g.f.f. state. Then it follows that $T_{2n+1} = 0$ $(n = 0, 1, 2 \ldots,)$ and T_{2n} is a sum of products of two point functions. In fact,

$$T_{2n}(x_1 \ldots, x_i, x_{i+1}, \ldots x_{2n}) = T_2(x_i, x_{i+1}) T_{2n-2}(x_1 \ldots, \hat{x}_i, \hat{x}_{i+1} \ldots x_{2n}) +$$

$$+ \sum_{\substack{\alpha < i \\ \beta < i \\ \alpha \neq \beta}} T_2(x_\alpha, x_i) T_2(x_\beta x_{i+1}) T_{2n-4}(x_1 \ldots, \hat{x}_\alpha, \hat{x}_\beta \hat{x}_i, \hat{x}_{i+1} \ldots x_{2n}) +$$

$$+ \sum_{\substack{\alpha < i \\ \beta > i+1}} T_2(x_\alpha, x_i) T_2(x_{i+1}, x_\beta) T_{2n-4}(\hat{x}_\alpha, \hat{x}_\beta, \hat{x}_i, \hat{x}_{i+1}) +$$

$$+ \sum_{\substack{\beta < i \\ \alpha > i+1}} T_2(x_i, x_\alpha) T_2(x_\beta, x_{i+1}) T_{2n-4}(\hat{x}_\alpha, \hat{x}_\beta, \hat{x}_i, \hat{x}_{i+1}) +$$

$$+ \sum_{\substack{\alpha > i+1 \\ \beta > i+1 \\ \alpha \neq \beta}} T_2(x_i, x_\alpha) T_2(x_{i+1}, x_\beta) T_{2n-4}(\hat{x}_\alpha, \hat{x}_\beta, \hat{x}_i, \hat{x}_{i+1}),$$

where $\hat{x}_k$ means that the variable x_k does not occur as argument. From this we get

$$(T_{2n}(x_1 \ldots x_2), f(x_i) g(x_{i+1}) - g(x_i) f(x_{i+1})) =$$

$$= (T_2, f \times g - g \times f) \cdot T_{2n-2}(x_1 \ldots \hat{x}_i, \hat{x}_{i+1}, \ldots x_2).$$

This implies equation (*).

We end this section by investigating g.f.f. Wightman states.

II.5.8. *Theorem*:

Let $T_2 \in \mathscr{S}'_2$. Then $e_{|s}^{T_2}$ defines a g.f.f. Wightman state exactly if T_2 fulfills the following conditions:

a) T_2 is positive, i.e. $(T_2, f^* \times f) \geqq 0$ for $f \in \mathscr{S}_1$;

b) T_2 is local, i.e. for $f, g \in \mathscr{S}_1$ with support f spacelike separated from support g we have the relation

$$(T_2, f \times g) = (T_2, g \times f);$$

c) T_2 is translation-invariant, i.e.

$$\alpha'_a T_2 = T_2, \qquad a \in \mathbb{R}^4;$$

d) T_2 satisfies the spectrum condition, i.e. for $f \in \mathscr{S}_1 \cap Sp$ we have

$$(T_2, f^* \times f) = 0.$$

Proof:

Condition a) has been treated in Theorem II.5.6. According to Theorem II.5.7 we get for $f, g \in \mathscr{S}_1$ and $h, l \in \mathscr{S}$

$$(T, h \times (f \times g - g \times f) \times l) = (T, h \times l)(T_2, f \times g - g \times f);$$

hence T is local if and only if T_2 is local. This proves condition b). Since we have $\alpha(e_{|s}^{T_2}) = e_{|s}^{\alpha T_2}$, we get condition c) if and only if $\alpha_a T_2 = T_2$.

Since T should annihilate the spectral ideal Sp, condition d) is evidently necessary.

Assume now that T is translation-invariant and fulfills condition d). In order to show that the spectrum condition is fulfilled, it suffices to show that for elements of the form

$$f = f^1 \times f^2 \times \ldots \times f^n, \qquad f^i \in \mathscr{S}_1$$

and every $h \in \mathscr{S}_1$ with $(\mathscr{F}^{-1}h)(p) = 0$ for $p \in V^+$, the relation

$$\int (T, (\alpha_a f)^* \times \alpha_b f) \, \overline{h(a)} \, h(p) \, da \, db = 0$$

holds.

T_2 is a sum over products $T_2(x_i, x_h)$ with $i < h$. Hence in $(T, \alpha_a f^* \times \alpha_b f)$ there appear products of the form

$$(T_2, \alpha_a f^{i*} \times \alpha_a f^{j*}) = (T_2, f^i \times f^j),$$

$$(T_2, \alpha_a f^{i*} \times \alpha_b f^j),$$

$$(T_2, \alpha_b f^h \times \alpha_b f^l) = (T_2, f^h \times f^l).$$

Hence the assertion is proved if we can show that

$$\int \prod_i (T_2, \alpha_a f^{i*} \times \alpha_b g^i) \, \overline{h(a)} \, h(b) \, da \, db = 0.$$

From the properties of T_2 it follows that

$$\mathrm{supp}\ (\mathcal{F}^{-1}T_2)(p_1, p_2) = \{p_1, p_2; p_1 + p_2 = 0, p_2 \in \bar{V}^+\}.$$

Since the support is a cone, it follows that

$$\mathrm{supp}\ \mathcal{F}^{-1}(T_2, \alpha_a f^{i*} \times \alpha_b g^i) = \{p_1, p_2; p_1 + p_2 = 0, p_2 \in \bar{V}^+\}$$

and hence the integral vanishes, since $(\mathcal{F}^{-1}h)(p) = 0$ for $p \in \bar{V}^+$.

This proves the theorem.

III. FURTHER RESULTS AND PROBLEMS

In the first two chapters we have introduced the algebra $\mathcal{S}$ and have shown its relation to Wightman field theory. We have introduced these notations in order to get away from the so-called linear program and to focus our attention on the non-linear part of the field theory, instead. By doing this, one hopes to get more insight into the existence problem. It should be possible, for instance, to find conditions on the two-, three- and four-point functions which imply the existence of a field theory with these given functions.

In this section I will list some results which have been obtained so far; I will also list some problems. It is not possible to indicate the proofs here. They will be published at a later time.

III/1. *Question*:

What is the structure of the positive cone $\mathcal{S}^+$?

Results:

    a) The cone $\mathcal{S}^+$ is closed.

    b) $\mathcal{S}^+ \cap - \mathcal{S}^+ = \{0\}$.

    c) $\mathcal{S}^+ - \mathcal{S}^+ = \mathcal{S}_h$.

    d) $\mathcal{S}^+$ is not a simplicial cone.

    e) $\mathcal{S}^+$ has no algebraic interior point and hence no topological interior point.

Unsolved:

    f) Which are the extremal rays of $\mathcal{S}^+$?

III/2. *Question*:

What do we know about the dual cone $\mathcal{S}'^+$?

    a) It is a closed cone with a base. The set of states $E(\mathcal{S})$ form a base of $\mathcal{S}'^+$.

    b) $\mathcal{S}'^+ \cap - \mathcal{S}'^+ = \{0\}$.

    c) $\mathcal{S}'^+ - \mathcal{S}'^+ \neq \mathcal{S}'_h$ but it is dense in $\mathcal{S}'_h$.

    d) Every positive linear functional on $\mathcal{S}^+$ is continuous [11].

e) There are enough elements in $\mathscr{S}'^{+}$, i.e. for $p_1, p_2 \in \mathscr{S}^{+}$ with $p_1 \neq p_2$ there exists an element $T \in \mathscr{S}'^{+}$ with $(T, p_1) \neq (Tp_2)$ [5].

f) The extremal points of $E(\mathscr{S})$ are elements T such that the representation $A_T(\mathscr{S})$ has a trivial weak commutant.

g) Every $T \in E(\mathscr{S})$ can be decomposed into extremal states

$$T = \int_0^1 T_\lambda \, d\mu(\lambda),$$

where T_λ is μ almost everywhere extremal.

h) $\mathscr{S}'^{+}$ is not a simplicial cone.

Unsolved:

i) Are the extremal Wightman states those fulfilling the cluster property? I conjecture that this problem has a negative answer.

III/3. *Question*:

What can we say about two-sided ideals and quotient algebras?

a) Since $\mathscr{S}$ contains only one invertible element, it can happen that the closure of a two-sided ideal is all of $\mathscr{S}$.

b) If I is a closed two-sided ideal, then $\mathscr{S}/I$ is again a complete algebra.

c) If I is a closed two-sided ideal and $\mathscr{A}$ a closed subalgebra of $\mathscr{S}$, then $\mathscr{A}/I \cap \mathscr{A}$ is a closed subalgebra of $\mathscr{S}/I$.

d) If I and $\mathscr{A}$ are as in c), then it follows that $\mathscr{A} + I$ is closed.

e) The set J of all closed two-sided ideals is closed under the following operations:
   i. taking the intersection of an arbitrary number of $I_\beta \in J$;
   ii. taking finite sums $\sum_{\text{finite}} I_\beta$.

f) Not every closed two-sided ideal is selfadjoint. Not every closed two-sided ideal is nice in the sense that the quotient algebra is free of zero divisors.

g) The good two-sided ideals are characterized by the following

Theorem:

Let I be a closed symmetric two-sided ideal. Then the following statements are equivalent.

1) $\mathscr{S}/I$ is a semisimple $*$-algebra, i.e.,

$$\overline{(\mathscr{S}/I)}^{+} \cap - \overline{(\mathscr{S}/I)}^{+} = \{0\}.$$

2) $I = \cap N(T)$, where $T \in \mathscr{S}'^{+} \cap I^{\perp}$ and $N(T)$ denotes the null space of T.

3) $I^{+} = I \cap \mathscr{S}^{+}$, where $I^{+} = \{ \sum i_n^{*} \times i_n; i_n \in I \}$.

4) I is an order ideal, i.e. if $a \geq 0$ and $a \in I$, then $0 \leq b \leq a$ implies $b \in I$.

Unsolved:

h) Does the locality ideal I_c have the properties listed in the above theorem?
   The following questions are all unsolved:

III/4. *Question*:

What is the structure of left ideals $\mathscr{L}$?

a) When do we have

$$\mathscr{L} = \cap\, L(T) \qquad L(T) > \mathscr{L} \quad ?$$

b) Let I be a closed two-sided ideal and $\mathscr{L}$ a closed left ideal. What can we say about $I + \mathscr{L}$?

III/5. *Question*?

What can we say about closed subalgebras of $\mathscr{S}$? In particular when does a state on such an algebra always have an extension to the whole algebra $\mathscr{S}$?

REFERENCES

A. Textbooks

[1] N. Bourbaki, "Espaces vectoriels topologiques". Hermann, Paris (1953/55).

[2] R. Jost, "The General Theory of Quantized Fields", Am. Math. Soc., Providence (1965).

[3] A. Pietsch, "Nukleare lokalkonvexe Räume", Akademie-Verlag, Berlin (1969).

[4] D. Ruelle, "Statistical Mechanics—Rigorous Results", Benjamin, New York–Amsterdam (1969).

[5] L. Schwartz, "Théorie des distributions", Hermann, Paris (1950/51).

[6] R.F. Streater and A.S. Wightman, "PCT, Spin and Statistics, and All That", Benjamin, New York (1964).

B. Papers

[1] H. Araki, "Einführung in die axiomatische Quantenfeldtheorie", Lecture Notes, Zürich (1961/62) (unpublished).

[2] H.J. Borchers, "On Structure of the Algebra of Field Operators", *Nuovo Cimento,* **24,** 214 (1962).

[3] O.W. Greenberg, "Generalized Free Fields and Models of Local Field Theory", *Ann. of Phys.,* **16,** 158 (1961).

[4] R. Haag and D. Kastler, "An Algebraic Approach to Quantum Field Theory", *J. Math. Phys.,* **5,** 848 (1964).

[5] G. Lassner and A. Uhlmann, "On Positive Functionals on Algebras of Test Functions for Quantum Fields", *Comm. Math. Phys.,* **7,** 152 (1968).

[6] H. Lehmann, K. Symanzik, and W. Zimmermann, "Zur Formulierung quantisierter Feldtheorien", *Nuovo Cimento,* **1,** 205 (1955).

[7] A.L. Licht and J.S. Toll, "Two-Point Function and Generalized Free Fields", *Nuovo Cimento,* **21,** 346 (1961).

[8] R.T. Powers, "Self-Adjoint Algebras of Unbounded Operators" (preprint) (1971).

[9] L. Schwartz, "Espaces de Fonctions Différentiables à Valeurs Vectorielles", *J. d'Analyse Mathématique (Jerusalem)*, **4**, 88 (1954).

[10] A.S. Wightman, "Quantum Field Theory in Terms of Vacuum Expectation Values", *Phys. Rev.*, **101**, 860 (1956).

[11] W. Wyss, "On Wightman's Theory of Quantized Fields", Boulder Lecture Notes (1968).

STATISTICAL MECHANICS AND FIELD THEORY
R. N. Sen and C. Weil, Editors

Pure Operations in Statistical Physical Theories

C. M. EDWARDS

The Queen's College, Oxford

Abstract

In the operational approach to the theory of statistical physical systems the set of states is represented by a norm-closed cone K in a complete base norm space (V, K, B) and the set of operations is represented by the set $\mathscr{P}$ of positive norm non-increasing linear operators on V. In physical experiments only certain subsets of K are available, and it is supposed that the set $\Gamma(K)$ of such subsets is the set of split faces of K. The properties of two classes of operations are studied. The first class $\mathbb{P}$ has the property that each member leaves every element of $\Gamma(K)$ invariant. The second class is a subset $\mathbb{P}_p$ of $\mathbb{P}$ such that each member sends pure states into pure states. A study is made, in terms of the structure of $\Gamma(K)$, of when such operations are physically relevant. Finally, an examination of $\Gamma(K)$, $\mathbb{P}$ and $\mathbb{P}_p$ is made in the von Neumann algebra model.

1. INTRODUCTION

The first appearance of the expression *pure operation* is to be found in the work of Haag and Kastler [12], in the consideration of systems whose set B of normalized states can be represented by a full family of states of a C^*-algebra, $\mathfrak{A}$, having identity e. They suggested that pure operations should be defined to be of the form

$$f \to \frac{a^*fa}{f(a^*a)}, \qquad \forall f \in B,$$

where $a \in \mathfrak{A}$, $\|a\| \leqq 1$ and for $b \in \mathfrak{A}$, $(a^*fa)(b) = f(a^*ba)$. The state obtained from f by the pure operation is supposed to be that obtained by some local effect on the system. It is implicit in this definition that B be invariant in the sense that $a^*fa/f(a^*a)$ lies in B, $\forall f \in B$, $a \in \mathfrak{A}$, $\|a\| \leqq 1$. This situation can be equivalently described in terms of the set $K = \{\alpha f : f \in B, \alpha \geqq 0\}$ of all states of the system. Then K is a weak* dense sub-cone of the cone $\mathfrak{A}^{*+}$ of positive linear functionals on $\mathfrak{A}$ and is invariant in the sense that $a^*fa \in K$, $\forall f \in K$, $a \in \mathfrak{A}$, $\|a\| \leqq 1$. The corresponding definition of pure operation is then $f \to a^*fa$.

Essentially, the operational approach to the theory of statistical physical systems is based upon the supposition that the set of states of the system is represented by an abstract cone K and that the algebraic approach only appears as an example when K is chosen to be a particular set of positive linear functionals on some C^*-algebra. The arguments for studying the more general situation stem from the difficulty of giving a physical interpretation of the multiplication operation. This is the point of view taken by Davies and Lewis [3], Mielnik [13] and others. Operations are then represented by positive linear mappings from the real vector space $K-K$ to itself. The main purpose of this article is to make a sensible definition of pure operation in the abstract setting and to examine its consequences when applied to algebraic models. In order to achieve this end, it is found necessary to examine the structure of a set $\Gamma(K)$ of subsets of K whose members correspond to subsystems of the original system. Section 2 is devoted to a resumé of the abstract operational approach the details of which may be found in [6, 7]. Section 3 studies the properties of the set $\Gamma(K)$ of split faces of K and is based upon results of Cunningham [1], Gerzon [11] and Wils [16]. In Section 4 the definition of pure operation is given, and in Sections 5 and 6 the preceding results are applied to algebraic models. Full details can be found in [8].

2. THE OPERATIONAL APPROACH

The set K of *states* of some statistical physical system is supposed to form a generating cone for a real vector space V. The set B of normalized states forms a base for K in the sense that every $f \in K$, $f \neq 0$ has a unique decomposition $f = \alpha g$, $g \in B$, $\alpha > 0$, and e is the uniquely defined strictly positive linear functional on V such that $B = e^{-1}(0) \cap K$.

If, for $f \in V$,

$$\|f\|_B = \inf\{\lambda \geqq 0 : f \in \lambda \, \text{conv}\,(B \cup - B)\},$$

then $\|\cdot\|_B$ is a semi-norm on V and it can be shown that if countable mixtures (sums) of states are supposed to exist, $\|\cdot\|_B$ is a norm with respect to which V is complete [9]. Without a great loss in generality it will also be supposed that K is closed in V for the norm topology. For $f \in K$, $e(f)$ represents the *strength* of the state f. An operation j on the system is a linear mapping from V to itself such that $j(K) \subset K$, $e(j(f)) \leqq e(f), \forall f \in K$, the latter condition ensuring that operations do not increase the strength of a state. The set $\mathscr{P}$ of operations on the system is the intersection of the unit ball in the space $\mathfrak{L}(V)$ of bounded linear operators from V to itself with the strongly closed cone $\mathfrak{L}(V)^+ = \{j : j \in \mathfrak{L}(V), j(K) \subset K\}$. The dual space V^* of V possesses a weak* closed generating cone $K^* = \{A : A \in V^*, A(f) \geqq 0, \forall f \in K\}$ and an order unit e in the sense that for $A \in V^*, \exists \lambda \geqq 0$ such that for the partial ordering defined by K^*, $-\lambda e \leqq A \leqq \lambda e$. Further, if for $A \in V^*$

$$\|A\|_e = \inf\{\lambda \geqq 0 : -\lambda e \leqq A \leqq \lambda e\},$$

$\|\cdot\|_e$ is a norm on V^* which coincides with its norm as a dual space. For $j \in \mathcal{P}$ there exists a unique element $T(j)$ in V^* such that $0 \leq T(j) \leq e$ and $T(j)(f) = = e(j(f))$, $\forall f \in K$. $T(j)$ is said to be the *simple observable* (or *effect*) corresponding to j, and $j \rightarrow T(j)$ maps $\mathcal{P}$ onto the set $\mathcal{Q} = \{A : A \in V^*, 0 \leqq A \leqq e\}$. The number

$$\frac{e(j(f))}{e(f)} = \frac{T(j)(f)}{e(f)}$$

can be interpreted as the transmission probability of the state f under any operation j designed to measure $T(j)$.

3. RESTRICTIONS

In the C^*-algebra approach, every representation is supposed to correspond to some restriction of the system under consideration, and quasi-equivalent representations are supposed to produce indistinguishable systems. The corresponding situation in the abstract approach is described below. It presents a strong physical motivation for the usual assumptions concerning representations.

The terminology of Section 2 is maintained. A subset $H \subset K$ describes the set of states of a restriction or, equivalently, a subsystem of the system, provided that:

(i) for $f, g \in H$, $\alpha \geqq 0$, $f + g, \alpha f \in H$,

(ii) for $f \in H$, $f = g + h$, $g, h \in K$, then $g, h \in H$,

(iii) there exists $H' \subset K$ satisfying (i)–(ii) above and such that every $f \in K$ has a unique decomposition $f = h + h'$, $h \in H$, $h' \in H'$.

Condition (i) says that H is a subcone of K, conditions (i)–(ii) that H is a face of K, and conditions (i)–(iii) that H is a split face of K with complementary face H'. If $\Gamma(K)$ denotes the set of split faces of K, the properties of $\Gamma(K)$ are important in the description of the system. If $H \in \Gamma(K)$, H, $H-H$ are norm-closed in V and the triple $(H-H,$ $H, H \cup B)$ has the same properties as the triple (V, K, B) and the norm $\|\cdot\|_{H \cap B}$ coincides with the norm $\|\cdot\|_B$ on $H-H$. There exists a bijection $H \rightarrow p_H$ from $\Gamma(K)$ onto the set $\mathscr{C}$ consisting of elements p such that $p^2 = p$, $0 \leqq p \leqq 1_V$, where 1_V is the identity operator on V and the partial ordering is defined by $\mathfrak{L}(V)^+$, defined by $p_H(f) = = h$, where for $f \in K$, $f = h + h'$ is the unique decomposition of f into $h \in H$, $h' \in H'$. The inverse mapping is defined by $p \rightarrow p(K)$. It follows that the subsystem with set H of states has as its simple observables the set $p_H^*(\mathcal{Q})$ where p_H^* is the adjoint of p_H.

$\Gamma(K)$ is a complete Boolean algebra when ordered by inclusion, where complementation is defined by $H \rightarrow H'$ and where for $\{H_i : i \in \Lambda\} \subset \Gamma(K)$,

$$\bigwedge_{i \in \Lambda} H_i = \bigcap_{i \in \Lambda} H_i, \qquad \bigvee_{i \in \Lambda} H_i = \left(\sum_{i \in \Lambda} H_i\right)^-,$$

the norm closure of the set

$$\sum_{i\in\Lambda} H_i$$

which consists of all finite sums $\sum_{i\in\Lambda'} f_i$, $f_i \in H_i$, $\Lambda' \subset \Lambda$. It is the structure of $\Gamma(K)$ as a Boolean algebra which is crucial for the description of the system. In particular, if $\Gamma(K) = \{\{0\}, K\}$, the system is said to be *primary*. For $H \in \Gamma(K)$, it is easily shown that $G \in \Gamma(H)$ if and only if $G \subset H$, $G \in \Gamma(K)$. Hence $H \in \Gamma(K)$ is the set of states of a primary system if and only if H satisfies the condition $G \in \Gamma(K)$, $G \subset H$ implies $G = \{0\}$ or $G = H$. In this case H is said to be a *primary* element of $\Gamma(K)$ and the set $P\Gamma(K)$ of primary elements of $\Gamma(K)$ is called the *quasi-spectrum* (or *quasi-dual*) of the system. For $G, H \in P\Gamma(K)$, either $G = H$ or $G \cap H = \{0\}$.

The complete Boolean algebra $\Gamma(K)$ is said to be *atomic* when

(i) $K = \vee \{H : H \in P\Gamma(K)\}$,

(ii) For $G \in \Gamma(K)$, $G = \vee \{H : H \in P\Gamma(K), H \subset G\}$.

It is easy to show that in this case (ii) is implied by (i) and the corresponding system is then said to be *atomic*.

One method of defining a restriction of a system is to specify an equilibrium or vacuum state $f \in K$, $f \neq 0$. In this case the corresponding restriction has as its set of states the smallest element H_f of $\Gamma(K)$ containing f. $f \in K$ is said to be *primary* if and only if $H_f \in P\Gamma(K)$ and $f, g \in K$ are said to be *quasi-equivalent* if and only if $H_f = H_g$. Notice that two primary states $f, g \neq 0$ are either quasi-equivalent or disjoint in the sense that $H_f \cap H_g = \{0\}$.

Let $E(K) = \{f : f \in K, 0 \leq g \leq f \Rightarrow g = \alpha f, \alpha \geq 0\}$ be the set of points on extreme rays of K, the set of *pure states* of the system. It is, of course, possible that $E(K) = \{0\}$. However, for $f \in E(K)$, $H_f \in P\Gamma(K)$ and hence every pure state is primary.

$H \in P\Gamma(K)$ is said to be *of Type* I when $H \cap E(K) \neq \{0\}$ and $P_{\rm I}\Gamma(K)$ denotes the set of all such elements. $Q \subset E(K)$ is said to be a *sector* if and only if $Q = H \cap E(K)$, for some $H \in P\Gamma(K)$. There exists a bijection $H \rightarrow H \cap E(K) = E(H)$ between $P_{\rm I}\Gamma(K)$ and the set of sectors either of which may be defined to be the *spectrum* (or *dual*) of the system. The set of sectors forms a disjoint covering of $E(K)$, and if $K_{\rm I} = \vee \{H : H \in P_{\rm I}\Gamma(K)\}$, $K_{\rm I}$ is the smallest element of $\Gamma(K)$ containing $E(K)$ and $E(K_{\rm I}) = E(K)$. An atomic system is said to be *of Type* I if and only if $P\Gamma(K) = P_{\rm I}\Gamma(K)$, or, equivalently, if and only if $K_{\rm I} = K$. It follows that for any system with set of states K, the subsystem described by $K_{\rm I}$ is atomic and of Type I.

4. PURE OPERATIONS

An operation $j \in \mathscr{P}$ on the system with the set of states K is said to be *pure*, provided that

$$\text{(i)} \quad j(H) \subset H, \ \forall H \in \Gamma(K) \Leftrightarrow jp = pj, \ \forall p \in \mathscr{C},$$

$$\text{(ii)} \quad j(E(K)) \subset E(K).$$

The first condition ensures that if only pure operations are supposed to be important, then every subsystem can be treated in isolation as a system in its own right, and adopting this procedure for any subsystem and its complement provides all possible information about the whole system. The second condition merely ensures that pure states are sent into pure states. Notice that this second condition imposes no extra requirement on elements of K_I' and therefore, in studying pure operations, it is sufficient to consider the case $K = K_I$. Let $\mathscr{P}_P(K)$ denote the set of pure operations on such an atomic system of Type I. The basic general result is the following:

Proposition 1.

Let K be the set of states of an atomic system of Type I. Then

(i) For $j \in \mathscr{P}_P(K)$, $H \in P\Gamma(K)$, $p_H j p_H = p_H j = j p_H \in \mathscr{P}_P(H)$ and j is the strong operator limit of the net $\{\sum_{i \in \Lambda} p_{H_i} j p_{H_i}\}$ where $\{H_i : i \in \Lambda\}$ ranges over finite subsets of $P\Gamma(K)$.

(ii) For each $H \in P\Gamma(K)$, let $j_H \in \mathscr{P}_P(H)$. Then there exists uniquely $j \in \mathscr{P}_P(K)$ such that $p_H j p_H = j_H$.

It follows from this result that the problem of identifying pure operations in particular models reduces to the Type I primary situation when $E(K) \neq \{0\}$, $\Gamma(K) = = \{\{0\}, K\}$.

5. THE VON NEUMANN ALGEBRA MODEL

For details of the theory of operator algebras see [4, 5, 10, 14].

Let $\mathfrak{B}$ be a von Neumann algebra acting on the Hilbert space X; let 1_X, the identity operator on X, be the identity in $\mathfrak{B}$, and let $\mathfrak{B}_*$ be the pre-dual of $\mathfrak{B}$. The von Neumann algebra model of the situation described above corresponding to $\mathfrak{B}$ is obtained by making the following identifications:

$$V \quad = \mathfrak{B}_*^h \text{ the self-adjoint part of } \mathfrak{B}_*;$$
$$K \quad = \mathfrak{B}_*^+ \text{ the positive part of } \mathfrak{B}_*;$$
$$B \quad = N(\mathfrak{B}) \text{ the set of normal states of } \mathfrak{B};$$
$$\|\cdot\|_B = \sup\{|\cdot(A)| : A \in \mathfrak{B}, \ \|A\| \leq 1\};$$
$$V^* \quad = \mathfrak{B}^h \text{ the self-adjoint part of } \mathfrak{B};$$
$$K^* \quad = \mathfrak{B}^+ \text{ the positive part of } \mathfrak{B};$$
$$e \quad = 1_X;$$
$$\|\cdot\|_e = \|\cdot\| \text{ the norm in } \mathfrak{B};$$
$$\Gamma(K) = \text{the set of all norm-closed invariant faces of } \mathfrak{B}_*^+. \text{ (A norm-closed face}$$
$$H \text{ of } \mathfrak{B}_*^+ \text{ is said to be invariant if } A^* f A \in H, \ \forall f \in H, \ A \in \mathfrak{B}.)$$

There exists a complete Boolean algebra isomorphism $H \to E_H$ from $\Gamma(K)$ onto the complete Boolean algebra $\mathscr{C}(\mathfrak{B})$ of central projections in $\mathfrak{B}$, defined by the condition that E_H is the smallest element of $\mathscr{C}(\mathfrak{B})$ such that $E_H f E_H = f$, $\forall f \in H$. The inverse mapping is defined by $E \to EKE$.

A von Neumann algebra is said to be atomic if its lattice of projections is atomic. The identifications above lead immediately to the following consequences:

(a) $H \in \Gamma(K)$ is primary if and only if $E_H \mathfrak{B} E_H$ is a factor.

(b) The system with set of states K is atomic if and only if $\mathfrak{B}$ has atomic center.

(c) $H \in P\Gamma(K)$ is of Type I if and only if $E_H \mathfrak{B} E_H$ is of Type I.

(d) An atomic system is of Type I if and only if $\mathfrak{B}$ is of Type I and has atomic center.

In Section 4 it was shown that in order to study pure operations it is sufficient to find all pure operations for Type I primary systems and then use Proposition 1 to extend to the only other relevant case, atomic systems of Type I. It follows from (c) that it is only necessary to examine pure operations for Type I factors. However, this situation has been solved by Davies [2] and Størmer [15]. Suppose $\mathfrak{B}$ is a Type I factor and let $j \in \mathscr{P}_P(K)$. Then for each $f \in K$ one of the following results holds:

(1) $j(f) = U^* f U$, $U \in \mathfrak{B}$, $U^* U = T(j) = j^*(1_X)$;

(2) $j(f) = c^* U^* f U c$, $U \in \mathfrak{B}$, $U^* U = T(j) = j^*(1_X)$, $c : X \to X^c$ is the conjugate mapping from X to its conjugate space X_c;

(3) $j(f) = f(T(j))\, \omega_x$, where $x \in X$, $\|x\| = 1$, and ω_x is a pure state of $\mathfrak{B}$ defined for $A \in \mathfrak{B}$ by $\omega_x(A) = \langle Ax, x \rangle$.

A combination of Proposition 1 and this result leads immediately to the following theorem:

Proposition 2.

Let $\mathfrak{B}$ be a Type I von Neumann algebra with atomic center $\mathfrak{C}$ acting on the Hilbert space X. For each $j \in \mathscr{P}_P(K)$ there exist mutually disjoint subsets $\Lambda_1, \Lambda_2, \Lambda_3$ of $P\Gamma(K)$ and mutually orthogonal projections E_1, E_2, E_3 in $\mathfrak{C}$ such that

$$\bigcup_{k=1}^{3} \Lambda_k = P\Gamma(K), \qquad \sum_{k=1}^{3} E_k = 1_X,$$

defined by

$$E_k = \sum_{H \in \Lambda_k} E_H, \qquad k = 1, 2, 3,$$

and, for $f \in K$,

(1) If $f = E_1 f E_1$, $j(f) = U^* f U$, $U \in E_1 \mathfrak{B}_1 E_1$, $U^* U = E_1 T(j) = j^*(E_1)$;

(2) If $f = E_2 f E_2$, $j(f) = c^* U^* f U c$, $U \in E_2 \mathfrak{B} E_2$, $U^* U = E_2 T(j) = j^*(E_2)$;

(3) If $f = E_3 f E_3$, $j(f) = \displaystyle\sum_{H \in \Lambda_3} f(E_H T(j))\, \omega_{x_H}$, $x_H \in E_H X$, $\|x_H\| = 1$ and ω_{x_H} is a pure state of $\mathfrak{B}$.

For arbitrary $f \in K$,

$$j(f) = j(E_1 f E_1) + j(E_2 f E_2) + j(E_3 f E_3).$$

This result solves the problem of describing pure operations for von Neumann algebras.

6. THE C^*-ALGEBRA MODEL

Let $\mathfrak{A}$ be a C^*-algebra with identity e, dual $\mathfrak{A}^*$ and von Neumann envelope $\mathfrak{A}^{**}$. If $\mathfrak{A}$ is identified with its canonical embedding in $\mathfrak{A}^{**}$, e is also the identity in the von Neumann algebra $\mathfrak{A}^{**}$ which has $\mathfrak{A}^*$ as its pre-dual. In this case there not only exists a bijection between $\Gamma(\mathfrak{A}^{*+})$ and the set of central projections in $\mathfrak{A}^{**}$ but also from $\Gamma(\mathfrak{A}^{*+})$ onto the set of quasi-equivalence classes of representations of $\mathfrak{A}$. To be precise, for $H \in \Gamma(\mathfrak{A}^{*+})$, the corresponding quasi-equivalence class $[\phi_H]$ is that containing the representation $\phi_H = \underset{f \in H}{\oplus} \phi_f$, where ϕ_f is the cyclic representation of $\mathfrak{A}$ obtained from f by the G.N.S. construction.

Alternatively, $[\phi_H]$ is that quasi-equivalence class which contains the representation $a \to E_H a E_H$, where E_H is the central projection in $\mathfrak{A}^{**}$ corresponding to H and $\mathfrak{A}$ is identified with its universal representation. The inverse mapping is defined as follows. If ϕ is any representation of $\mathfrak{A}$, then $K_\phi = \{\omega \circ \phi : \omega \in (\phi(\mathfrak{A})'')_*^+ \}$ lies in $\Gamma(\mathfrak{A}^{*+})$, where $\phi(\mathfrak{A})''$ is the von Neumann algebra generated by $\phi(\mathfrak{A})$. Two representations ϕ_1, ϕ_2 of $\mathfrak{A}$ are quasi-equivalent if and only if $K_{\phi_1} = K_{\phi_2}$ and therefore K_ϕ depends only on the quasi-equivalence class $[\phi]$. When K is identified with $\mathfrak{A}^{*+}$, the concepts associated with $\Gamma(K)$ in Section 3 have the following analogues in representation theory:

(a) $H \in \Gamma(K)$ is primary if and only if ϕ_H is primary or, equivalently, $\phi_H(\mathfrak{A})''$ is a factor.

(b) There exists a bijection $H \to [\phi_H]$ from $P\Gamma(K)$ onto the quasi-spectrum (or quasi-dual) $\hat{\mathfrak{A}}$ of $\mathfrak{A}$.

(c) $H \in P\Gamma(K)$ is of Type I if and only if $\phi_H(\mathfrak{A})''$ is a Type I factor or, equivalently, if and only if ϕ_H is quasi-equivalent to an irreducible representation which may be taken to be ϕ_f, where $f \in E(H) = H \cap E(K)$.

(d) There exists a bijection $H \to [\phi_H]$ from $P_1\Gamma(K)$ onto the spectrum (or dual) $\hat{\mathfrak{A}}$ of $\mathfrak{A}$.

(e) $f \in K$ is primary if and only if ϕ_f is primary.

(f) $f, g \in K$ are quasi-equivalent if and only if ϕ_f, ϕ_g are quasi-equivalent.

(g) K_I is weak* dense in K (by the Krein–Milman Theorem) and ϕ_{K_I} is quasi-equivalent to the reduced atomic representation of $\mathfrak{A}$ (the direct sum of irreducible representations, one taken from each unitary equivalence class).

The usual C^*-algebra approach requires not that $\mathfrak{A}^{*+}$ be regarded as the set of states of the system but that the set of states be represented by some weak* dense sub-cone C of $\mathfrak{A}^{*+}$. The weak* denseness of C in $\mathfrak{A}^{*+}$ ensures that the representation $\phi_C = \underset{f \in C}{\oplus} \phi_f$ is faithful. The local simple observables are then identified with the set $\{\phi_C(a) : a \in \mathfrak{A}, 0 \leq a \leq e\}$, and all simple observables with the set $\{A : A \in \phi_C(\mathfrak{A})'', 0 \leq A \leq 1_X\}$ where 1_X denotes the identity operator on the Hilbert space on which ϕ_C is defined. In order that this agree with the von Neumann algebra model it is necessary that the set K of all states of the system be identifiable with $(\phi(\mathfrak{A})'')_*^+$ or, equivalently, with the smallest element of $\Gamma(\mathfrak{A}^{*+})$ containing C. In this case it would seem reasonable to require at the outset that C be an element of $\Gamma(\mathfrak{A}^{*+})$ in which case $K = C$. Therefore it follows that each element of $\Gamma(\mathfrak{A}^{*+})$ weak* dense in $\mathfrak{A}^{*+}$ defines a possible set of states of some system. Note that all elements of $\Gamma(\mathfrak{A}^{*+})$ are weak* dense in $\mathfrak{A}^{*+}$ if and only if $\mathfrak{A}$ is simple. If it is supposed that only the local simple observables are important, then the elements of $\Gamma(\mathfrak{A}^{*+})$ which are weak* dense in $\mathfrak{A}^{*+}$ form sets of states of systems which are "physically equivalent" in the sense of Haag and Kastler [12]. If it is not required to separate physically equivalent systems, it follows from (g) above that it is sufficient to choose $K = = (\mathfrak{A}^{*+})_1$ or, what is equivalent, choose the von Neumann algebra model defined by $\mathfrak{B} = \phi(\mathfrak{A})''$ where ϕ is the reduced atomic representation of $\mathfrak{A}$. Then $\mathfrak{B}$ is of Type I and has an atomic center in which case the results of Section 5 apply. From Proposition 2 it follows that a pure operation j is in general a sum of three components, one being described by a sequence of pure states, one from each sector, and the other two by a bounded linear and a bounded conjugate linear operator, respectively, on the Hilbert space X on which ϕ is defined. It is not clear precisely when such operators may be chosen to lie in $\phi(\mathfrak{A})$ or its conjugate, or when the result for the last two terms will reduce to the definition of Haag and Kastler.

REFERENCES

[1] F. Cunningham, "*L*-structure in *L*-spaces", *Trans. Am. Math. Soc.*, **95**, 274–299 (1960).

[2] E. B. Davies, "Quantum stochastic processes", *Commun. math. Phys.*, **15**, 277–304 (1969).

[3] E. B. Davies and J. T. Lewis, "An operational approach to quantum probability", *Commun. math. Phys.*, **17**, 239–260 (1970).

[4] J. Dixmier, "Les *C**-algèbres et leurs representations". Gauthier–Villars, Paris (1964).

[5] J. Dixmier, "Les algèbres d'operateurs dans l'espace hilbertien". Gauthier–Villars, Paris (1969).

[6] C. M. Edwards, "The operational approach to algebraic quantum theory, I", *Commun. math. Phys.*, **16**, 207–230 (1970).

[7] C. M. Edwards, "Classes of operations in quantum theory", *Commun. math. Phys.*, **20**, 26–56 (1971).

[8] C. M. Edwards, "Theory of pure operations", *Commun. math. Phys.*, **24**, 260–288 (1972).

[9] C. M. Edwards and M. A. Gerzon, "Monotone convergence in partially ordered vector spaces", *Ann. Inst. Henri Poincaré*, **A 12**, 323–328 (1970).

[10] E. G. Effros, "Order ideals in a C^*-algebra and its dual", *Duke Math. J.* **30**, 391–412 (1963).

[11] M. A. Gerzon, "Split faces of convex sets" (In preparation).

[12] R. Haag and D. Kastler, "An algebraic approach to quantum field theory", *J. math. Phys.*, **5**, 846–861 (1964).

[13] B. Mielnik, "Theory of filters", *Commun. math. Phys.*, **15**, 1–46 (1969).

[14] R. T. Prosser, "On the ideal structure of operator algebras", *Mem. Am. Math. Soc.*, 45 (1963).

[15] E. Størmer, "Positive linear maps of operator algebras", *Acta math.*, **110**, 233–278 (1963).

[16] W. Wils, "The ideal center of a partially ordered vector space", *Acta math.*, **127**, 41–79 (1971).

Two-Fluid Aspects of Condensed Matter

CHARLES P. ENZ*

IBM Zürich Research Laboratory, 8803 Rüschlikon (Switzerland)

Abstract

The close analogy in the hydrodynamics of a wide class of many-body systems characterized by a state with a condensed phase is stressed. These systems are neutral superfluids (liquid helium), superconductors, dielectric crystals, magnetic crystals and nematic liquid crystals. The condensed phase gives rise to a two-fluid description which is physically characterized by its modes of excitation, typically first and second sound.

This article is intended to balance against the more formal aspects of this book and, therefore, will stress the phenomenological properties of statistical systems, that is, of condensed matter.

It has been realized in recent times that a number of systems can be phenomenologically described by two-fluid hydrodynamic equations of astonishing similarity [1]. The class of these systems is characterized by a phase transition into a state in which a condensed phase can be defined. By analogy with superfluid helium this condensed phase then gives rise to a two-fluid description with its characteristic hydrodynamic modes of propagation: first and second sound.

Quite generally, a phase transition can be characterized by a change of order. In the gas–liquid transition this change goes from no order to short-range order and in the liquid–solid transition from short-range to long-range order. A change from short-range or no order to some type of long-range order also occurs in the magnetic and electric transitions of solids, in the nematic transition of liquid crystals and, in the sense of off-diagonal long-range order, in the superfluid and superconductive transitions. A definite change of long-range order occurs in the structural phase transitions of solids, while order–disorder transitions of solids are similar to the solid–liquid case. Finally, the metal–insulator transition, which has been much discussed recently, is analogous to the structural type but with respect to the electron

* On partial leave from Institut de Physique Théorique, Université de Genève, Geneva, Switzerland.

system. With the exception of the latter case and of the superconducting transition which also occurs in the electron system (this may also be said of itinerant magnetism), phase transitions occur in systems of atoms or molecules.

In putting up this classification one should, of course, bear in mind that Nature pops out of any scheme forced upon her! This is even more true when one wants to invoke broken symmetries, Goldstone bosons, soft modes or order parameters to characterize phase transitions, which works only in a more restricted class of systems. In our case this means restriction to phase transitions leading from short-range or no order to some kind of long-range order. The principal systems having this property are:

1)   Neutral Superfluids (He II)
2)   Superconductors
3)   Dielectric Crystals
4)   Magnetic Crystals
5)   Nematic Liquid Crystals.

The respective broken symmetry groups are the gauge group in cases 1 and 2, the translation (and rotation) group in case 3, and the rotation group in cases 4 and 5. The breaking of these symmetry groups is related by Noether's theorem to non-conservation of the physical quantity acting as generator of this group, which is respectively particle number, momentum and angular momentum. From the statistical ensemble point of view this means that the system is described by a density matrix ρ which, although commuting with the Hamiltonian H of the system, does not commute with this group generator G. The generator, however, commutes with H:

$$[\rho, G] \neq 0, \qquad [\rho, H] = 0, \qquad [G, H] = 0. \tag{1}$$

There exists a physical density $\eta(r)$ playing the role of an order parameter, the time dependence of which is given by the Heisenberg representation

$$\eta(r, t) = e^{i(H - \lambda G)t} \eta(r) e^{-i(H - \lambda G)t}, \tag{2}$$

where $-\lambda G$ is a potential energy of thermodynamic origin $(T \neq 0)$. Because of (1), the average of η still depends on time:

$$\langle \eta(r, t) \rangle \equiv \mathrm{Tr}(\rho \eta(r, t)) = \langle e^{-i\lambda Gt} \eta(r) e^{i\lambda Gt} \rangle. \tag{3}$$

Above the phase transition $(T > T_c)$, $[\rho, G] = 0$ and $\langle \eta \rangle = 0$. In a two-fluid $(T < T_c)$ description, $\langle \eta \rangle$ determines the hydrodynamic motion of the second fluid or superfluid. A ρ for which $[\rho, G] \neq 0$ can be obtained from the equilibrium density matrix $\rho_e = \exp[\beta(F - H + \lambda G)]$ by adding to the Hamiltonian H a coupling to an external source ϕ_{ext},

$$H' = \int d^3 r M(r) \phi_{ext}(r, t) \eta(r) + h \cdot c. \tag{4}$$

Then $\langle \eta \rangle$ is just given, in linear response to H', by a Kubo formula. If both M and ϕ_{ext} are independent of r, the above formulas hold with an r-independent $\langle \eta \rangle$. If, however, M or ϕ_{ext} depend on r, then the system will be close to local equilibrium, as described by the density matrix $\rho_{\text{l.e.}} = \exp\left[\int \beta(r) \{ f(r) - h(r) + \lambda(r) g(r) \} d^3 r \right]$, where $f(r), h(r), g(r)$ are the densities associated with the free energy F, the Hamiltonian H, and the group generator G, respectively. This local equilibrium is perturbed by time-dependent variations of the external source coupling H'. Then $\langle \eta \rangle$ also becomes a function of r.

In the superfluid, η is the creation operator Ψ^+, and the external source ϕ_{ext} is the external annihilation operator Ψ_{ext}. Then (4), with an appropriate matrix element M, is just a tunneling Hamiltonian coupling the external particle source to the system. We start with an infinitely long wave coupling H'. Since Ψ is complex we can write

$$\langle \Psi^+(t) \rangle = \sqrt{n_0} \exp\left[-i\phi(t) \right], \tag{5}$$

where n_0 is the condensate density, G is the number operator N, and λ is the chemical potential $m\mu$, where m is the atomic mass. Since Ψ^+ raises the particle number by one, we conclude that

$$e^{-im\mu Nt} \Psi^+ e^{im\mu Nt} = \Psi^+ e^{im\mu t}, \tag{6}$$

so that

$$\dot{\phi} = -m\mu. \tag{7}$$

If the coupling H' varies with r, then n_0, ϕ and μ depend on the position r and there is a nonvanishing superfluid velocity defined by

$$v_s = \frac{\hbar}{m} \nabla\phi. \tag{8}$$

From (7) and (8) follows the hydrodynamic equation of motion for the second fluid or superfluid

$$\dot{v}_s = -\nabla\mu. \tag{9}$$

This is of course well known. What is less well-known is that the other systems under consideration are closely analogous.

In a superconductor, η is the creation operator of a Cooper pair S^+, and ϕ_{ext} the external annihilation operator S_{ext}. H' is the tunneling Hamiltonian of a Josephson junction, and

$$\langle S^+(t) \rangle = \sqrt{n_0} e^{-i\phi(t)}. \tag{10}$$

Since S^+ raises the particle number by two and $G = N$, $\lambda = m\mu$ as before, but m is now the electron mass,

$$e^{-im\mu Nt} S^+ e^{im\mu Nt} = S^+ e^{i2m\mu t}, \tag{11}$$

so that

$$\dot{\phi} = -2m\mu. \tag{12}$$

With

$$v_s = \frac{\hbar}{2m}\nabla\phi, \tag{13}$$

the former equation of motion (9) follows. This analogy between superconductor and neutral superfluid was first stressed by Ginzburg [2].

In the case of a (planar) ferromagnet, η is the spin raising operator at site i, $S_i^+ = S_i^x + iS_i^y$, ϕ_{ext} is the external magnetic field, and $-M$ is the Bohr magneton μ_B. Then

$$\frac{1}{v}\langle S_i^+(t)\rangle = M_\perp\, e^{i\phi(t)}, \tag{14}$$

where v is the volume of the unit cell, $M_\perp$ is the magnetization in the x, y-plane, and ϕ is the precession angle. Here $G = -\sum_i S_i^z$ and $\lambda = m\mu = -\mu_B(H_z - h_z)$, where H_z is the external field and $h_z(M_z T)$ is the equilibrium field at given magnetization M_z and temperature T.

Since S_i^+ raises the spin by one, we have

$$\exp\left(-im\mu\sum_i S_i^z t\right) S_i^+ \exp\left(im\mu\sum_i S_i^z t\right) = S_i^+ \exp\left(-im\mu t\right), \tag{15}$$

so that equation (7) follows again, and with the definition (8), also equation (9). This analogy with superfluids was first pointed out by Halperin and Hohenberg [3].

In the dielectric crystal, η is the displacement operator $d(R)$ at lattice position R, G is the momentum operator, and $-\lambda$ a velocity parameter v. With these definitions, equation (3) leads to the elastic displacement field

$$\langle d(r + vt)\rangle = u(r + vt) \tag{16}$$

which satisfies the equation of motion of elasticity theory,

$$\ddot{u} = \frac{1}{\rho}\nabla\sigma, \tag{17}$$

where σ is the stress tensor and ρ the mass density. Defining a lattice velocity by

$$v_L = \dot{u}, \tag{18}$$

equation (17) takes a form analogous to equation (9),

$$\dot{v}_L = \frac{1}{\rho}\nabla\sigma, \tag{19}$$

so that v_L here plays the role of the superfluid velocity and the elastic displacement field represents the condensed phase. This analogy with superfluidity was first emphasized by Goetze and Michel [4].

In the nematic phase of a liquid crystal, the order parameter η is to be associated with the quadrupole moment Q_{ij} of the molecules, as was pointed out by Martin et al. [5]. If we assume that the molecules have an axis of symmetry, then the average

$$\langle Q_{ij} \rangle = q(n_i n_j - \tfrac{1}{3}\delta_{ij}) \qquad (20)$$

defines the director field $n(r)$ which is an axial unit vector. Here the group generator G is the orbital angular momentum operator and $-\lambda$ is an angular velocity ω. The external source ϕ_{ext} is an electric field E. For a homogeneous field E, equation (4) has the form

$$H' = \sum_{ij} Q_{ij} E_i E_j. \qquad (21)$$

From rotational invariance of this expression it follows for an infinitesimal rotation $\delta E_i = (\omega \times E)_i \delta t = \sum_{kl} \varepsilon_{ikl}\omega_k E_l \delta t$ that

$$\delta Q_{ij} = \sum_{kl} (Q_{il}\varepsilon_{jkl} + Q_{jl}\varepsilon_{ikl})\,\omega_k \delta t, \qquad (22)$$

and by comparison with (20), that

$$\delta n = (\omega \times n)\,\delta t. \qquad (23)$$

In an inhomogeneous electric field $E(r)$ a drift velocity $v(r)$ develops in the liquid, and $\omega = \tfrac{1}{2}(\nabla \times v)$. Then $\dot{n}$ is not given by the rotational contribution $\omega \times n$ alone, but is also influenced by this drift. The result is again an expression analogous to equation (9) [5]:

$$\dot{n} = -\nabla\chi, \qquad (24)$$

which means that the director field n represents the condensed phase.

The two-fluid description is completed by hydrodynamic equations for the transport density ρ_T, the momentum density j, and the energy density per unit mass e. The equation for ρ_T is a local conservation law

$$\dot{\rho}_T + \nabla \cdot j_T = 0, \qquad (25)$$

where j_T is the transport current density. j satisfies in general only a balance law

$$\dot{j} + \nabla\pi = -\frac{1}{\tau_J}j, \qquad (26)$$

where π is the momentum flux and the right-hand side describes momentum dissipation in the lattice (by Umklapp or imperfections). This term is evidently missing in fluid systems ($\tau_J = \infty$). e satisfies the local conservation law

$$(\rho e)^{\cdot} + \nabla \cdot J_e = 0, \qquad (27)$$

where ρ is the mass density of the system and $\boldsymbol{J}_e$ the energy flux.

In the superfluid and superconductor, the translational motion of the total mass density ρ with $\boldsymbol{v}_s$ defines the supercurrent

$$\boldsymbol{j}_s = \rho\boldsymbol{v}_s, \tag{28}$$

and the motion relative to $\boldsymbol{v}_s$ defines the normal current of excitations

$$\boldsymbol{j}_n = \rho_n(\boldsymbol{v}_n - \boldsymbol{v}_s), \tag{29}$$

ρ_n being the excitation mass density. The total current is

$$\boldsymbol{j} = \boldsymbol{j}_n + \boldsymbol{j}_s = \rho_n\boldsymbol{v}_n + \rho_s\boldsymbol{v}_s, \tag{30}$$

where

$$\rho_s = \rho - \rho_n \tag{31}$$

is the superfluid density. Furthermore,

$$\rho_T = \rho, \qquad \boldsymbol{j}_T = \boldsymbol{j}. \tag{32}$$

In the dielectric crystal, mass conservation in the lattice is expressed by equation (25) with

$$\rho_T = \rho, \qquad \boldsymbol{j}_T = \boldsymbol{j}_L \equiv \rho\boldsymbol{v}_L. \tag{33}$$

But the momentum balance (26) holds for the phonon fluid

$$\boldsymbol{j} = \boldsymbol{j}_p \equiv \rho_p\boldsymbol{v}_p, \tag{34}$$

where $\boldsymbol{v}_p$ is the phonon drift velocity and ρ_p is the associated excitation mass density.

In the planar ferromagnetic crystal, equation (25) expresses conservation of magnetization M_z with the identification

$$\rho_T = mM_z, \qquad \boldsymbol{j}_T = \boldsymbol{j}_s \equiv \rho_s\boldsymbol{v}_s, \tag{35}$$

where ρ_s is a stiffness parameter [3]. By analogy with the dielectric crystal, the momentum balance holds for the magnon fluid

$$\boldsymbol{j} = \boldsymbol{j}_M \equiv \rho_M\boldsymbol{v}_M \tag{36}$$

where $\boldsymbol{v}_M$ is the magnon drift velocity (which exists only for a linear dispersion law $\omega_q = cq$ and hence not for the isotropic ferromagnet) and ρ_M is the associated excitation mass density.

Finally, in the nematic liquid crystal

$$\rho_T = \rho, \qquad \boldsymbol{j}_T = \boldsymbol{j} \equiv \rho\boldsymbol{v}. \tag{37}$$

This case is exceptional in that the "normal" motion described by equations (25) and (26) is completely decoupled from the "superfluid" motion described by equation (24).

It is useful to transform the energy conservation (27) into an entropy balance

$$(\rho s)^{\cdot} + \mathbf{V} \cdot \mathbf{J}_s = \sigma, \tag{38}$$

where s is the entropy per unit mass, $\mathbf{J}_s$ is the entropy flux, and σ the entropy production density. This is achieved with the aid of the thermodynamic differential form for $d(\rho e) - Td(\rho s)$. This must be a linear form in the differentials of the free variables, that is, those which are governed by a hydrodynamic equation of motion. Hence

$$d(\rho e) - Td(\rho s) = \alpha d\rho_T + \boldsymbol{\beta} \cdot d\mathbf{j} + \boldsymbol{\gamma} \cdot d\mathbf{v}_s. \tag{39}$$

Substituting from equations (9), (25), (26), (27), we find

$$T(\rho s)^{\cdot} = - \mathbf{V} \cdot \mathbf{J}_e + \alpha \mathbf{V} \cdot \mathbf{j}_T + \boldsymbol{\beta} \cdot \left(\mathbf{V}\pi + \frac{1}{\tau_J}\mathbf{j} \right) + \boldsymbol{\gamma} \cdot \mathbf{V}\mu. \tag{40}$$

Here the generalized fluxes, i.e. the quantities on the right-hand side on which the gradient acts, consist of reactive and dissipative parts, the latter (together with the dissipative term $1/(\tau_J)\mathbf{j}$) give rise to the entropy production σ. Thus leaving out the dissipative parts, $\sigma = 0$ and equation (38) expresses entropy conservation or adiabaticity. It then follows that the reactive part of the right-hand side of (40) divided by T must be the divergence of a vector. This condition, together with expressions for the reactive parts of $\mathbf{V} \cdot \mathbf{j}_T, \mathbf{V}\pi$ and $\mathbf{V}\mu$, determines the reactive parts of $\mathbf{J}_e$ and $\mathbf{J}_s$ as well as the coefficient γ, while α and β are determined from established physical laws.

In the linear response approximation, the dissipative parts of the fluxes $\mathbf{J}_e, \mathbf{j}_T, \pi$ and μ are linear functions of the generalized forces, i.e. of odd powers of the gradients of the coefficients in the expression (39) for $d(\rho e)$. Invariance against time reversal and rotations restrict these linear functions. Marking the dissipative parts with a prime, we then have to lowest order in the gradients:

$$\begin{aligned}
\mathbf{J}'_e &= - \kappa \mathbf{V}T \\
\mathbf{j}'_T &= - \xi \mathbf{V}\alpha \\
\pi' &= - (\zeta_1 \mathbf{V} \cdot \boldsymbol{\gamma} + \zeta_2 \mathbf{V} \cdot \boldsymbol{\beta})\, \mathbb{1} - (\eta_1 \mathbf{V}) \cdot \boldsymbol{\beta} - (\eta_2 \mathbf{V}) \cdot \boldsymbol{\gamma} \\
\mu' &= - (\zeta_3 \mathbf{V}) \cdot \boldsymbol{\gamma} - (\zeta_4 \mathbf{V}) \cdot \boldsymbol{\beta}.
\end{aligned} \tag{41}$$

Here κ is the heat conductivity, ξ a diffusion coefficient and the ζ's and η's are viscosities. These quantities may be tensors of various rank. They lead to damping of the hydrodynamic modes.

In global equilibrium the vector quantities in equations (39) and (40) all vanish. Thus for small excitation of the system we can treat them to first order as well as derivatives of all quantities. To this order then there is no entropy production,

$\sigma = 0$. Eliminating j and v_s with the help of equations (9) and (26), we are left with the two equations of motion (25) and (38) for the transport density ρ_T and the entropy s. The first leads to a nonthermal mode of propagation, the second to a thermal mode. The nonthermal mode is an isothermal first sound or phonon in the superfluid, the dielectric crystal and the nematic liquid crystal, a plasmon in the superconductor (due to the Coulomb forces) and an isothermal spin-wave or magnon in the magnetic crystal. The thermal mode is an incompressible second sound or second phonon in the superfluid, the superconductor, the dielectric crystal, a second magnon in the magnetic crystal with linear magnon dispersion and a heat diffusion in the nematic liquid crystal. The fact that the latter does not admit a second sound is due to the lack of coupling between the "normal" and "superfluid" motions mentioned above. While in the superfluid second sound is easily observable at arbitrarily long wave length and at all but very low temperatures below the λ-point, this is not so in crystalline systems. This is due to momentum dissipation through the term $1/(\tau_J)j$ in equation (26), which is missing in the superfluid as mentioned above. The consequence of this term is the existence of a window condition [6] for the frequency,

$$\frac{1}{\tau_J} \ll \omega \ll \frac{1}{\tau_{eq}}, \tag{42}$$

where τ_{eq} is the relaxation time responsible for local thermal equilibrium. This window condition was the reason for the considerable difficulty in observing second sound in dielectric crystals [7]. The same difficulty exists for the not yet observed second magnon for which, however, the chances of detection in particular systems are not unrealistic [8]. In the superconductor the window condition practically excludes observation of second sound because electron–phonon interaction reduces τ_J too severely [1].

In reality the nonthermal and thermal modes are coupled through the temperature dependence of the nonthermal quantity α. This coupling transforms the nonthermal mode from isothermal to adiabatic but does not appreciably alter the thermal mode.

In addition to these modes there are others, more particular ones. One has fourth sound in the superfluid and its analogue in the magnetic case under conditions where $j = 0$ [1]. In the superfluid this is achieved by enclosing it in a porous substrate so that the normal component does not flow. In magnetic crystals $j = 0$ holds, e.g., for a quadratic magnon dispersion. The geometry-dependent mode of Poisenille flow exists in the systems that allow for a second sound and has been observed in liquid [9] and solid helium [10]. In nematic liquid crystals, particular diffusive modes of the director field exist and have been observed [5].

In closing, I want to make the link to the soft-mode concept which has recently been applied with remarkable success to a large number of phase transitions [11]. In liquid helium both the second-sound velocity and damping go to zero at the λ-point [12], while the first-sound velocity and absorption do not vanish. This means that second sound becomes soft, i.e. its complex frequency goes to zero. Above the λ-point it moves away from zero along the imaginary axis, which means that it has become a thermal diffusion mode. The same is true for the superconductor, where the second-sound mode is overdamped below the transition temperature and diffusive above.

Finally I want to draw attention to the analogy between the superfluid and external fields at $T = 0$ (field theory). Indeed, the latter can be described as coherent states which are the $T = 0$ anaolgue of a condensed phase. Thus quantized fields in the presence of external fields can be viewed as a two-fluid description of field theory.

REFERENCES

[1] C. P. Enz, "Unified Hydrodynamic Description of Ordered Systems" (to be published).

[2] V. L. Ginzburg, *J. Exp. Theor. Phys. USSR,* **14**, 177 (1944) and *J. Phys. USSR,* **8**, 148 (1944).

[3] B. I. Halperin and P. C. Hohenberg, *Phys. Rev.,* **188**, 898 (1969).

[4] W. Götze and K. H. Michel, *Phys. Rev.,* **156**, 963 (1967).

[5] D. Forster, T. C. Lubensky, P. C. Martin, J. Swift, and P. S. Pershan, *Phys. Rev. Letters,* **26**, 1016 (1971).

[6] E. W. Prohowsky and J. A. Krumhansl, *Phys. Rev.,* **133**, A1403 (1964);
R. A. Guyer and J. A. Krumhansl, *Phys. Rev.,* **133**, A1411 (1964).

[7] C. C. Ackerman, B. Bertman, H. A. Fairbank, and R. A. Guyer, *Phys. Rev. Letters,* **16**, 789 (1966);
C. C. Ackerman and W. C. Overton, Jr., *Phys. Rev. Letters,* **22**, 764 (1969);
T. F. McNelly, S. J. Rogers, D. J. Channin, R. J. Rollefson, W. M. Goubou, G. E. Schmidt, J. A. Krumhansl, and R. O. Pohl, *Phys. Rev. Letters,* **24**, 100 (1970);
H. E. Jackson, C. T. Walker, and T. F. McNelly, *Phys. Rev. Letters,* **25**, 26 (1970).

[8] K. H. Michel and F. Schwabl, *Phys. Rev. Letters,* **26**, 1568 (1971).

[9] R. W. Whitworth, *Proc. Roy. Soc.,* **A 246**, 390 (1958).

[10] L. P. Mezov-Deglin, *J. Exp. Theor. Phys. USSR,* **46**, 1926 (1964) (English transl.: *Soviet Phys. JETP,* **19**, 1297 (1964)).

[11] T. Schneider, G. Srinivasan, and C. P. Enz, *Phys. Rev.,* **A5**, 1528 (1972).

[12] J. Wilks, "The Properties of Liquid and Solid Helium", Fig. 22, p. 71 and Fig. 18, p. 209. Oxford (1967).

Momentum-Like Constants of Motion

A. GROSSMANN

Centre de Physique Théorique, C.N.R.S., Marseille (France)

Abstract

We study consequences of invariance under an arbitrary closed subgroup of the additive group of phase space (for a finite number of degrees of freedom). The motivation comes mainly from the one-electron theory of solids. In some physically interesting cases (magnetic translations) the algebra of constants of motion can be of type II.

Contents

1. INTRODUCTION

We are all familiar with the separation of center-of-mass motion in nonrelativistic quantum mechanics of finitely many particles. The main point is this: if a function on phase space (the classical Hamiltonian) is invariant with respect to a continuous family of translations in x-space, then the corresponding quantum-mechanical operator commutes with a "conserved momentum".

Related but more general constants of motion appear in the one-electron theory of solids. First of all, periodic lattice potentials bring in discrete translations in x-space. Secondly, a superimposed external magnetic field, constant in space, introduces linear combinations of the x's and the p's so that, geometrically, the kinetic energy part of the Hamiltonian is invariant under "slanted" phase–space translations (see Section 6).

So essentially every closed subgroup of (the additive group of) phase space occurs in some physical setting. It is natural then to attempt a general study of the corresponding reduction process. One cannot expect it to be always entirely elementary, since infinite, discrete, non-abelian groups of quantum-mechanical translation operators may appear. Such groups have a notorious tendency not to be of type I.

As we shall see, algebras of type II do occur. This means that the familiar procedure for exploiting symmetries (with the help of tensor product decompositions) can break down within elementary quantum mechanics.

2. FOURIER ANALYSIS ON PHASE SPACE

Our notations will be as follows:

E denotes an even-dimensional vector space over reals, with its natural topology $(E \simeq R^{2\nu}, 0 < \nu < \infty)$.

On E, we consider a fixed symplectic (i.e. bilinear, nondegenerate, antisymmetric) form σ;

$$\sigma(a, v) = -\sigma(v, a) \qquad (a \in E, v \in E).$$

$S'(E)$ is the space of all tempered distributions on E with its usual topology.

The "symplectic" Fourier transform of a distribution $f \in S'(E)$ will be denoted by $\tilde{f}$. If f is a testing function, then $\tilde{f}$ is defined by

$$\tilde{f}(v) = \int \exp\left[2i\pi\sigma(v, v_1)\right] f(v_1)\, dv_1.$$

The Lebesgue measure dv_1 will be normalized so that $(\tilde{f})\tilde{\ } = f$.

If $f \in S'(E)$ and if $a \in E$, then $T^a f$ denotes the distribution displaced by a. For functions, this means $(T^a f)(v) = f(v - a)$.

By Λ we denote an arbitrary closed subgroup of (the additive group of) E. For instance, Λ can be any vector subspace, or a discrete subgroup of E. The most general Λ can be written as a sum of a vector space (the connected component of $\{0\}$ in Λ) and of a discrete subgroup.

If Λ is purely discrete and E/Λ is compact, then Λ is called a lattice.

Given $\Lambda \subseteq E$, denote by Λ' the set of all $v \in E$ such that $2\sigma(v, a)$ is an integer for all $a \in \Lambda$. Thus Λ' is always a closed subgroup of E. If Λ is a closed subgroup of E, one has

$$(\Lambda')' = \Lambda.$$

A distribution $f \in S'(E)$ is said to be invariant under Λ if $T^a f = f$ for all $a \in \Lambda$. The set of all such f will be denoted by $P(\Lambda)$.

$\Sigma(\Lambda)$ stands for the set of all tempered distributions that can be obtained from measures with support in Λ by finitely many directional derivatives along the connected component of $\{0\}$ in Λ. In other words, the support of any $f \in \Sigma(\Lambda)$ is contained in Λ, and the evaluation of f does not involve derivatives transversal to Λ. It is natural to say that a $f \in \Sigma(\Lambda)$ is a distribution *in* Λ.

Theorem 1 (Fourier Expansion Theorem):

$$f \in P(2\Lambda) \quad \text{if and only if} \quad \tilde{f} \in \Sigma(\Lambda').$$

Here 2Λ is the set of all $2v$ $(v \in \Lambda)$.

Examples:

If Λ is a lattice, then Λ' is also a lattice and Theorem 1 says that the Fourier transform of a periodic function is a sum of Dirac measures.

If Λ is a vector subspace of E, then Λ' is also a vector subspace of E, and Theorem 1 describes δ-functions due to conservation of momentum.

3. CANONICAL COMMUTATION RELATIONS AND WEYL CORRESPONDENCE

A *representation of canonical commutation relations* is defined as a strongly continuous map $v \to U(v)$ $(v \in E)$ from E into unitary operators in a separable Hilbert space $H(U)$, satisfying

$$U(v) U(a) = \exp\left[2i\pi\sigma(v, a)\right] U(v + a) \qquad (v, a \in E). \tag{1}$$

One can think of $U(v)$ as

$$U(v) = \exp\left[i\sum (x_j P_j - p_j X_j)\right],$$

where $v = \{x_1, \dots x_v, p_1, \dots p_v\}$ and where the X_j, P_j are operators satisfying the Heisenberg commutation relations.

If the contrary is not stated, we shall assume that U is irreducible, i.e. that there is no closed subspace of $H(U)$ stable under all the $U(v)$. This determines U up to unitary equivalence.

An immediate consequence of (1), independent of the irreducibility assumption, is

Proposition:

Let $\Lambda \subseteq E$. Then the conditions

(a) $v \in \Lambda'$

(b) $U(v)$ commutes with all $U(a)$ $(a \in \Lambda)$

are equivalent.

Let $f \in S(E)$ be a testing function. Consider in $H(U)$ the operator $[f]$ defined by

$$[f] = \int U(v)\, \tilde{f}(v)\, dv. \tag{2}$$

The operator $[f]$ is the quantum-mechanical "observable" corresponding to the classical observable f.

The operator $[f]$ is of trace class for every $f \in S(E)$. The correspondence (2) can be inverted with the help of

$$\tilde{f}(v) = k\, \mathrm{tr}\,(U(-v)[f]), \tag{3}$$

where k is a suitable constant.

The correspondence (3) can be extended as follows: To every bounded operator A acting in $H(U)$ (that is, $A \in L(H(U))$), associate the tempered distribution g^A defined by

$$\langle g^A, \varphi \rangle = k\, \mathrm{tr}\,(A^*[\varphi]) \qquad (\varphi \in S(E)). \tag{4}$$

Here A^* is the adjoint of A. Denote by $(S'(E))_B$ the set of all g^A $(A \in L(H(U)))$. Then $(S'(E))_B \subset S'(E)$ (proper inclusion). The correspondence $A \to g^A$ is continuous if $L(H(U))$ is given the ultraweak topology and $(S'(E))_B$ the topology inherited from $S'(E)$. It follows that, for every $\varphi \in S(E)$, the linear functional $A = [f] \to \langle f, \varphi \rangle^*$ is ultraweakly continuous.

Remarks:

If $f(v) = \lambda$ is a constant, then $[f] = \lambda I$ where I is the identity in $H(U)$.

If $f, g \in S(E)$, then the operator product $[f][g]$ can be written as $[f][g] = [f \circ g]$, where

$$(f \circ g)(v) = \int\!\!\int \exp\left[2i\pi(\sigma(v, v_1) + \sigma(v_1, v_2) + \sigma(v_2, v))\right] f(v_2)\, g(v_1)\, dv_1\, dv_2.$$

It is easy to verify that, for every $a \in E$, one has

$$T^a(f \circ g) = (T^a f) \circ (T^a g).$$

4. CONSEQUENCES OF TRANSLATIONAL INVARIANCE

Define $P_B(\Lambda)$ as

$$P_B(\Lambda) = P(\Lambda) \cap (S'(E))_B.$$

In other words, $P_B(\Lambda)$ is the set of distributions invariant under Λ and giving rise to bounded operators in $H(U)$.

If $f \in P_B(2\Lambda')$, then, by Theorem 1, $\tilde{f}$ is a distribution *in* Λ. A look at (2) suggests then that $[f]$ commutes with all $U(a)$ $(a \in \Lambda')$. Actually more can be proved: denote by $R(\Lambda) = \{U(\Lambda)\}''$ the von Neumann algebra generated by the restriction of U to Λ.

Theorem 2:

The conditions

$$\text{(a)} \qquad f \in P_B(\Lambda)$$

$$\text{(b)} \qquad [f] \in R(\Lambda)$$

$$\text{(c)} \qquad [f] \in (R(\Lambda'))'$$

are equivalent.

Here $(R(\Lambda'))'$ is the commutant of $R(\Lambda')$, i.e. the set of all bounded operators in $H(U)$ that commute with every element of $R(\Lambda')$. The irreducibility of U is of course necessary for the validity of Theorem 2.

A more intuitive way of stating Theorem 2 is the following: Suppose that we know that some classical observables, say Hamiltonians, are functions on phase space invariant under (phase space) translations belonging to a group $2\Lambda' \subseteq E$. We are interested in quantum-mechanical constants of motion that appear as a consequence of this invariance. Theorem 2 tells us that the algebra of these constants of motion coincides with the algebra $R(\Lambda)$ generated by the restriction of U to Λ. This motivates our interest in the algebras $R(\Lambda)$.

5. THE ALGEBRAS $R(\Lambda)$

We can now settle some natural questions about $R(\Lambda)$. If $\Lambda \neq \{0\}$, then $R(\Lambda)$, considered as a vector space, is infinite-dimensional, since any finite family of $U(v_i)$ is linearly independent.

Theorem 3:

The von Neumann algebra $R(\Lambda)$ is finite (in the sense that it admits a nontrivial finite normal trace) if and only if the orbits of Λ in E/Λ' have compact closure.

Theorem 4:

The von Neumann algebra $R(\Lambda)$ is a factor (that is, the intersection of $R(\Lambda)$ and of its commutant $R(\Lambda)'$ contains only multiples of the identity) if and only if the closure of $\Lambda + \Lambda'$ in E is all of E.

For terminology on orbits, see e.g. lectures of L. Michel in this volume.

Let us illustrate the above theorems by three examples where the answers are known beforehand.

1) If $\Lambda = E, \Lambda' = \{0\}$, then $R(\Lambda) = L(H(U))$ (the set of all bounded operators

in $H(U)$). The orbit of Λ in $E/\Lambda' = E$ is all of E. It is closed and not compact, so that $R(\Lambda)$ is not finite. $R(\Lambda)$ is a factor, since $\Lambda + \Lambda' = E$.

2) If $\Lambda = \{0\}$, then $\Lambda' = E$. The algebra $R(\Lambda)$ consists of multiples of the identity. The orbit of Λ in E/Λ' is a single point, so $R(\Lambda)$ is finite.

3) Assume that $\Lambda' \supseteq \Lambda$ (This means that $R(\Lambda)$ is abelian). Then the orbits of Λ in E/Λ' are points, and so $R(\Lambda)$ is finite. The algebra $R(\Lambda)$ is not a factor unless $\Lambda' = = E$ (i.e. $\Lambda = \{0\}$), since $\Lambda + \Lambda' = \Lambda'$ which is closed.

It should be noted that the assumptions of Theorem 4, making sure that $R(\Lambda)$ is a factor, can be satisfied by a purely discrete Λ. For example, let $v = 1$, so that dim $E = 2$. Let $a \in E$ and $b \in E$ be such that $\sigma(a, b)$ is irrational. Let Λ be the (discrete) subgroup of E generated (additively) by a and b. Then Λ' is also purely discrete, but the closure of $\Lambda + \Lambda'$ is all of E. On the other hand, E/Λ' is compact so that one is sure that the closure of any orbit of Λ in E/Λ' is compact (the orbits turn out to be dense in E/Λ'). It follows that $R(\Lambda)$ is a factor of type II_1.

This example can be immediately generalized:

Corollary:

If E/Λ' is compact (in particular if Λ' is a lattice), then $R(\Lambda)$ is finite. If E/Λ' is compact and if the closure of $\Lambda + \Lambda'$ is all of E, then $R(\Lambda)$ is a factor of type II_1.

Sketch of proofs:

Theorem 5:

Consider the space $S(E/2\Lambda')$ of C^∞-functions defined on $E/2\Lambda'$ and decreasing fast in (possible) directions of non-compactness of $E/2\Lambda'$. The problem of finding a finite normal trace on $R(\Lambda)$ is solved if we find a nonzero testing function $\varphi \in S(E/2\Lambda')$ which is invariant under the action of 2Λ on $E/2\Lambda'$. The invariance requirement reflects the commutativity of the trace. Testing functions with the required properties exist if and only if the orbits of Λ in E/Λ' have compact closure.

Theorem 6:

Let $[f] \in R(\Lambda) \cap R(\Lambda)' = R(\Lambda) \cap R(\Lambda')$. Then the distribution f is invariant under 2Λ and under $2\Lambda'$, and consequently under the closed group generated by 2Λ and $2\Lambda'$. This is all of E, so f is a constant.

6. MAGNETIC TRANSLATIONS

There are practically important physical situations where $R(\Lambda)$ can be of type II. Consider a spinless nonrelativistic particle ("electron") subjected to a periodic ("lattice") potential and to a magnetic field constant in space. In two space dimensions and with an appropriate choice of gauge and of units, the classical Hamiltonian is the function

$$h(v) = h_\beta(x_1, p_1 ; x_2, p_2) = (p_1 - \beta x_2)^2 + (p_2 + \beta x_1)^2 + V(x_1, x_2)$$

defined on four-dimensional phase space.

Here β is proportional to the magnetic field and $V(x_1, x_2)$ is periodic:

$$V(x_1 + 1, x_2) = V(x_1, x_2 + 1) = V(x_1, x_2).$$

The function h_β is invariant under the discrete group of phase–space translations, generated by the vectors

$$m_1 = \{1, 0, 0, -\beta\}$$
$$m_2 = \{0, 1, \beta, 0, 0\},$$

that is: $h_\beta(v + m_1) = h_\beta(v + m_2) = h_\beta(v)$. The theory of the preceding sections may be applied to describe the algebras of constants of motion that appear as a consequence of this symmetry. This algebra is of type I if and only if a certain condition "of rational flux" holds.

7. INDUCED REPRESENTATION OF CANONICAL COMMUTATION RELATIONS

Let us return to the general problem of exploiting invariance under an arbitrary closed subgroup of E. The separation of constants of motion, if possible, is greatly facilitated by an appropriate explicit realization of $U(v)$ (within the unique irreducible equivalence class).

By a standard procedure, we first recast (1) into a problem of group representation. Let T_1 be the unit circle on the complex plane (the set of all $\tau \in C$ with $|\tau| = 1$). For any subset $F \subseteq E$, write $\underline{F}$ to denote the cartesian product $\underline{F} = F \times T_1$. Make $\underline{E}$ into a (non-abelian) group by defining

$$\{v_1, \tau_1\} \{v_2, \tau_2\} = \{v_1 + v_2, \tau_1 \tau_2 \exp [2i\pi\sigma(v_1, v_2)]\}.$$

Let $\{v, \tau\} \to U(v, \tau)$ be a (strongly continuous, unitary) representation of $\underline{E}$ in a separable $H(U)$, such that $U(0, \tau) = \tau I$. Then $U(v, 1) = U(v)$ is a representation of canonical commutation relations.

The maximal abelian subgroups of $\underline{E}$ are exactly sets of the form $\underline{W}$, where $W \subset E$ satisfies

$$W' = W. \tag{5}$$

Examples of sets satisfying (5) are: (a) Maximal isotropic vector subspaces of E, for instance the x-space or the p-space. (b) Certain lattices in E, for example the direct sum of a lattice in x-space and of the corresponding reciprocal lattice in p-space. There are of course other examples, e.g. "mixtures" of (a) and of (b).

Representations induced from a suitable character of $\underline{W}$ can be described as follows:

Let η be a measurable map from W into T_1, satisfying

$$\eta(w_1 + w_2) = \exp\left[2i\pi\sigma(w_1, w_2)\right]\eta(w_1)\eta(w_2) \qquad (w_1, w_2 \in W). \tag{6}$$

Note that the exponential on the right-hand side of (6) can take only the values ± 1. Consider the Hilbert space $H(U^{W;\eta})$ of complex-valued, locally square integrable functions $\Psi(v)$, defined on phase space and satisfying the "covariance condition"

$$\Psi(v - w) = \eta(w)\exp\left[2i\pi\sigma(w, v)\right]\Psi(v) \qquad (v \in E,\ w \in W).$$

The norm squared of $\Psi \in H(U^{W;\eta})$ is defined as the integral of $|\Psi|^2$ over E/W.

Define in $H(U^{W;\eta})$ the unitary operators $U^{W;\eta}(v)$ $(v \in E)$, by

$$(U^{W;\eta}(v)\,\Psi)(v_1) = \exp\left[-2i\pi\sigma(v, v_1)\right]\Psi(v + v_1).$$

Then $U^{W;\eta}$ is an irreducible representation of canonical commutation relations. Note that the $U^{W;\eta}(w)$ $(w \in W)$ are multiplication operators.

8. REDUCTION OF $R(\Lambda)$ IN $H(U^{W;\eta})$

The steps are as follows:

1) Find $W \subset E$ such that $W' = W$ and that $W + \Lambda$ is closed.

2) Identify the representation space $H(U^{W;\eta})$ with $L^2(E/W)$ (there is some ambiguity here).

3) Write $L^2(E/W)$ as a Hilbert bundle based on $B = E/(W + \Lambda)$ (that is, the set of orbits of Λ in E/W), with fibers isomorphic to $L^2(\Lambda/(W \cap \Lambda))$ (Hilbert space of complex-valued functions square integrable over an orbit of Λ in E/W).

Then the elements of $R(\Lambda)$ act "within the fibers" and we have achieved a reduction. In the case that the orbits of Λ in E/W are points (i.e. that $\Lambda \subseteq W$) the algebra $R(\Lambda)$ is diagonalized.

It is clearly advantageous to choose W so that the orbits of Λ in E/W are "as small as possible".

The procedure breaks down if $W + \Lambda$ is not closed; the action of Λ on E/W can be strictly ergodic, and we are again faced, as in Section 5, with the impossibility of a clean separation of constants of motion.

9. SPECTRAL PROPERTIES OF HAMILTONIANS OF BLOCH TYPE

This is an example where reduction is easy and where the spectral theory of "reduced operators" is made trivial by an appropriate choice of W.

Let $E = E_1 \oplus E_2$, where E_1 and E_2 are maximal isotropic vector subspaces of E (x-space and p-space, respectively), and dim $E_1 = $ dim $E_2 = 3$. Elements of E/E_2 may be denoted by x, since they are in a natural 1-to-1 correspondence with elements of E_1. So $V(x)$ denotes a function defined on phase space and invariant under E_2. Similarly, $T(p)$ is a function invariant under E_1.

Let $L \subset E_1$ be purely discrete and such that E_1/L is compact. That is, L is a lattice in x-space. Denote by $L^\perp$ the corresponding reciprocal lattice in p-space.

We are interested in operators $[h(v)]$, where the function $h(v)$ is of the form

$$h(v) = T(p) + V(x)$$

and where $V(x)$ is invariant under $E_2 + L$. It follows that $h(v)$ is invariant under L and that $[h]$, if bounded, belongs to the algebra $R(\Lambda)$ where $\Lambda = E_1 + L^\perp$.

A natural choice of W is $W = E_1$ (the p-representation). The orbits of Λ in E/W are $\simeq L^\perp$. We are thus led to the natural decomposition of $H(U^{W;\eta}) = L^2(E_2)$ into a Hilbert bundle based on the abstract Brillouin zone $B = E_2/L^\perp$ and having $l^2(L^\perp)$ as a fiber. (This is essentially the "crystal momentum representation").

The operators $[h]$ are then written as direct integrals

$$[h] = \int_B^\oplus h_k \, dk.$$

Spectral properties of the "reduced Hamiltonians" h_k become now trivial, since the "reduced kinetic energies" T_k are multiplication operators on a Hilbert space of sequences, namely $l^2(L^\perp)$. One has, for instance

Theorem 7:

Let $T(p)$ be continuous and tending to $+\infty$ as $|p| \to \infty$. Let $|V(x)|^2$ be integrable over E_1/L. Then, for every $k \in B$, the operator h_k has compact resolvent, i.e. a purely discrete spectrum of finite multiplicity and without finite accumulation points.

This theorem covers the familiar case $T(p) = |p|^2$. It can be refined to give information about the dependence on k of the eigenvalues $\varepsilon_n(k)$ and to allow for nonlocal potentials.

Another suitable choice of W is $W = L + L^\perp$ (the lattice representation, also called the k,q-representation). Now E/W is the abstract unit cell in phase space. The orbits of Λ in E/W are $\simeq E_1/L$ (the abstract unit cell in x-space). The Hilbert space $L^2(E/W)$ is decomposed into a bundle based on $E/(E_1 + L)$ (again the abstract Brillouin zone) with fiber $\simeq L^2(E_1/L)$. Suitable choices of $T(p)$ give rise to elliptic reduced Hamiltonians, and the conclusions are the same as above, because of the compactness of E_1/L.

10. COMMENTS

Section 3 is a sketch of quantum mechanics over phase space. The subject has a long history, and I cannot give here an adequate bibliography; see an earlier paper [1]. Araki [5] treats $R(\Lambda)$ (Λ a vector subspace), allowing dim $E = \infty$.

About magnetic translations, see Zak [2], Brown [3], Opechowski and Tam [4], and Boon [6]. The occurrence of situations not of type I has been recognized independently by Boon.

Induced representations of canonical commutation relations were studied by Cartier [7]. Lattice representations have been introduced into solid-state physics by Zak [8].

For spectral properties of reduced Hamiltonians, compare Mackey [9] and, in the "rational magnetic" case, Bentosela [10].

Mathematically related material can be found in Auslander–Moore [11, chap. X], Pukanski [12], Slawny [13], Sirugue, Testard and Verbeure [14].

I have not said anything about momentum-like constants of motion in Schrödinger N-particle theory. See Combes [15].

REFERENCES

[1] A. Grossmann, G. Loupias, and E. M. Stein, *Ann. Inst. Fourier,* p. 343, **18**, 2 (1968).

[2] J. Zak, *Phys. Rev.,* **134**, A 1602 (1964); **134**, A 1607 (1964).

[3] E. Brown, *Solid State Physics,* **22**, 313.

[4] W. Opechowski and W. G. Tam, *Physica,* **42**, 529 (1969); W. G. Tam, *Physica,* **42**, 557 (1969).

[5] H. Araki, *Journ. Math. Phys.,* **4**, 1343 (1963); *Prog. Th. Phys.,* **32**, 956 (1964).

[6] M. Boon, *Journ. Math. Phys.,* **13**, 1268 (1972).

[7] P. Cartier, *Am. Math. Soc. Symposia on Pure Mathematics,* p. 361, Vol. 9 (1966).

[8] J. Zak, *Phys. Rev.,* **168**, 686 (1968); **177**, 1151 (1969).

[9] G. W. Mackey, in "Functional Analysis and Related Fields", F. Browder, Ed. Springer, Chicago (1968).

[10] F. Bentosela, "Magnetic Bloch Functions" (to appear in *Nuovo Cimento*).

[11] L. Auslander and C. C. Moore, *Memoirs Am. Math. Soc.,* No. 62.

[12] M. Pukanski, *Ann. Ec. Norm. Sup.* (1971).

[13] J. Slawny, *Commun. Math. Phys.,* **24**, 151 (1972).

[14] J. Manuceau, M. Sirugue, D. Testard, and A. Verbeure, Marseille preprint 71/P397.

[15] J. M. Combes, *Journ. Math. Phys.,* **12**, 1719 (1971).

Haag's Theorem

MARCEL GUENIN

Department of Theoretical Physics, University of Geneva (Switzerland)

Abstract

Haag's theorem and its consequences are reviewed. As a way out, time evolution of field operators is described without using the Hamiltonian. Although this enables one to bypass Haag's theorem, it brings one into contact with unsolved problems concerning the vacuum state, some of which are pointed out.

1. INTRODUCTION

Haag's theorem essentially states that a quantum field theory (QFT) which obeys the canonical commutation rules (CCR) at a sharp time, possesses a vacuum state and is invariant under a sufficiently large group, is trivial.

Later on, we shall, of course, state the theorem precisely in its different versions. But the importance of Haag's theorem arises mainly because it imposes in QFT the use of inequivalent representations of the CCR (inequivalent meaning not unitarily equivalent to the Fock representation of the CCR). This is also the reason why one has to introduce the use of spatial cutoffs in the constructive methods of QFT.

We shall see that it is now possible to say that we understand how to bypass Haag's theorem, at least in principle, and that therefore the problems it poses can be considered as solved. One has to realize, however, that the problem of Haag's theorem is only one of the difficulties inherent in all quantum field theories which, following Wightman, we can list as follows:

— Problems of the vacuum (i.e. existence, stability, uniqueness);
— Haag's theorem;
— Divergences (infrared and ultraviolet).

We refer to the article of B. Simon (this volume) for more information on some of these topics. We shall leave aside infrared problems, which can be considered as at least partly solved, after the thesis of Blanchard [1]. We can say that for the simplest

111

cases, all three problems are solved, as well as the uniqueness of the vacuum. The problems arising from the most severe ultraviolet divergences, however, are still far from being solved.

Before coming to the subject itself, we specially want to attract your attention to one fact. Although we shall formulate most of our remarks and propositions in the language of QFT, they also apply, *mutatis mutandis*, to numerous problems of statistical mechanics, in fact to all systems with an infinite number of degrees of freedom.

I finally want to apologize to the experts because this talk will bring them essentially nothing new, since it is aimed mainly at the large audience of younger physicists.

2. THE VARIOUS FORMULATIONS OF HAAG'S THEOREM

We shall first discuss the formulation given by Wightman [2], because it is the form which shows most clearly the different aspects and also the possible generalizations. This version is the following:

Theorem:

In a Euclidean invariant QFT which uses the Fock representation of the CCR, the no-particle state is Euclidean invariant.

Proof:

We have the CCR

$$[\phi(x, t), \pi(y, t)] = i\delta(x - y)$$

and

$$[\phi(x, t), \phi(y, t)] = [\pi(x, t), \pi(y, t)] = 0.$$

Let us suppose $|0\rangle$ is the no-particle state, and also that it is unique (this is the meaning of having the Fock representation). The no-particle state can be characterized by the fact that an annihilation operator maps it on the zero-vector. An annihilation operator may be written as

$$b(k) = \frac{1}{\sqrt{2}(2\pi)^{s/2}} \int \{(-\Delta + m^2)^{1/4}\phi(x, t) + i(-\Delta + m^2)^{-1/4}\pi(x, t)\}e^{ikx}d^sx$$

where one can show that the right-hand side is independent of t. Demanding $b(k)$ to annihilate the vacuum amounts to putting

$$\{(-\Delta + m^2)^{1/4}\phi(x, t) + i(-\Delta + m^2)^{-1/4}\pi(x, t)\}|0\rangle = 0$$

for almost all x.

Now if we have a unitary representation of the Euclidean group $U(a, \mathscr{R})$ with

$$U(a, \mathscr{R})\phi(x, t)U(a, \mathscr{R})^{-1} = \phi(\mathscr{R}x + a, t)$$

and

$$U(a, \mathcal{R})\pi(y, t)U(a, \mathcal{R})^{-1} = \pi(\mathcal{R}y + a, t),$$

it follows from the above equation and the uniqueness of the vacuum that

$$U(a, \mathcal{R})\,|\,0\,\rangle = \omega(a, \mathcal{R})\,|\,0\,\rangle,$$

where $|\,\omega(a, \mathcal{R})\,| = 1$.

It is furthermore clear that $\{a, \mathcal{R}\} \to \omega(a, \mathcal{R})$ is a continuous one-dimensional representation of the Euclidean group. If the number of space dimensions is greater than one, there is none, so that

$$U(a, \mathcal{R})|0\rangle = |0\rangle.$$

Remarks:

1) The case of one space dimension can be reduced to the preceding; one defines a new representation of the Euclidean group by multiplying the previous one with a phase factor (cf. [3]).

2) As far as the group is concerned, we see that we do not use explicitly the particular nature of the Euclidean group. What is important is that the group should not possess one-dimensional representations. That can be formulated by saying that if G is the invariance group, Haag's theorem will be true whenever there exists no invariant subgroup K of G such that G/K is abelian.

The reason why the preceding theorem is troublesome is that if the no-particle state is Euclidean invariant, it cannot be interpreted in any other way except as representing the physical vacuum. Now, this again implies that this no-particle state should be left invariant under time evolution (a phase can always be absorbed by adding a constant to the Hamiltonian). This in turn implies that the vacuum will be an eigenvector of the Hamiltonian with eigenvalue zero (as we have already absorbed the constant).

Now this means that we must have an interaction which does not polarize the vacuum. Stated otherwise, the Wick-ordered interaction part must be the sum of terms each of which contains at least one annihilation operator. Theories of this kind (which one calls theories with persistent vacuum) are generally not very interesting in QFT, although they play a significant role in solid state physics. The trouble is that such an interaction term is non-local and thus unsuited for a relativistic theory.

At this point there is an important remark to be made. We shall see below that the unavoidable consequence of Haag's theorem will be that one is obliged to introduce inequivalent representations of the CCR. However, it is not true in general that one can get along with the Fock representation of the CCR in the cases where Haag's theorem does not apply. For this there are two different groups of reasons. The first is tied up with the fact that an uncountable set of possible states exists in the thermodynamic limit (i.e. the limit of infinite volume; the meaning is clear in N-body theory and is applied by extension to QFT for the limit of no space cutoff) and that different boundary conditions (for instance different temperatures) will select in general different inequivalent representations. This is the case in solid state physics, for instance for spin systems. The second reason is that in theories with ultraviolet divergences, one is performing a "dressing" transformation to get a good domain for the Hamiltonian, which also defines an inequivalent representation of the CCR.

At this stage, it is also useful to consider the relativistic forms of Haag's theorem. If we assume more about the invariance group, then we shall also get stronger results.

Let us therefore assume that the fields ϕ_1, ϕ_2 satisfy Wightman's axioms (irreducibility, locality, spectrum condition) (we shall not write the smearing with test functions, to simplify writing). Furthermore, the fields are supposed to exist when smeared out in space only, otherwise it would not make sense to speak of "equal-time commutation relations". Let $\mathscr{H}_1$ and $\mathscr{H}_2$ be the corresponding Hilbert spaces, and let $\{a, \Lambda\} \to U_j(a, \Lambda)$ be a representation of the inhomogeneous $SL(2, \mathbb{C})$, such that

$$U_j(a, \Lambda)\phi_j(x)U_j(a, \Lambda)^{-1} = \phi_j(\Lambda x + a).$$

Suppose further that

$$U_j(a, \Lambda)\psi_{0j} = \psi_{0j} \qquad (\psi_{0j} \in \mathscr{H}_j, \quad \text{vacuum state}).$$

Suppose finally that $\exists\, V$

$$\phi_2(x, t) = V\phi_1(x, t)V^{-1}, \qquad \forall\, x, t \text{ fixed}.$$

Then:

1) $U_2(a, \Lambda) = V U_1(a, \Lambda) V^{-1}$;

2) $c\psi_{02} = V\psi_{01}, \quad |c| = 1$.

3) The equal-time Wightman functions are identical:

$$(\psi_{01}, \phi_1(x_1, t) \dots \phi_1(x_n, t)\psi_{01}) = (\psi_{02}, \phi_2(x_1, t) \dots \phi_2(x_n, t)\psi_{02}).$$

4) The Wightman functions of up to 4 arguments are the same for arbitrary arguments (generalized Haag's theorem).

5) If ϕ_1 is a free field, then ϕ_2 is a free field (Haag's theorem).

For the proof, see Streater and Wightman [4].

3. CONSEQUENCES AND THE WAY OUT

We shall first quickly discuss the consequences for the ordinary approaches, using the interaction picture. In the interaction picture, one supposes that at a certain time, say $t = 0$, the Heisenberg, Schrödinger and the interaction or Dirac pictures coincide. The idea is to subtract from the total evolution, as it would be given by the Schrödinger picture, the part which is known and which we call the free part. If H_0 is the Hamiltonian corresponding to this free part, and H is the total Hamiltonian, we have that

$$|\psi(t)\rangle_D = e^{iH_0 t}\, e^{-iHt}|\psi(0)\rangle_D \equiv W(t)|\psi(0)\rangle_D; \qquad |\psi(0)\rangle_D = |\psi(0)\rangle_H = |\psi(0)\rangle_S.$$

One supposes, further, that at this time $t = 0$, the field operators satisfy the canonical commutation relations (CCR) in the Fock representation. Since $W(t)$ is obviously unitary, this implies that these properties will hold as well at every time.

If we apply now Haag's theorem and thus assume enough invariance properties, we get that the theory must be trivial, or that at least one of the assumptions does not hold. The essential assumptions are three:

— Euclidean invariance;
— Fock representation of the CCR;
— Unitarity of $W(t)$.

The last point deserves some comments, since it is not usually considered as a separate postulate. Indeed, one usually considers in any quantum system the time evolution as being generated by a continuous unitary representation of the translation group, and the application of Stone's theorem then yields the existence of a Hamiltonian. But we shall see that it is necessary to enlarge the standard way of looking at general quantum mechanics, yet without preventing us from getting back on our feet at the end.

The first idea to bypass Haag's theorem is, therefore, to break Euclidean invariance. This is usually done in putting the system in a box. The finiteness of the box imposes the choice of new boundary conditions, and one usually takes the periodic ones. One then has the problem of showing the existence of the limit for suitable objects as the box is removed. This may eventually be done in some cases, but the trouble is that one does not have any indication of what to look for, or for which physical reasons should some objects have a limit and some others none.

The second possibility is to work with the inequivalent representations of the CCR which must enter into the theory. From the generalized Haag's theorem, however, we learn that the choice of the representation of the CCR essentially determines the Hamiltonian, at least up to modifications which leave the 1-, 2-, 3- and 4-point functions invariant. A more powerful result along this line has been obtained by Araki [5] without using Haag's theorem. He shows that for a relativistically invariant scalar theory in which the representation of the CCR is given, the Hamiltonian is uniquely defined on a dense domain; there may, however, exist different extensions. But for practical purposes the path should go the other way round. We are usually given a formal expression as Hamiltonian, and not an inequivalent representation of the CCR. We know how to construct an infinity of mutually inequivalent representations, but how are we to choose the right one for a given model? The constructive approach is constructive in the sense that it does define a domain for the spatially (and also eventually energetically) cutoff Hamiltonian; in case ultraviolet divergences are present, this is obtained using a so-called "dressing" transformation and it indeed defines an inequivalent representation of the CCR. But this is still not the end of the game, since one has to remove the space cutoff, and this again leads to still another representation of the CCR: a representation of which we can, as of now, only assert the existence, but not construct it explicitly.

As we have seen that—to say the least—it is not particularly easy to start directly with the inequivalent representations, we shall ask ourselves whether there is no other way open that could give us some insight as to where to look for the physical reasons which could make the limits of no cutoff exist. The best intuitive approach is to start using a perturbative expansion in the Heisenberg picture, or in a form of the interaction picture that is particularly well-suited for this purpose [6]. We shall work in the Heisenberg picture. The time evolution of a field operator $\phi(f, t)$ can be given formally by the expansion of Dyson and Schwinger,

$$\phi(f, t) = \phi_0(f, t) + i \int_0^t \left[H_I(t - t_1), \phi_0(f, t) \right] dt_1 +$$

$$+ i^2 \int_0^t dt_1 \int_0^{t_1} dt_2 \left[H_I(t - t_1), \left[H_I(t - t_2), \phi_0(f, t) \right] \right] + \ldots,$$

where $\phi_0(f, t)$ is the free field at time t and $H_I(\tau) = e^{iH_0\tau} H_I e^{-iH_0\tau}$.

Suppose that we don't need to worry about the ultraviolet problems. By this I mean that either they are not present, as is the case in the $:\phi^4:$ model in 2 space-time dimensions, or that we have been able to get rid of them using a good dressing transformation which provides a large enough invariant set of analytic vectors for the field operator and the Hamiltonian, whenever the latter is cut off in space. That is, the Hamiltonian is supposed to be the integral over a space-like surface of a density $\mathscr{H}_I(x)$, and the spatially cut-off Hamiltonian is defined by

$$H_I(g) = \int \mathscr{H}_I(x) g(x) dx, \qquad H(g) = H_0 + H_I(g),$$

where $g \in \mathscr{D}$, $g = 1$ for $|x| < R$, $g = 0$ for $|x| > 2R$ and $0 \leq g \leq 1$.

If these hypotheses are satisfied, it is easy to see that the perturbative expansion of Dyson and Schwinger is perfectly well defined term by term. We have now to understand why $\phi(f, t)$ should be independent of g for large enough R. The basic reason for this is the local nature of the Hamiltonian. If we look at the first term in the perturbative expansion, $\int_0^t \left[H_I(g)(t - t_1), \phi_0(f, t) \right]$, we see that in representation space we can picture the relevant domains as in the following figure:

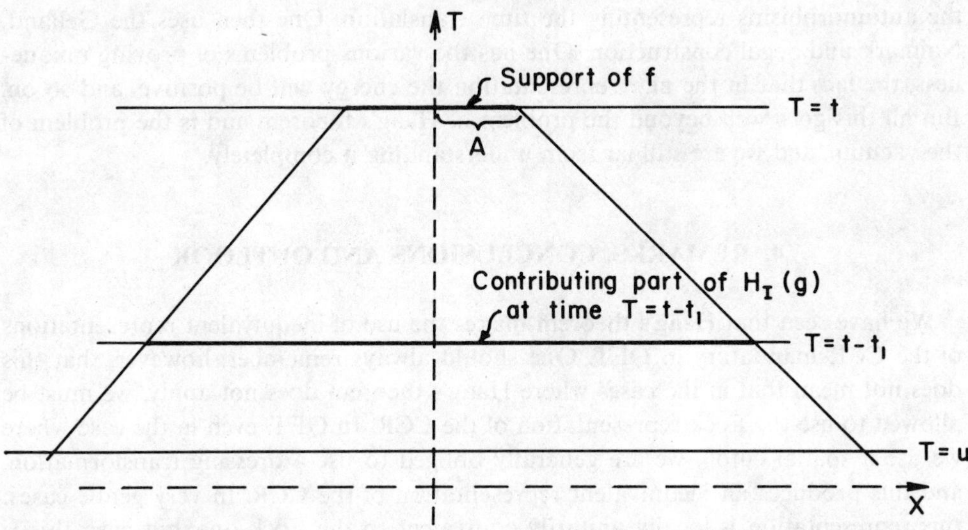

$H_I(g)(t - t_1)$ is the integral of a local density over space at time $T = t - t_1$. Since we take the commutator of $H_I(g)(t - t_1)$ with $\phi_0(f, t)$, only the part of $H_I(g)(t - t_1)$ which comes from the integration of $\mathscr{H}_I(x, t - t_1)$ in the region intersected by the backward cone spanned by the support of f at time t, will give a nonvanishing contribution. Therefore, if $R > t + A$ (where A is the radius of a sphere enclosing the support of f and centered at the same spatial point as the center of the support of g), then the first term in the perturbative expansion becomes independent of g. This will obviously also be true for all other terms in the expansion [6].

The perturbative expansion can give us an idea as to what is reasonable to expect. But it is not a proof unless we can say something about the convergence of the series. This can be done in the cases where $\mathscr{H}_I(x, t)$ is a bounded operator and quasi-local, in the sense that $\mathscr{H}_I(x, t)$ will commute with $\mathscr{H}_I(y, t)$ and $\phi_0(y, t)$ if the distance between x and y is greater than some fixed constant Λ. For proofs, see [6] and [7]. In the general case, one prefers to use results based on Trotter's product formula [8, 9, 10].

What one gets, therefore, is that we have a well-defined mapping which defines the time evolution of the field operators, without having a Hamiltonian, since this was forbidden to us by Haag's theorem. In this sense Haag's theorem is bypassed. There remains, however, still a long way to go. In fact, we have got a time evolution, but not in the sense that is usual in quantum mechanics. We can show that this time evolution is given by a group of automorphisms of a suitable C^*-algebra of local observables. In order to get the standard form of quantum mechanics, one has to get a new representation of this C^*-algebra in which the time evolution will be represented by a strongly continuous family of unitary operators. This is done by constructing a positive linear functional on the C^*-algebra, which is invariant under the automorphisms representing the time translation. One then uses the Gelfand, Naimark and Segal construction. One has the various problems of proving uniqueness, the fact that in the new representation the energy will be positive, and so on. But all this goes well beyond the problem of Haag's theorem and is the problem of the vacuum, and we are still far from understanding it completely.

4. REMARKS, CONCLUSIONS AND OUTLOOK

We have seen that Haag's theorem makes the use of inequivalent representations of the CCR mandatory in QFT. One should always remember, however, that this does not mean that in the cases where Haag's theorem does not apply, we must be allowed to use the Fock representation of the CCR. In QFT, even in the case where we use a spatial cutoff, we are generally obliged to use a dressing transformation, and this produces an inequivalent representation of the CCR. In very gentle cases, this representation is locally unitarily equivalent to the Fock one, but even this is not the rule. In all these cases, Araki's theorem [5] does not hold, but there should be a theorem saying under which conditions two different dressing transformations corresponding to different spatially cut-off interactions will lead to inequivalent representations. One would guess that this is the case whenever the difference between the two Hamiltonians is not "relatively" small, in a sense to be defined.

For the N-body problem, there exists a theorem by Takesaki [11] which tells us that two KMS states corresponding to different temperatures lead to inequivalent representations. It is, therefore, a theorem telling us to change the representation of the CCR if we change the boundary conditions. There does not yet, however, exist any theorem exactly corresponding to Haag's theorem, that is a theorem which would assert that if two theories are unitarily equivalent at a given time and possess a sufficiently large invariance group, then all Green functions will be identical. Such a statement is probably too strong, but a weaker form will certainly soon be found to be valid.

Haag's theorem has historically been very important, since it has been the first theorem forcing the use of inequivalent representations of the CCR. The younger physicists present here are certainly going to wonder about the fuss which is made about this theorem, since for them the use of inequivalent representations of the CCR has become an everyday exercise. They should not forget, however, that they represent only a tiny minority of the physicists of our days, and that we have still a very long way to go until we can convince a majority of our colleagues that the inequivalent representations of the CCR are something more than mere mathematical curiosities.

REFERENCES

[1] P. Blanchard, Thesis, ETH Zurich (1969).

[2] A. S. Wightman, "Cargèse Lectures," 1964. Gordon and Breach, New York.

[3] M. Guenin, "Boulder Lectures", 1966, 1969. Univ. of Colorado Press.

[4] R. F. Streater and A. S. Wightman, "PCT, Spin and Statistics and All That". Benjamin, New York (1964).

[5] H. Araki, *J. Math. Phys.*, **1**, 492 (1960).

[6] M. Guenin, *Commun. Math. Phys.*, **3**, 120 (1966).

[7] R. F. Streater and I. Wilde, *Commun. Math. Phys.*, **17**, 21 (1970).

[8] I. Segal, *Proc. N.A.S.*, **57**, 1178 (1967).

[9] J. Glimm and A. Jaffe, *Phys. Rev.*, **176**, 1945 (1968).

[10] K. Hepp, "Théorie de la renormalisation", Springer Verlag (1969).

[11] M. Takesaki, *Commun. Math. Phys.*, **17**, 33 (1970).

Combinatorial Analysis of the Lie Identities
in Field Theory

A. JOSEPH

Department of Physics and Astronomy, Tel-Aviv University, Ramat-Aviv (Israel)

Abstract

The Lie bracket structure of the commutator between fields (or currents) taken at different space-time points is investigated. Calculations performed at a formal algebraic level are given a rigorous setting and their combinatorial structure is examined. The modification to this structure in the original field theoretic framework is discussed by way of examining the divergences which arise.

The specification of a model in quantum field theory lies almost entirely in the form given to the commutator of the fields taken at different space-time points. A prime requirement of the commutator is that it satisfy the Lie identities. This may be studied at various levels of sophistication, and even at its most elementary—that is when formal algebraic manipulations are allowed—it is a condition which has a nontrivial content. Here we report on some calculations carried out at this formal level, which we justify through the observation that each component can be given a rigorous meaning without essentially affecting the algebraic structure we attempt to examine.

Generally, a field theoretic model is specified through the set of equal-time commutation relations and dynamical equations for the fields. Formally the dynamical equations determine a set of differential equations for the commutator with boundary conditions given by the equal-time relations. In principle, this system admits a solution which generally takes the following form:

$$[\phi(q), \phi(p)] = \sum_{n=0}^{\infty} \kappa^n \int \cdots \int \phi^n(q_1, q_2, \ldots q_n) f_n(q, p, q_1, q_2, \ldots q_n) \, dq. \tag{1}$$

In (1) the momentum space formalism has been adopted. In the left-hand side ϕ denotes the field, q, p its momentum coordinates and any additional indices. In the right-hand side κ is an arbitrary parameter which generally represents an interaction strength and enables one to keep track of orders. The symbol ϕ^n denotes a polynomial of degree n in the field ϕ taken at the given momentum coordinates, with f_n a real or complex valued function. Finally, dq denotes invariant integration (or summation) over the dummy variables.

Of the properties of (1) implied by its commutator structure, the Jacobi identity represents a highly nontrivial requirement. In this we compute the double commutator $[[\phi(q), \phi(p)], \phi(k)]$ through (1) taking $[\ldots, \phi(k)]$ to act like a derivation on the polynomial algebra of the fields. Setting the cyclic sum of double commutators to zero, order by order identification of terms implies conditions on the ϕ^n and the f_n which have the general structure of an infinite coupled set of nonlinear integral equations. In principle this (and antisymmetry) determines the most general form of (1). Though this is hardly of use computationally, it does illustrate the essential nonlinearity of the constraint imposed by the Jacobi identity and may be of use in establishing other general properties. It is no less easy to develop (1) through expressing the field in terms of objects having simpler commutation relations, since to obtain the right-hand side one must still re-express the computed commutator in terms of the fields.

The abstraction of (1) from a model field theory and the consequent attempt at formal verification of the Lie identities involves some rather delicate questions which go beyond the mere imposition of mathematical rigor. This is because if (1) is taken from field theory, then a number of constraints which stem from its field-theoretic origin are imposed on ϕ. These constraints may be essential to the validity of the Lie identities. This does not rule out the possibility that (1) may be modified within its general form to recover the Lie identities without these constraints, but rather that the Lie identities must be examined afresh.

All such constraints fall essentially into the two following classes.

1. THE DYNAMICAL CONSTRAINT

The field ϕ satisfies a set of dynamical equations which, as explained above, are themselves used in the derivation of (1). Hence there is no *a priori* reason for supposing that the commutator so derived will satisfy the Lie identities as identities, even though its boundary values (i.e., equal-time values) do have this property. (We remark that if the boundary is taken to be the light cone, as is becoming popular [1], then the Lie identities for the boundary values are no longer trivial. This provides a further motivation for examining the Lie structure of (1).) Rather, (1) may satisfy the Lie identities as a dynamical relation, that is, one whose verification requires the use of the dynamical equations satisfied by the fields. This circumstance is also apparent from the conditions implied on the ϕ^n and the f_n discussed previously. This difficulty should not be considered a serious one. Thus it is very probable that one can always rearrange the right-hand side of (1) through the dynamical equations to a new expression which satisfies the Lie identities as identities. On the other hand, the specification of such a procedure is not a trivial matter and it must be remembered that in a given development of (1) through the dynamical equations, the Lie identities need not be contained.

2. ANALYTICAL CONSTRAINTS

In order to avoid the unwanted singularities of field theory (some singularities do have a physical meaning!), it is necessary to make certain reinterpretations of the formal expressions given in the bare theory. In general, such procedures upset the simple algebraic structure upon which our present considerations have implicitly been based. We give two examples of this circumstance, though there are certainly others.

a) Wick Ordering

This procedure consists in decomposing the field at a fixed time as a linear combination of creation and annihilation operators and then reordering such expressions as $\phi^n(x)$ so that the creation operators appear to the right and the annihilation operators to the left. The Wick-ordered product $:\phi^n(x):$ so obtained is not equal to $(:\phi(x):)^n$. As such expressions appear in the dynamical equations for the field, the ϕ^n in the derived commutator can no longer be assumed to have a simple polynomial form. This in turn vitiates the assumption that $[\ldots, \phi(k)]$ acts like a derivation on ϕ^n in the verification of the Jacobi identity.

b) Renormalization [2]

Even after Wick ordering, unwanted singularities still persist in the prediction of physical quantities through the perturbation expansion. The renormalization procedure systematically eliminates these divergences through the introduction of counter terms which in turn modify both the equal-time commutation relations and the dynamical equations for the fields. Evidently the general commutator changes and consequently it cannot be assumed that the original expression contained the Lie identities. Apparently the commutator derived from the bare theory also requires the addition of counter terms [3], a circumstance which indicates a delicate interplay between algebraic and analytical requirements.

It might appear from the foregoing that formal algebraic manipulation is entirely inappropriate for the examination of Lie bracket structure. Our belief that this is not so is founded on two observations which enable one to translate to a framework in which the analytic difficulties are suppressed. (On the other hand, one must be prepared for the possibility that the algebraic complexity may thereby be enhanced). These are as follows:

a) The Structure of the Propagator

The function f_n together with the specification of the ϕ^n must be such as to ensure convergence of integration. When (1) is taken from field theory these quantities are given by polynomials in the forward Δ^+ and backward Δ^- free field propagators. In general $\Delta^\pm$ have either too singular (infrared divergence) or else insufficient fall-off (ultraviolet divergence) to ensure convergence [2]. However, at least in simple cases (e.g. ϕ^3, ϕ^4 models) the validity of the Lie identities does not require the explicit form of the $\Delta^\pm$ but merely some extremely mild requirements, such as

$$\Delta^+(q) = \Delta^-(-q). \tag{2}$$

This circumstance allows convergence to be imposed without the algebraic structure being upset. Since (2) is also required of the true propagator, it further allows the latter to be reached in a suitable limiting sense.

b) The Role of Integration

In (1) integration is extended over the whole of momentum space and as such may give rise to ultraviolet divergences. However, we maintain that this integration plays only one role with respect to the Lie identities (an assertion easily justified in a wide variety of cases); that is, it defines a convolution algebra which on functions is commutative and associative. Such a structure is recovered through invariant integration over any compact abelian group. In particular we can choose this group to be finite and thus make convergence trivial. Such a procedure is less appealing in that the connection with space-time structure becomes blurred. Yet it is of significance that the algebraic structure can be preserved whilst giving the verification of the Lie identities a rigorous setting.

Thus far we have suppressed the rather unpleasant fact that the coefficients of the expansion are not uniquely determined. This is so because the interpretation of the commutator permits a reordering of a given polynomial in the fields appearing in the right-hand side through substitution from the left. Applied to a given term, no lower-order terms are affected and to given order only finitely many different expressions are possible, since a polynomial has only finitely many ordered forms. Hence it still makes sense to verify the Lie identities order by order, though reordering at each step may be required. This latter possibility greatly increases the complexity. We have found in the ϕ^3 model (and we suspect it to be generally true) that one cannot recover the Lie identities if reorderings are not allowed. Apart from manipulative difficulties there are also analytical implications of this circumstance. These arise because a reordering generally creates terms containing integrations involving only the propagators. Such integrations are generally found to be divergent when substitution is made for the known propagators of the model. Hence reorderings may be excluded on analytical grounds even though they are required algebraically. Obviously, a more complete analysis must recover the Lie identities without invoking such infinities. Even if this were possible in principle it is certainly impractical for the present. On the other hand, formal manipulation does at least uncover the basic combinatorial structure of the given model that in many ways is its most important content. Moreover, such calculations can be performed in a rigorous framework. On a higher level we believe the implicit infinities of the formal manipulations have an underlying connection with the appearance of renormalization counter terms and as such their explicit description is of some importance.

In order to separate out the algebraic and analytical difficulties associated with reordering, we decompose the overall problem into two stages. In the first stage we identify expressions which differ only in order of the field operators in polynomial terms. To give this a well-defined meaning we require the fields to commute, thereby generating a commutative algebra $\mathscr{A}$. In $\mathscr{A}$ we replace the commutator which is trivial by a bilinear map $X, Y \to \{X, Y\}$ of $\mathscr{A} \times \mathscr{A}$ into $\mathscr{A}$ such that for each $X \in \mathscr{A}$ the maps $\{X, \}: Y \to \{X, Y\}$ and $\{, X\}: Y \to \{Y, X\}$ of $\mathscr{A}$ into $\mathscr{A}$ are derivations. We define this map on the fields through (1) and extend through linearity and derivation to the whole of $\mathscr{A}$. Then we verify the Lie identities for (1) in the sense that we show that $\{, \}$ defines a Lie bracket relation on $\mathscr{A}$. This completes the first stage. The second stage is a return to the original problem. Of course, it is important to realize that stage one is an integral part of stage two and indeed represents the cancellation of a certain class of terms. Moreover, we shall see that the commutative algebra $\mathscr{A}$ arises in a very natural fashion.

Stage one itself needs not be trivial. We have given [3] this calculation for a current algebra model which has both internal and Poincaré symmetry, an analysis which is of some complexity. Here we describe a much simpler calculation which pertains to the ϕ^3 model. For stage one the same argument applies directly to ϕ^4 and indeed to any polynomial self-interaction model.

For stage one the field $\phi(q)$ is an element of the commutative algebra $\mathscr{A}$. An important observation is that in $\mathscr{A}$ two types of multiplication arise naturally. These are pointwise multiplication and convolution defined respectively by

$$(XY)(q) = X(q)\,Y(q) \tag{3}$$

$$(X \circ Y)(q) = \int X(q')\,Y(q - q')dq' \tag{4}$$

for all $X, Y \in \mathscr{A}$. Both are separately commutative and associative. This derives from the commutativity of $\mathscr{A}$ and, for convolution, the invariance of integration. As we noted, for our purpose this is the only role integration plays. Associativity fails on mixed products, so it is a key observation that (2) translates to the identity

$$((\Delta^+ X) \circ Y)(0) = (X \circ (\Delta^- Y))(0) \tag{5}$$

for all $X, Y \in \mathscr{A}$. Further, if we define clothed propagators $\underline{\Delta}^\pm$ through

$$\underline{\Delta}^\pm X = \Delta^\pm X + \kappa \Delta^\pm(\phi \circ (\Delta^\pm X)) + \kappa^2 \Delta^\pm(\phi \circ (\Delta^\pm(\phi \circ (\Delta^\pm X)))) + \cdots, \tag{6}$$

then from (5) and associativity of convolution

$$((\underline{\Delta}^+ X) \circ Y)(0) = (X \circ (\underline{\Delta}^- Y))(0) \tag{7}$$

for all $X, Y \in \mathscr{A}$. This is a central result. For the ϕ^3 model (1) takes the form [3]

$$\{\phi(q), \phi(p)\} = ((\underline{\Delta}^+ \delta_p) \circ \delta_q)(0) - ((\underline{\Delta}^- \delta_p) \circ \delta_q)(0), \tag{8}$$

where δ_p is the function given by

$$\delta_p(q) = \delta(q + p).$$

Antisymmetry of (8) is an immediate consequence of (7). Again using (7) to transpose the propagator, it can be shown [3] that the double commutator takes the form

$$\{\{\phi(q), \phi(p)\}, \phi(k)\} = [(\underline{\Delta}^+ \delta_k) - (\underline{\Delta}^- \delta_k)] \circ [(\underline{\Delta}^+ \delta_p) \circ (\underline{\Delta}^- \delta_q) - (\underline{\Delta}^- \delta_p) \circ (\underline{\Delta}^+ \delta_q)]. \tag{9}$$

Using the commutativity and associativity of convolution, it is easily verified that the right-hand side of (9) vanishes upon cyclic summation. Hence the Lie identities are established. We remark that the only property required of the propagators is (2). Finally, expanding (8) through (6) and (4) gives the explicit κ expansion for the bracket relation which is seen to have the form given by (1).

The momentum formalism used above is generally unsatisfactory for such calculations, as within this framework the fields are distribution-valued. However, since we have not invoked the distributional character of the propagators (and hence of the fields) this formalism has not been invalidated here. Yet to give the above result a wider context, we should still like to recast this computation in terms of smooth fields. Starting from (8) this is quite straightforward, whereas from the equal-time commutation relation and the equation of motion the various terms (such as $\phi^2(x)$) require a more careful interpretation. We should add that the propagator cannot be a a completely arbitrary distribution, nor does smoothing of itself eliminate the infinite terms which arise from field reorderings encountered in stage two.

Let $\mathscr{S}$ denote the Schwartz space of complex-valued smooth rapidly decreasing test functions on configuration space R^4. Let $\mathscr{S}'$ be the dual space of tempered distributions. For $f \in \mathscr{S}$ and $T \in \mathscr{S}'$ we define $f^- \in \mathscr{S}$ through

$$f^-(x) = f(-x); \qquad T^-(f) = T(f^-). \tag{10}$$

Given $f \in \mathscr{S}$, $T \in \mathscr{S}'$ their convolution $T \circ f \in \mathscr{S}'$ is defined

$$(T \circ f)(g) = T(f^- \circ g) \tag{11}$$

for all $g \in \mathscr{S}$. From (10) and (11) by easy computation we have

$$(T \circ f)(g) = (T^- \circ g)(f) \tag{12}$$

for all f, $g \in \mathscr{S}$.

We derive an expression for the Lie bracket $\{\phi(f), \phi(g)\} : f, g \in \mathscr{S}$ from (8) by formal computation. We then investigate under what conditions the Lie bracket relations are satisfied in a well-defined sense. Towards this aim let $\mathscr{H}$ denote the Hilbert space of complex-valued square integrable functions on R^4 with Lebesgue measure dx. Let $\{a_n \in \mathscr{S}; n = 0, 1, 2, \ldots\}$ be a complete set of orthonormal vectors for $\mathscr{H}$. Then by formal computation from (8) we obtain (dropping Fourier transform symbols)

$$\{\phi(f), \phi(g)\} = \sum_n \kappa^n \sum_{m_i} \left\{ \Delta^+ \circ (a_{m_1}^* (\Delta^+ \circ (a_{m_2}^* \ldots a_{m_n}^* (\Delta^+ \circ g) \ldots)(f)\, \phi(a_{m_1})\, \phi(a_{m_2}) \ldots \right.$$
$$\left. \cdots \phi(a_{m_n}) - (+) \to (-) \right\}. \tag{13}$$

In (13), $*$ denotes complex conjugation and $(+) \to (-)$ indicates that Δ^+ is replaced by Δ^-. The propagators $\Delta^\pm$ are interrelated through (10) which replaces (2).

A formal computation similar to the above shows that (13) satisfies the Lie identities. Its justification derives from the lemma below. First let $\mathscr{S}'_2$ denote the subspace of $\mathscr{S}'$ given by

$$\mathscr{S}'_2 = \{T \in \mathscr{S}' : (T \circ f) \in \mathscr{H} \quad \text{for all} \quad f \in \mathscr{S}\}.$$

Lemma:

For all $f, g, f_1, f_2, \ldots f_n \in \mathscr{S}$ and $T_1, T_2, \ldots T_n \in \mathscr{S}'$,

a) $f_1(T_1 \circ (f_2(T_2 \circ \ldots f_n(T_n \circ g)) \ldots) \in \mathscr{S}$ and

b) $(T_1 \circ (f_1(T_2 \circ \ldots f_n(T_n \circ g)) \ldots)(f) = (T_n^- \circ (f_n(T_{n-1}^- \circ \ldots f_1(T_1^- \circ f)) \ldots)(g).$

For all $f, g, h, f_1, f_2, \ldots f_k, g_1, g_2, \ldots g_1 h_1, \ldots h_m \in \mathscr{S}$ and $T \in \mathscr{S}'_2$, with $\{a_n\}$ as above,

c) $\sum_{n=0}^{\infty} (T \circ (f_1(T \circ \ldots f_k(T \circ (a_n^* (T \circ (g_1(T \circ (g_2 \ldots g_l(T \circ g)) \ldots)))(f) \times$

$$\times (T \circ (h_1(T \circ (h_2(T \circ \ldots h_m(T \circ h)) \ldots)(a_n) =$$
$$= \{T^- \circ (f_k(T^- \circ (f_{k-1} \ldots f_1(T^- \circ f)) \ldots)(T \circ (g_1(T \circ \ldots g_l(T \circ g)) \ldots)\} \times$$
$$\times (T \circ (h_1(T \circ \ldots h_m(T \circ h)) \ldots).$$

Proof:

a) This is an elementary consequence of the fact [4] that $T \circ f$ is a complex-valued function which is infinitely differentiable and polynomially bounded for all $f \in \mathscr{S}$ and $T \in \mathscr{S}'$.

b) This follows from a) and (12).

c) As in b), we use a) and (12) to transpose the f_j in the left-hand side. This gives an expression of the form

$$\sum_{n=0}^{\infty} (ab)\,(a_n^*)\,c(a_n)\,.$$

Since $T \in \mathscr{S}'_2$, we have from a) that $a, b, c \in \mathscr{H}$. Hence by the completeness of $\{a_n\}$ the above summation reduces to $(ab)(c)$ which is the right-hand side. The lemma is proved.

Given $\Delta^+ \in \mathscr{S}'$, then a) and b) of the lemma validate the antisymmetry of (13). Given $\Delta^+ \in \mathscr{S}'_2$, then c) of the lemma validates the Jacobi identity. We should add that if Δ^+ is computed from the equal-time commutation relations then $\Delta^+ \in \mathscr{S}'$; but $\Delta^+ \notin \mathscr{S}'_2$. Indeed we pick up an infinite contribution from the integration over time. This circumstance deserves comparison with the fact that renormalization effects a (formal) modification of the equal-time commutation relation [2].

We remark that for the triple and higher brackets to be well-defined within the above framework an even stronger condition on Δ^+ is required. That Δ^+ have compact support suffices. This (and also the previous condition) excludes Lorentz covariance, since the latter requires Δ^+ to be constant on the orbits of the homogeneous Lorentz group acting on R^4, which are not compact. One may restrict integration to the coset space; but then some of the structure of the theory is lost.

We conclude with some remarks concerning stage two. We first note that in the following sense the Lie identities cannot be recovered from the ϕ^3 model without field reordering. In (1) take for each n the homogeneous polynomial of degree n in ϕ^n, multiply by $\kappa^n f_n$ and sum over n. To within a reordering of each homogeneous polynomial this expression is independent of the original ordering in (1) and, moreover, every ordering of the homogeneous polynomials can be recovered.

Put in another way, there is a single equivalence class of such expressions and within this class field orderings can be ignored. We identify this class with (8). Now suppose that (1) has a form from which the Lie identities can be recovered without reorderings. Then antisymmetry implies that for each n the homogeneous polynomial defined by (8) as above must be symmetric with respect to the interchange of dummy variables given by $q_1, q_2, \ldots q_n \rightarrow q_n, q_{n-1}, \ldots q_1$. When field reorderings are excluded this can be shown [3] to be inconsistent with the Jacobi identity. Hence reordering is necessary.

The reordering procedure can be described in a language of operator-valued Feynman diagrams. We define these for the ϕ^3 model. To each term in (1) we assign a diagram given as follows. Each diagram has plane lines, wiggly lines and vertices. Three lines, of which at least two must be plane, meet at a vertex. Lines joining vertices are called internal and external otherwise. All wiggly lines, and for the n-fold commutator $(n + 1)$ plane lines, are external. Each line carries a 4-momentum directed by an arrow with momentum conservation at each vertex. To each wiggly line is assigned the field ϕ at the given momentum and to each plane line the propagator Δ^+ (through (2), Δ^- terms appear by arrow reversal). All momenta, except those carried by the external plane lines which define the arguments of the n-fold commutator, are integrated out.

Unlike ordinary Feynman diagrams, the ordering of the wiggly lines must be specified since these correspond to non-commuting operators—namely the fields at the given momenta. This is a difficult problem which we have not yet been able to solve in a satisfactory manner.

Let us call the set of all diagrams appearing in the commutator the commutator line. This has two external plane lines. The action of $[.., \phi(k)]$ on any diagram is given by replacing one of the wiggly lines by the commutator line and taking the sum over all possible replacements. For a given diagram the interchange of two field operators is given by replacing the corresponding wiggly lines by a commutator line joining the vertices. This produces a loop. To a given loop there is an integration over the internal propagator lines. As noted above, this is a possible divergence associated with reordering.

Diagrams with no loops (so-called tree diagrams) cannot be a result of field reordering. Hence to within the ordering of the wiggly lines, a tree diagram is invariant under the reordering of fields in (1). Moreover, identifying the different wiggly line orderings, it is clear that the class of all tree diagrams must separately satisfy the Lie identities. This essentially describes stage one. We note in particular that (8) is given by tree diagrams alone [5]. Yet it is wrong to conclude that this is the only class of diagrams which can be made to satisfy the Lie identities without reordering. For example, if we include all possible 1-fold, 2-fold, ... replacements of

$\longrightarrow$ by $\rightarrow\!\bigcirc\!\rightarrow$ in each tree diagram of (8) then the Lie identities still

hold. Momentum conservation implies that the inclusion of this diagram represents a modification of the propagator Δ^+ to $\Delta^+\Delta^+(\Delta^+ \circ \Delta^+)$, so that we recover essentially the same set of tree diagrams, but with two sets of propagators. Again we can

take a combination of diagrams of the form $\longrightarrow\!\bigcirc\!\longrightarrow$ such that, ignoring external

legs, the overall contribution is symmetric in the external momentum. This generalizes the vertex, and a similar inclusion in (8) recovers the Lie identities. Since loops can only be produced by reordering, each component of a diagram having two external plane lines and any number of wiggly lines must be totally symmetric in its external momentum to recover the Lie identities. In the sense of the generalized vertex so defined, only tree diagrams admit the Lie identities without reorderings.

Though we are not yet able to satisfactorily define the total ordering of the wiggly lines required for stage two, we can introduce a partial ordering in a natural fashion, which we believe is an important step. We proceed as follows. Note that (8) has exactly two components which are interrelated by ($\pm$) interchange. We shall call these the forward ($+$) and backward ($-$) commutator lines. Observe that in the forward line the arrows on each plane line point in the same direction for every tree. This defines a natural ordering of the attached wiggly lines. On any such tree eliminate any pair of ordered wiggly lines and join the corresponding vertices by the forward commutator line so that its arrows define the same ordering. This gives a loop for which the arrows on plane lines define a partial ordering of wiggly lines. Joining totally ordered vertices in the above fashion inductively generates a class of loop diagrams with this partial ordering. With a factor $-\frac{1}{2}$ for each internal loop momentum, this is exactly the class which derives by solution of the dynamical equations, though it remains to lift the partial ordering to a total ordering.

The above inductive procedure corresponds to the reordering of fields on totally ordered wiggly lines, given that the backward component of the commutator line does not contribute. The condition for this can be shown to be

$$\int \Delta^+(q)\,\Delta^+(q-q_1)\,\Delta^+(q-q_1-q_2)\cdots\Delta^+(q-q_1-q_2-\ldots-q_n)\,dq = 0 \quad (14)$$

for all positive integers n and all $q_1, q_2, \ldots q_n$. This holds, given that the supports of the Fourier-transformed functions $\hat{\Delta}^+$ and $\hat{\Delta}^-$ are separated by a three plane which by (2) can always be chosen to pass through the origin. This holds in field theory since $\hat{\Delta}^+$ has support in the forward light cone. In this case the partial ordering defined above can be identified as a time ordering.

Acknowledgements

The work described here was carried out at Tel-Aviv University with the support of a Royal Society–Israel Academy Research Fellowship.

I should like to thank Professor Yuval Ne'eman for the hospitality of the Physics Department and Dr. Joseph Slawny for discussions on distributions.

REFERENCES

[1] See for example: J. M. Cornwall and R. Jackiw, "Canonical Light Cone Commutators in Photon–Hadron Interactions", and references therein. International Summer School in Theoretical Physics, Hamburg, DESY, 1971.

[2] S.S. Schweber, "Relativistic Quantum Field Theory", Chap. 16. Harper and Row, New York (1964).

[3] A. Joseph, "On the Dynamical Solution to the Sugawara Model". II, "The Lie Identities", *Ann. Phys.* (in press).

[4] L. Gårding and J. Lyons, *Nuovo Cimento, Suppl.,* **14**, 9 (1959).

[5] Compare with a similar result for ordinary Feynman diagrams: D.G. Boulware and L.S. Brown, *Phys. Rev.,* **172**, 1628 (1968).

Nonlinear Group Action.
Smooth Action of Compact Lie Groups on Manifolds

LOUIS MICHEL

Institut des Hautes Etudes Scientifiques, 91-Bures-sur-Yvette (France)

Abstract

Linear representations of groups is a particular case of group action which is well known to physicists. The aim of these two lectures is to give the basic concepts for general group action.

In section 1 we consider the action of (abstract) groups on sets. In section 2 we study the homogeneous spaces of a given group. Section 3 is devoted to continuous actions of topological groups on topological spaces. The last section gives some results on the important particular case of smooth action of compact Lie groups on manifolds. For lack of time, no physical applications are given, but some references are provided.

Contents

1. GROUP ACTION ON SETS

1.1. Definitions

Let $\mathscr{P}(E)$ be the group of permutations of the elements of a set E. Each action of a group G on E is given by a group homomorphism

$$G \xrightarrow{\ f\ } \mathscr{P}(E). \tag{1}$$

Often, instead of $f(g)[x]$, we shall simply denote by $g.x$ the transform of $x \in E$ by $g \in G$.

An action of G on E can also be given by a map

$$G \times E \xrightarrow{\ \Phi\ } E \tag{2}$$

which satisfies

$$\Phi(e, x) = x \qquad \Phi(g_1, \Phi(g_2, x)) = \Phi(g_1 g_2, x). \tag{2$'$}$$

Indeed, $x \xrightarrow{\ \varphi_g\ } \Phi(g, x)$ is a permutation of E whose inverse is $\varphi_{g^{-1}}$ and $g \rightsquigarrow \varphi_g$ is the homomorphism f in (1).

Let us introduce some vocabulary.

G acts *effectively* on E if $\operatorname{Ker} f = \{e\}$ (trivial) $\Leftrightarrow g \neq e$ implies $g.x \neq x$ for some $x \in E$.

G acts *freely* on E if $g \neq e$ implies $g.x \neq x$ for all $x \in E$.

G acts *transitively* on E if for all $x, y \in E$, $\exists\, g \in G$ such that $y = g.x$. In this case E is called a homogeneous space of G.

The set of fixed points, which we denote by E^G, is defined by

$$E^G = \{x \in E,\ \forall g \in G,\ g.x = x\}.$$

Of course, if $E^G = E$, G acts *trivially* on $E (\Leftrightarrow \operatorname{Ker} f = G)$.

1.2. Natural Transfer of Actions

Given G actions $f_1, \ldots, f_n$ on sets $E_1, \ldots, E_n$, then there are well-defined "natural" actions of G on the sets one can form from the E_i:

(a) G acts on $E_1 \times E_2$

$$g \cdot (x_1, x_2) = (g.x_1, g.x_2). \tag{3}$$

(b) Let $\text{Maps}(E_1, E_2)$ be the set of maps from E_1 to E_2. Then for any $\theta \in \text{Maps}(E_1, E_2)$ and $g \in G$, the transform $g.\theta$ must satisfy the commutative diagram of maps

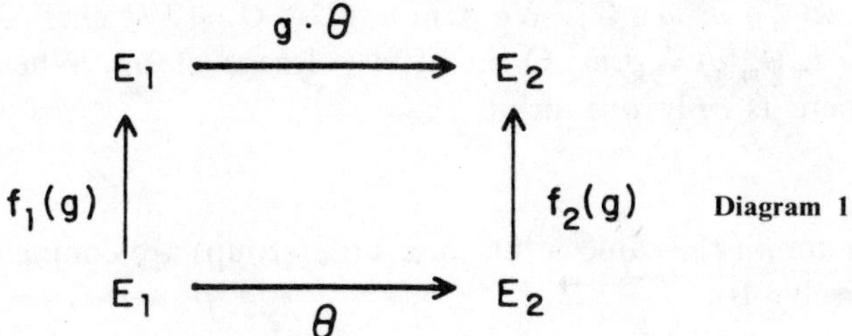

Thus it is defined by

$$(g.\theta)(x_1) = g.(\theta(g^{-1}.x_1)). \tag{3'}$$

(c) If G acts on disjoint (i.e. $E_i \cap E_j = \varnothing$) sets, its natural action on the union $\cup_i E_i$ is obvious.

1.3. Equivariant and Equivalent G-Actions

Given G, we define the "morphisms" between its actions and, as a particular case, the isomorphisms, and consider isomorphic actions as equivalent. The action of G on E_1 is equivariant to the action of G on E_2 if $\text{Maps}(E_1, E_2)^G$ is not empty, i.e. if there exists $\theta \in \text{Maps}(E_1, E_2)$ invariant (such that $g.\theta = \theta$) for any $g \in G$. Then Diagram 2 of maps is commutative for all $g \in G$:

$$f_2(g) \circ \theta = \theta \circ f_1(g). \tag{4}$$

We say that θ is an equivariant map. If furthermore θ is a bijective map (i.e. one-to-one and onto) the two actions of G are isomorphic. We shall also say that E_1 and E_2 are isomorphic G-spaces.

1.4. Little Group $=$ Isotropy Group $=$ Stabilizer of $m \in E$

This is the set $\{g \in G, g.m = m\}$. It is a subgroup of G which we denote by G_m.

1.5. The Orbit of $m \in E$

This is the set $\{g.m, \; g \in G\}$. We denote it by $G.m$. We shall denote by ψ_m the map $G \xrightarrow{\psi_m} E, \psi_m(g) = g.m$: Orbit of $m =$ Image of ψ_m. When the group acts transitively there is only one orbit.

Lemma

If m' and m are on the same orbit, their little groups are conjugate.
More precisely: If

$$m' = g.m, \quad \text{then} \quad G_{m'} = gG_mg^{-1}. \tag{5}$$

1.6. Strata

If two points m' and m have conjugate little groups, they need not be on the same orbit. By definition they are on the same *stratum*. When the group acts freely, there is only one stratum ($\forall \, m \in E, G_m = \{e\}$). When E^G is not empty it is a stratum. We denote the stratum of m by $S(m)$.

1.7. Orbit Space

Being on the same orbit is an equivalence relation between the elements of E. Thus E is partitioned into orbits. The set of orbits is called the orbit space. It is usually denoted by E/G. We shall use π for the canonical map $E \xrightarrow{\pi} E/G = \pi(E)$.

1.8. Space of Strata

Having conjugate little groups is an equivalence relation for the elements of E. Thus E is partitioned into strata. We denote by σ the canonical map of E into the space of strata. This map factorizes:

$$\sigma = \tau \circ \pi. \tag{6}$$

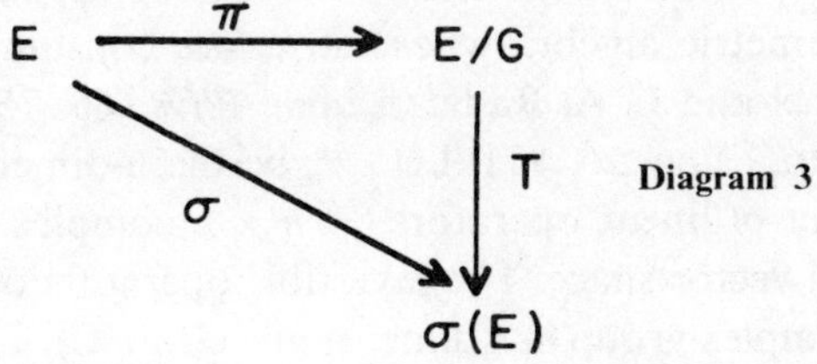

1.9. Examples

(a) $SO(2)$ *action on* S_2 (The 2-dimensional sphere)

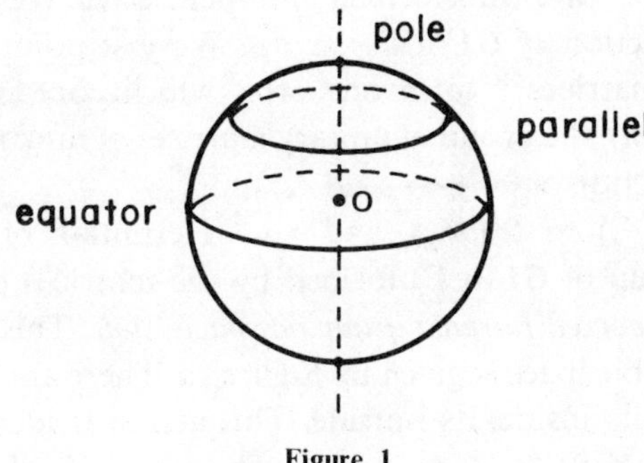

Figure 1

The orbits are the parallel circles (little group $= \{1\}$) and the two poles ($=$ fixed point, little group $= SO(2)$). So there are two strata.

(a') $SO(2) \times Z_2$ *action on* S_2. The group of (a) is enlarged with the symmetry through the center 0. There are no fixed points. There are now three strata:

(i)   the generic one, open, dense: each orbit consists of the two parallels with the same N and S latitude.

(ii) one with one orbit: the 2 poles (little group $SO(2)$).

(iii) one with one orbit: the equator (the little group has two elements).

(a") $SO(2)$ *action on* P_2. The real projective plane: action obtained by the equivariant map $S_2 \longrightarrow P_2$ which identifies points symmetric through 0. 3 strata, images of the 3 strata of (a') (not of (a)).

(b) $SU(3)$ *action on* S_7. The unit sphere of the adjoint representation ($=$ octet space). Note that $SU(3)$ does not act effectively, but the adjoint group $SU(3)/Z_3$ does. Same image as (a) but the 2 poles are four-dimensional orbits (little group $U(2)$) and the generic stratum has 6-dimensional orbits (little group $U_1 \times U_1$).

(b') $\mathrm{Aut}\,(SU(3)/Z_3)$ *action on* S_7. $SU(3)$ or $SU(3)/Z_3$ has only one class of outer automorphism (it corresponds in physics to charge conjugation). This example is similar to (a'); the "equator" becomes the 6-dimensional orbit of the "roots" of the Lie algebra. The poles are replaced by the 4-dimensional set of "pseudo-roots" (= roots of the symmetric algebra whose structure constants are the d_{ijk} of Gell–Mann); see L. Michel and L. A. Radicati, *Ann. Phys.*, **66**, 758 (1971).

(c) *Action of* $GL(n, \mathbb{C})$ *on* $\mathcal{L}(\mathcal{H}_n)$. Let $\mathcal{H}_n$ be the n-dimensional complex vector space, $\mathcal{L}(\mathcal{H}_n)$ the set of linear operators (= $n \times n$ complex matrices) on $\mathcal{H}_n$: that is an n^2-dimensional vector space. The invertible operators of $\mathcal{L}(\mathcal{H}_n)$ form a group, the general linear complex group in n dimensions: $GL(n, \mathbb{C})$. This group acts (linearly) on the vector space $\mathcal{L}(\mathcal{H}_n)$ according to the law

$$g \in GL(n, \mathbb{C}), \quad \mathcal{L}(\mathcal{H}_n) \ni x \rightsquigarrow gxg^*$$

(where * is the Hermitian conjugation of matrices). We leave as an exercise the study of the strata. We note that the Hermitian operators $x = x^*$, which form a real vector space $\mathscr{E}_{n^2}$ of dimension n^2, are transformed into themselves. We also leave as an exercise the study of the action of $GL(n, \mathbb{C})$ on $\mathscr{E}_{n^2}$. We just point out that the strictly positive (Hermitian) matrices x form one orbit which contains the unit matrix I, whose little group is $U(n)$, the group of unitary matrices of rank n (i.e. $n \times n$ matrices). Another interesting action is $x \rightsquigarrow gxg^{-1}$.

(c') *Action of* $SL(n, \mathbb{C})$ *on* $\mathcal{L}(\mathcal{H}_n)$, and $\mathscr{E}_{n^2} =$ Hermitian of $\mathcal{L}(\mathcal{H}_n)$. The group $SL(n, \mathbb{C})$ is the subgroup of $GL(n, \mathbb{C})$ formed by the matrices of determinant unity.

(d) *Action of the connected Lorentz group on space–time.* This action is well known to the audience. The orbit space is given by Figure 2a. There are four strata, the origin 0, the light cone $-O$, its inside, its outside. This action is identical to that studied in (c') for $n = 2$: action of $SL(2, \mathbb{C})$ on $\mathscr{E}_{n^2} = \mathscr{E}$. The center Z_2 of $SL(2, \mathbb{C})$ acts trivially and the quotient group $SL(2, \mathbb{C})/Z_2$, which acts effectively, is isomorphic to the connected Lorentz group.

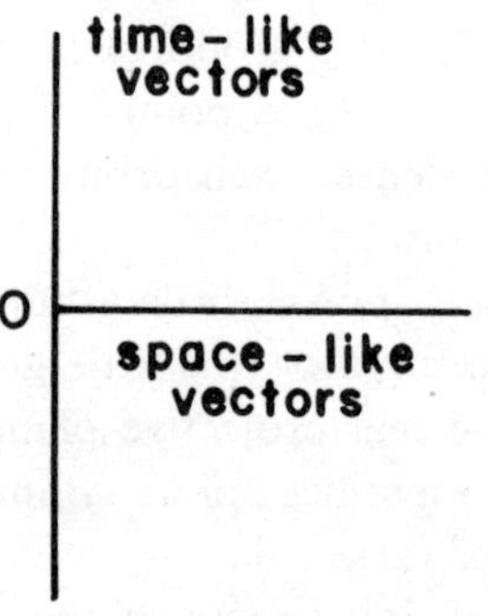

Figure 2a

The strictly positive and strictly negative matrices form the stratum of time-like vectors. The corresponding little groups are conjugates of $SU(2)$.

(d') *Action of the complete group* $\mathcal{L}$ *on space–time.* The same four strata, but the orbit space is reduced to that of Figure 2b. Note that the little group $\mathcal{L}(p)$ of a time-like vector p is conjugate to $O(3)$, the three-dimensional orthogonal group, which is the little group for the time axis.

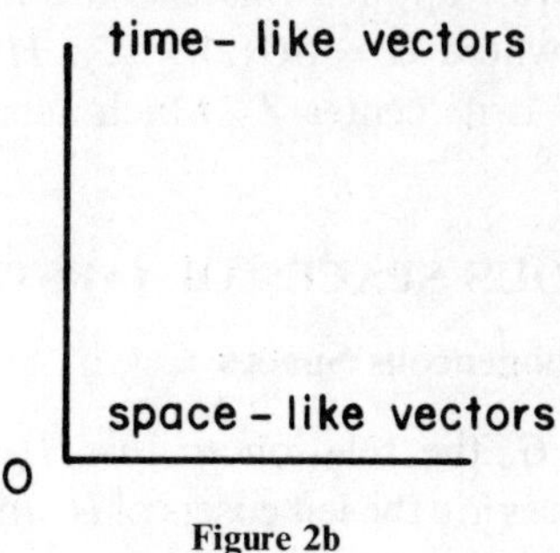

Figure 2b

(e) *Action of the connected Poincaré group of space–time.* The group generated by the translations and the connected Lorentz group transformations is the connected Poincaré group. The space–time is one orbit of this group. The stabilizer of any point 0 is the Lorentz group leaving 0 fixed.

(f) *Action of* $\mathscr{L}(p)$, *the little group of* p *on three-particle phase space* $p = p_1 + p_2 + p_3$. As we have seen in (d'), $\mathscr{L}(p)$ is conjugate to $O(3)$. There are two strata: The generic stratum S; p_1, p_2, p_3 are linearly independent, the 2-element little group is generated by the symmetry through the plane defined by p_1, p_2, p_3. The exceptional stratum S': when p_1, p_2, p_3 are linearly dependent; the little group is conjugate to $O(2)$. The orbit space is the Dalitz plot; $\pi(S')$ is its boundary.

We leave the case when two (or three) particles are identical to be handled as an exercise.

(g) *Any group G acts on itself* (= its set of points) *by different actions*:

(i)   by left translations $x \xrightarrow{g} gx$, so G acts freely and transitively on itself.

(ii)   by conjugation $x \xrightarrow{g} gxg^{-1}$. In this case the orbits are the conjugation classes. This action of G on itself is by automorphisms: $G \longrightarrow \mathrm{Aut}\, G$, where $\mathrm{Ker}\, f$ is the center of G (it is also the stratum of fixed points) and image of f is the group of inner automorphisms.

(g') *H, subgroup of G acts on G by left translation.* The orbits are the right cosets of H. There is only one stratum and H acts freely on G.

1.10. Exercises

(i) If G does not act effectively on E, then $\mathrm{Ker}\, f = \bigcap_{m \in E} G_m$. It is the largest invariant subgroup contained in any little group G_m.

(ii) If H is a subgroup of G, the action of G on E defines an action of H on E. If, furthermore, H is an invariant subgroup of G, this defines an action of G on the orbit space E/H and also of the quotient group G/H on the orbit space E/H. Note that $\pi_H : E \xrightarrow{\;\pi_H\;} E/H$ defined by the action of H on E is an equivariant map of G actions.

The reader has already twice applied this exercise in very simple cases: to pass from example (a′) to (a″), where $G = SO(2) \times Z_2$, $H = Z_2$, $E = S_2$, $E/H = P_2$; and in (d), $G = SL(2, \mathbb{C})$, H is its center Z_2 which acts trivially on space–time.

2. HOMOGENEOUS SPACES OF (ABSTRACT) GROUPS

2.1. Classification of G-Homogeneous Spaces

Given a subgroup H of G, the relation $a^{-1}b \in H (\Leftrightarrow b \in aH)$ is an equivalence relation, the equivalence classes are the left cosets of H, and we will denote the quotient space ($=$ coset space) by $[G:H]$.

By left translation G acts on $[G:H] \ni gH \rightsquigarrow gaH$. This action is transitive, so $[G:H]$ is an orbit of G: the little group of $H \in [G:H]$ is H itself.

Let E be a G-homogeneous space with little-group conjugate to $H, \ni m \in E$, $G_m = = H$. Let $m' = g_1.m = g_2.m$; then $g_2^{-1}g_1 \in H$, so g_2 and g_1 are in the same left coset of H, i.e. $g_1 H = g_2 H$. The correspondence $g.m \rightsquigarrow gH$ is therefore a map $E \xrightarrow{\;\theta\;} [G:H]$;

it is equivariant: $g_1\theta(g_2.m) = g_1 g_2 H = \theta(g_1 g_2.m)$;

it is one-to-one: $\theta(g_1 m) = \theta(g_2 m) \Leftrightarrow g_1 H = g_2 H \Leftrightarrow g_2^{-1}g_1 \in H \Leftrightarrow g_2^{-1}g_1 m = m$, i.e. $g_1 m = g_2 m$;

it is onto: every gH is the image of a $g.m$.

So the G-homogeneous space E is isomorphic to the "standard" G-homogeneous space $[G:H]$ with G-action by translation. It is easy to see that $[G:H]$ and $[G:gHg^{-1}]$ are isomorphic G-homogeneous spaces. Hence we have proved the

Theorem:

There is a natural bijective map between the classes of isomorphic G-homogeneous spaces and the classes of conjugate subgroups of G.

A class of isomorphic G-homogeneous spaces is also called a *type* of G-orbits.

2.2. Partial Order on the Set $\mathcal{O}_G$ of Types of G-Orbits

We denote by (H) the class of subgroups of G conjugate to H.

With the inclusion $(H_1 \subset H_2)$, the subgroups of a group form a lattice. The classes of conjugate subgroups also form a lattice. This lattice structure can also be applied to the set $\mathcal{O}_G$, but it is done in the usual language by reversing the order: the larger a subgroup H of G, the smaller is the orbit $[G:H]$; indeed, for finite groups (Card H) . (Card $[G:H]$) = Card G.

In an action of G on E, a stratum is the union of all orbits of the same type. So the stratum space (see Section 1.8) $\sigma_G(E)$ can be identified with a subset of $\mathcal{O}_G$. One can therefore speak of maximal or minimal strata. For example, if there are fixed points, they form the unique minimal stratum.

2.3. Morphisms of Homogeneous Spaces. Imprimitivity

$$E_1 \xrightarrow{\ \theta\ } E_2$$

is an equivariant map from E_1 to E_2. θ is always surjective ($=$ onto). If it is not injective ($=$ one-to-one) $(H_1) \leqq (H_2)$, e.g.

$$\theta^{-1}(H_2) = [H_2 : H_1] \subset [G : H_1].$$

If H_2 is a maximal proper subgroup of G (i.e. $H_2 \subset H \subset G$ implies for the subgroup H either $H = H_2$ or $H = G$), the action of G on E_2 is said to be *imprimitive*. If (H_1) is strictly smaller than (H_2), H_2 a maximal subgroup, the inverse images $\theta^{-1}(m_2)$ of elements of E_2 are called imprimitivity classes of E_1.

Example: The connected Lorentz group action on a light cone (without summit). The generatrices are imprimitivity classes. The little groups are isomorphic to $E(2)$, the 2-dimensional Euclidean group. They are not maximal in Lorentz, the maximal group $TR(2)$ is isomorphic to $E(2)$ and dilations

$$\left(TR_2 = \left\{ \text{triangular matrices} \begin{pmatrix} a & b \\ 0 & a^{-1} \end{pmatrix} \text{ in } SL(2, \mathbb{C}) \right\} \right).$$

The set of points at infinity for the light cone (or any—real or imaginary—mass hyperboloid $=$ asymptotia...) can be identified with the orbit $[\mathscr{L} : TR(2)]$; it is homeomorphic to the sphere S_2.

2.4. Remark on the Isomorphy of G-Spaces and G-Orbits

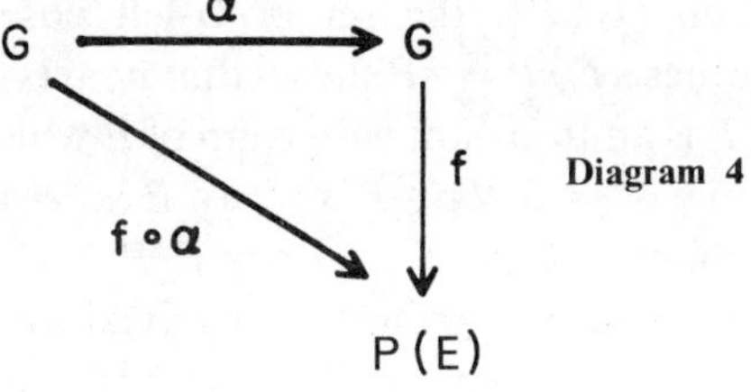

Let $\alpha \in \text{Aut } G$. If f of Diagram 4 defines an action of G on E, $f \circ \alpha$ defines another action. Are the two actions equivalent? The answer is yes if α is an inner automorphism of G; let $a \in G$ be an element which realizes α, i.e.

$$\forall g \in G \qquad aga^{-1} = \alpha(g). \tag{7}$$

Then $f(a)$ is an equivariant map on E which establishes the equivalence of the two actions. Indeed, for any $g \in G$, using (7) and the fact that f is a group homorphism:

$$f(\alpha(g)) \circ f(a) = f(aga^{-1}) \circ f(a) = f(a) \circ f(g) \tag{8}$$

If α is not an inner automorphism of G, it may happen that the two G-actions are inequivalent although they produce on E the same orbits and the same strata. But some orbits may be nonisomorphic G-homogeneous spaces. Indeed, consider the case when E is one orbit, isomorphic to $[G:H]$ for the action f. For the action $f \circ \alpha$, it is isomorphic to $[G:\alpha^{-1}(H)]$. If $\alpha^{-1}(H)$ and H are not conjugate in G, these two homogeneous spaces are nonisomorphic. Indeed, there is no $\varphi \in \mathscr{P}(E)$ such that

$$\forall\, g \in G \qquad f(\alpha(g)) \circ \varphi = \varphi \circ f(g). \tag{9}$$

Let $m \in E$ such that $G_m = H$ for the action f. Its little group for the action $f \circ \alpha$ is $\alpha^{-1}(H)$. If there is a φ satisfying (9), apply both sides of this equation to m for all $h \in H$; we obtain:

$$\alpha f(a(h))\, \varphi(m) = \varphi(m),$$

so that H is also the little group of $\varphi(m)$. Since points on the same orbit have conjugate little groups, H has to be conjugate to $\alpha^{-1}(H)$, the little group of m; this is absurd.

2.5. Physical Application

Generalization of the McGlinn Theorem. McGlinn (*Phys. Rev. Lett.*, **12**, 467 (1964)) proved that if $G = S.P$ (i.e. $\forall\, g \in G$ can be decomposed uniquely into a product $g = sp, s \in S, p \in P$) where P is the Poincaré group ($P = T.L$) and S a semi-simple Lie group, and if $\forall\, s \in S, \forall\, l \in L, sl = ls$, then $G = S \times P$, the direct product. Michel (*Phys. Rev.*, **137B**, 405 (1965)) extended this result to the case when S can also be an arbitrary simple group or direct product of simple groups, under the weaker hypothesis: there is a $p_0 \in P, p_0 \notin T$ such that $\forall\, s \in S, s^{-1}p_0 s \in P$. As we saw in Section 2.1, G acts on $[G:P]$, the set of G-left cosets of P, by the action: $sP \overset{f(g)}{\rightsquigarrow} gsP$. The hypothesis $p_0 sP = sP$ shows that p_0 acts trivially, so that $p_0 \in K = {} = \operatorname{Ker} f \subset G$. And $K \cap P$ is an invariant subgroup of P which contains p_0. It must be P itself, so that $P \subset K$; i.e. $\forall\, s \in S, \forall\, p \in P, s^{-1}ps \in P$ so P is an invariant subgroup of G, and there is a homomorphism $S \overset{\varphi}{\longrightarrow} \operatorname{Aut} P/P = R$. Since S is a product of simple groups (or a semisimple Lie group), φ is trivial and G is isomorphic to the direct product $S \times P$. Hint: $G = S' \times P, s' = sq(s^{-1})$ where $sps^{-1} = q(s)\, pq(s^{-1})$, and $q: E \overset{q}{\longrightarrow} \operatorname{Inner} \operatorname{Aut} P = P$.

3. CONTINUOUS ACTION OF A TOPOLOGICAL GROUP G
ON A TOPOLOGICAL SPACE X

3.1. Introductory Remarks

If E has a mathematical structure, it has a group of automorphisms Aut E for this structure. The elements of Aut E have various names, e.g.:

E	Name of elements of Aut E
set	permutation = bijective map
ordered set	order preserving map
topological space	homeomorphism
metric space	isometry
vector space	invertible linear operator
complex Hilbert space	unitary operator
real Hilbert space	orthogonal operator
differentiable manifold	diffeomorphism

and so on.

One considers only the actions of G given by the group homomorphisms $G \xrightarrow{f} \text{Aut } E$.

Example: If E is a vector space, f is a linear representation of G. One should distinguish the different structures on the same set, for example:

If E is the space–time *vector space*, i.e. the space of energy momenta, the vector zero must be invariant and the group which preserves the Minkowski scalar product is the Lorentz group. If E is the affine space–time (i.e. of x and t coordinate), then translation belongs also to Aut E, which is the full Poincaré group. If the space–time E is defined just as a set with the partial order relation due to causality:

$$x < y \Leftrightarrow y - x \in V^+,$$

the inside of the future light cone, it is then quite remarkable that (without linear and even topological structure for E) Aut E is the orthochronous Poincaré group with the dilations added (E. C. Zeeman, *J. Math. Phys.*, **5**, 490 (1964)).

This point of view is too general for our purpose, because we wish also to take into account a richer structure on the group G than that of the group law. For instance, G can be a topological group; then we do not want to consider as possible actions of G on the topological space X all homomorphisms of G into the group of homeomorphisms of X. We shall use the alternative way (eqns. (2) and (2′)) of defining the action of G and require, for example, that the map Φ preserve the structure on G and X.

3.2. Definition and General Remarks

A continuous action of a topological group G on a topological space X is defined by a continuous map

$$G \times X \xrightarrow{\Phi} X$$

which satisfies (2').

If G is a topological group, the homogeneous space $[G:H]$ is a topological space with the quotient topology: i.e. the open sets of $[G:H]$ are all the subsets S of $[G:H]$ such that $\rho^{-1}(S)$ is an open set of G, where ρ is the canonical map $G \xrightarrow{\rho} [G:H]$. By definition, the topology on $[G:H]$ is the finest one such that ρ be continuous.

Consider the continuous action $G \times X \longrightarrow X$. On any orbit $G \cdot x (x \in X)$, one has to distinguish two topologies: That of the homogeneous space $[G:G_x]$ and that of topological subspace $G.x \subset X$ (the open sets of $G.x$ are then the intersections of $G.x$ and open sets of X).

If the canonical bijective map $[G:G_x] \xrightarrow{l} G(x)$ is not continuous (i.e. if the topology of $[G:G_x]$ is not equal to or finer than that of $G.x$), then the G-action is not continuous.

The orbit space X/G is also a topological space with the quotient topology, defined by the canonical map $\pi: X \xrightarrow{\pi} X/G$.

Exercise

Let $K \subset H \subset G$, K an invariant subgroup of G. Then $[G:K]$ is also a topological group G/K and $[G:H]$ is homeomorphic to $[G/K:H/K]$.

3.3. Basic Facts on Continuous Group Action

$\Phi(g,.) = \varphi_g$ is a homeomorphism of X.

$\Phi(.,x) = \psi_x$ is a continuous map from G to X.

Let L be a subset of G, S a subset of X, $LS = \{l.s, l \in L, s \in S\}$.

	G	X	L	S	LS
a)	topological	topological	arbitrary	open	open
b)	Hausdorff	Hausdorff	compact	compact	compact
c)	Hausdorff	Hausdorff	compact	closed	closed
d)	compact	Hausdorff	closed	closed	closed.

Note 1. Definition of X, Hausdorff: $\forall\, x, y \in X$, $\exists$ open neighborhood of x, V_x and open neighborhood V_y of y, $V_x \cap V_y = \varnothing$. Moreover, to be compact or to have compact subsets, X must be Hausdorff. Remark: a point of a Hausdorff space is a closed set.

Proofs

(a) Φ_e = homeomorphism $\to l.S$ open, $L.S = \bigcup_{l \in G}$ open sets = open set.

(b) $L.S$ is the continuous image of a compact set, product of two compact sets.

(c) Choose sets g'_α in L, s_α in $S, g_\alpha s_\alpha \to x \in \overline{LS}$ (closure of LS) L compact $\Leftrightarrow \exists$ a subset g_α in $L, g_\alpha \to l, \lim s_\alpha = \lim g_\alpha^{-1}(g_\alpha s_\alpha) = l^{-1}, x \in S$ since S closed, so $x \in LS$ and $\overline{LS} = LS$.

(d) L closed in G compact $\Rightarrow L$ compact $\Rightarrow$ (c).

Then for $L = G, S = \{x\}$ (one-point set, which is closed, $LS = G.x$, an orbit, so in case (d) the *orbits are closed*). They are also compact, since $G.x = \psi_x(G)$, the continuous image of a compact set.

In the general case (a), if S open, $G.S = \pi^{-1}(\pi(S))$ open, hence $\pi(S)$ open (by definition of the quotient topology), so that π is an open map. In case (d), S closed $\Rightarrow$ $\Rightarrow G.S = \pi^{-1}(\pi(S))$ closed, so that π is then also a closed map.

3.4. Particular Case when G Is Locally Compact and Compact and X Is Metrizable

A locally compact group G has a left-invariant Haar measure dg. If, furthermore, G is compact, then dg (which is also right-invariant) can be normalized by

$$\int_G dg = 1. \tag{10}$$

Let $\Delta(x, y)$ be a distance in X. It can be averaged by the Haar measure of the compact G into

$$\tilde{\Delta}(x, y) = \int_G \Delta(gx, gy)\, dg. \tag{11}$$

This is a G-invariant metric on X and it yields the same topology as Δ. The orbit space X/G is also metrizable, with the metric:

$$d(\pi(G.x), \pi(G.y)) = \underset{x' \in G.x;\, y' \in G.y}{\text{Min}} \tilde{\Delta}(x', y'). \tag{12}$$

The orbit space X/G is indeed a generalization of the meridian section of an $SO(2)$-invariant domain in our 3-dimensional space.

3.5. Physical Applications. Crystallography

Our space $\mathscr{E}$ as an affine space has for its automorphism group $IL(3, \mathbb{R})$, the inhomogeneous general linear group in 3 dimensions; it is the semidirect product of the translations T by $GL(3, \mathbb{R})$. We denote by $GL^+(3, \mathbb{R})$ the subgroup of index 2 of $GL(3, \mathbb{R})$ formed by the 3×3 real matrices of positive determinant; we denote by $IL^+(3, \mathbb{R})$ the semidirect product $T \wedge GL^+(3, \mathbb{R})$.

The Euclidean group $E(3)$ is a subgroup of $IL(3, \mathbb{R})$. A crystal is represented as a space lattice C of $\mathscr{E}$, and the crystal symmetry group is the subgroup $K \subset E(3)$ which transforms C into itself. An equivalent but more abstract definition of the crystallographic groups K is: a discrete subgroup of $E(3)$ such that the homogeneous space $[E(3):K]$ is compact. Two crystallographic groups K_1 and K_2 are considered as equivalent if they are isomorphic and their actions on $\mathscr{E}$ are equivalent. However, in order to distinguish between left and right, crystallographers have restricted the equivalence map θ of diagram 2 to be in $IL^+(3, \mathbb{R})$ and not in $IL(3, \mathbb{R})$. So there are 230 classes of equivalent crystallographic groups in 3 dimensions; among them 11 pairs are isomorphic: they are conjugate in $IL(3, \mathbb{R})$ but not in $IL^+(3, \mathbb{R})$. In 2 dimensions there are 17 classes of crystallographic groups. It was one of the famous Hilbert problems to know if the number of crystallographic classes is finite for any dimension n. The positive answer was given by Bieberbach in 1911.

4. SMOOTH ACTION OF A COMPACT LIE GROUP G ON A C^∞-MANIFOLD M

Smooth $= C^\infty =$ infinitely differentiable.

Then G and M are both (smooth) manifolds and the map defining the action $G \times M \xrightarrow{\Phi} M$, which satisfies (2′), is a morphism (= smooth map) of smooth manifolds.

4.1. General Results

Let $\Delta(x, y)$ be a Riemann metric on M. By averaging with the Haar measure of G one obtains a G-invariant Riemann metric and, as we have seen, M/G is metrizable. If in V_m, neighborhood of m, we choose geodesic coordinates, G_m transforms geodesics through m into geodesics through m, so G_m acts linearly in V_m.

However, there are stronger results:

(i) R. C. Palais (*Amer. J. Math.*, **92**, 748 (1970)): C^1 (= continuous action) $\Rightarrow C^\infty$ action when M is compact.

(ii) Myers and Steenrod (*Ann. Math.*, **40**, 400 (1939)): G acts isometrically $\Rightarrow G$ acts differentially.

For C^1-action: Mostow proved (*Ann. Math.*, **65**, 513 and 432 (1957)) *Theorem* 1: If M is compact, the number of strata is finite. *Theorem* 2: If the number of strata is finite, the G-action on M is equivariant, by an injective map, to a linear (orthogonal) representation of G on a finite dimensional vector space.

(iii) Palais (1961): When M is compact, the number $N(G, M)$ of nonisomorphic G-actions on M is at most countable, and there exists a k such that on the k-dimensional sphere S_k, $N(G, S_k)$ is infinite (countable).

(iv) Montgomery and Yang (1961) have proved for C^∞-action that there is a stratum which is open dense. More precisely, let $M_{r,c}$ be the set of $m \in M$ such that $\dim G_m = r$, number of connected components of $G_m = c$. Any connected component of $M_{r,c}$ is in one stratum only. The open dense stratum is $M_{r_{\max}, c_{\min}}$; it is the maximal stratum (i.e. the little group is *the* minimal one among all little groups which appear in the G-action on M; see Section 2.2). Let $t < r_{\max}$. Then

$$\bigcup_{r \leq t, c} M_{r,c}$$

is closed and its dimension is $\leq n - r_{\max} + t - 1 \leq n - 2$.

References for (iii) and (iv) can be found in the review paper of D. Montgomery, "Compact Groups of Transformations", p. 43 of *Differential Analysis*, Bombay Colloquium (1964). A good reading for an introduction to the subject is Palais, "The Classification of G-Spaces", *Memoirs Amer. Math. Soc.*, No. 36 (1960).

(v) Consider the set $\mathscr{F}$ of real-valued, G-invariant smooth functions on M ($\forall\, g \in G, \forall\, m \in M, f(g.m) = f(m)$). We call an orbit $\mathcal{O}$ *critical* if for *all* $f \in \mathscr{F}$, $m \in \mathcal{O} \Rightarrow (df)_m = 0$. Then (L. Michel, *C. R. Acad. Sc., Paris*, **272**, 433 (1971)):

Theorem 1

An orbit is critical if and only if it is isolated in its stratum. (I.e., any connected component of this orbit is a connected component of its stratum). If a stratum has a finite number of orbits, it is closed and all its orbits are critical.

The importance of this theorem for physical problems blending an invariance by G and a variational principle is obvious. This theorem is a generalization of the intuitive result: A C^1 real even function $f(x)$ of the real numbers $\mathbb{R}$, $f(x) = f(-x)$, satisfies $df/dx|_{x=0} = 0$.

We shall introduce some essential tools for the proof of this theorem.

4.2. The Equivariant Retraction

Let $d(x, y)$ be the G-invariant Riemann metric on M. Let Q be a G-invariant sub-manifold of M. There exists a neighborhood V_Q such that

$$x \in V_Q, \qquad \operatorname*{Inf}_{y \in Q} d(x, y)$$

is well defined and unique. We denote by $r_Q(x)$ the point of Q for which this minimum is reached: this point is the foot of the geodesic passing through x and normal to Q. Since it is defined in terms of the metric and G acts isometrically, the *retraction* $V_Q \xrightarrow{\;r_Q\;} Q$ is equivariant, i.e.

$$\forall\, g \in G \qquad \Phi_g \circ r_Q = r_Q \circ \Phi_g. \tag{13}$$

Applying this equation to x, we see for $g \in G_x$ that

$$g \in G_x \Rightarrow g \in G_{r_Q(x)}. \tag{14}$$

So, if Q is the orbit $G.m$,

$$x \in V_{G.m} \Rightarrow G_x \leqq G_m \tag{15}$$

that is, for any $m \in M$, there is a neighborhood V_m such that for any $x \in V_m$, the little group G_x is equal or smaller (i.e. $G_x \subset G_m$ up to a conjugation). The submanifold $r_Q^{-1}(m)$ is called the (local) slice at m. We denote it by $N(m)$. In geodesic coordinates it is a linear manifold. As a particular case of (15),

$$x \in N(m) \Rightarrow G_x \subset G_m. \tag{16}$$

The set of points of $N(m)$ such that $G_m = G_x$ will be denoted by $F(m)$:

$$F(m) = N(m)^{G_m}. \tag{17}$$

Equation (16) implies:

$$S(m) \cap N(m) = F(m). \tag{18}$$

If $F(m) = m$, this is true also for all $m' \in G.m$ and we shall say that the orbit $G(m)$ is isolated in its stratum $S(m)$.

4.3. The Local Action of G_m

As we have seen, in a geodesic coordinate system G_m acts linearly on a neighborhood V_m of m, by an orthogonal ($=$ real unitary) representation. So it leaves invariant a Euclidean scalar product (corresponding to the invariant metric) and the representation space decomposes into the direct sum of orthogonal subspaces.

$$V_m = \underset{\substack{\text{tangent plane to} \\ \text{the orbit}}}{T_m(G.m)} \oplus \underset{\text{slice}}{N(m)} \tag{19}$$

$$= \underbrace{T_m(G.m) \oplus F(m)}_{\substack{\text{tangent plane to} \\ \text{the stratum}}} \oplus K(m) \tag{19'}$$

($N(m)$, linear manifold is identical to $T_m(G_m)^\perp \subset T_m(M)$ in the tangent plane of M at m).

The linear representation of G_m on $T_m(G.m)$ depends only on G (It is the restriction of the adjoint representation of G to G_m, for the subspace of the Lie algebra $\perp$ to $\mathscr{G}_m$ for the Cartan–Killing metric).

The representation on $F(m)$ is trivial.

The representation on $K(m)$ does not contain the trivial representation.

For $m' \in S(m)$, $G_{m'} \sim G_m$ and the representation is equivalent. So (19) and (19') depends only on the stratum (at least $S(m)$ is connected).

4.4. Theorem on Critical Orbits of *G*-Invariant Functions

We now sketch the proof of Theorem 1. Let $f \in \mathcal{F}$, i.e. f is a *G*-invariant real-valued smooth function on M. Its differential at m, df_m, is in the cotangent plane to M at m. With the Euclidean local metric at m we use the dual notion, the gradient of f at m. The *G*-invariance of f implies that

$$(\operatorname{grad} f)_m \in T_m(G.m)^{\perp} = N(m) \tag{20}$$

and it has to be invariant under G_m, so that

$$(\operatorname{grad} f)_m \in F(m) \subset N(m). \tag{21}$$

Hence if $G.m$ is isolated in its stratum,

$$\forall f \in \mathcal{F}, \qquad \forall m \in G(m), \qquad (\operatorname{grad} f)_m = 0.$$

This proves the "if" of Theorem 1.

Conversely, if $F(m)$ has points other than m, i.e. $\dim F(m) > 0$ (since $F(m)$ is a linear manifold!), then one can explicitly build in V_m an invariant smooth function on M with compact support on $F(m)$ and a nonvanishing gradient (which is in $F(m)$ at m).

4.5. *G*-Invariant Gradient Vector Fields

Note that for any $f \in \mathcal{F}$, $(\operatorname{grad} f)_m \in T_m(S(m))$, the tangent plane to the stratum of m.

Therefore f and $f|_{S(m)}$, its restriction to the stratum $S(m)$, have the same gradient. Consider a minimal stratum (in the order defined in Section 2.2). Such a stratum is closed. If M *is compact*, a closed stratum $S(m)$ is compact so it has a finite number of (closed, compact) connected components. Let $S_0(m)$ be that of m. Either $S_0(m) \subset G.m$, and the orbit is isolated in its stratum, or the points of $S_0(m)$ belong to an infinite number of orbits. Then on the compact $S_0(m)$, for any $f \in \mathcal{F}, f|_{S_0(m)}$ reaches its maximum and its minimum on orbits on which its gradient vanishes. Since f has same gradient, we have (Michel, *loc. cit.*)*

Theorem 2

If $S(m)$ is compact and has an infinite number of orbits, every $f \in \mathcal{F}$ has at least $\operatorname{grad} f = 0$ on two different orbits of $S(m)$.

---

* As an application of this to Example 1.9(f), every $O(3)$-invariant smooth function on the 3-particle phase space Ω has at least two extrema, on the boundary of Ω. This is the case for smooth functions f defined on the Dalitz plot. Then $f_0\pi$ is defined on Ω.

However, these orbits generally depend on f. One can give more results on extrema of $f \in \mathscr{F}$ with an equivariant Morse theory (see A. G. Wassermann, *Topology*, **8**, 127 (1969)).

4.6. *G*-Invariant Vector Fields

It is easy to generalize to any *G*-invariant vector fields what we have done for gradient fields.

Decompose the vector field into two $\perp$ components, one normal to $T_m(G.m)$, hence in $F(m)$, the other in $T_m(G.m)$.

If dim $F(m) = 0$, and if the representation of G_m on $T_m(G(m))$ does not contain the trivial one, then all *G*-invariant vector fields vanish on $G(m)$. If $T_m(G(m))$ contains a trivial representation of G_m, but the Euler characteristic of $G(m)$ is $\neq 0$, then each *G*-invariant vector field has to vanish on some orbit of $G(m)$.

4.7. Physical Applications

The directions of breaking of the hadronic internal symmetry by the electro-magnetic, semileptonic and nonleptonic weak, and *CP*-violating interactions are on four critical orbits of the adjoint action of $(SU(3) \times SU(3)) \wedge (I, C, P, CP)$ on the unit sphere S_{15}. In this action there are twelve strata, and five critical orbits (each one forms a stratum). For lack of time we just refer to: L. Michel and L. A. Radicati, *Ann. Phys.*, **66**, 758 (1971). Let us end these two lectures by a question: what is the use of the fifth critical orbit?

Symmetries, Currents and Infinitesimal Generators

HELMUT REEH

Max-Planck-Institut für Physik und Astrophysik, 8 München 23 (Germany)

Abstract

We review recent work concerning spontaneous breakdown of symmetries and the connection between conserved local currents and infinitesimal generators of symmetry transformations, including dilatation and conformal transformations.

1. INTRODUCTION

In this article I shall not go into the history of the subject, I shall also not give a complete list of references. Furthermore, I shall not go into all technical details. Those who want to study the subject more closely are advised to study the review articles of ref. [1], as well as the references I am going to give in the following.

To begin with, let me remind you of the notion of a spontaneously broken symmetry within the Wightman framework [2] of relativistic quantum field theory. (Alternatively, one can use the Araki–Haag–Kastler framework [3] of local observables which sometimes has advantages since the operators are bounded.)

Let $\mathfrak{R}$ denote the algebra of unbounded local fields, i.e. polynomials in Wightman fields smeared with test functions having compact support ($\in \mathscr{D}(\mathbb{R}^4)$). For simplicity we assume that there are no spinor fields. $\mathscr{H}$ denotes the Hilbert space on which $A \in \mathfrak{R}$ acts, and Ω is the unique vacuum vector.

A symmetry transformation is a map

$$A \leftrightarrow \alpha_\tau(A) \equiv A_\tau, \qquad \forall A \in \mathfrak{R}, \qquad -\infty < \tau < \infty \tag{1}$$

of $\mathfrak{R}$ onto $\mathfrak{R}$ preserving the algebraic structure (including $(\alpha_\tau(A))^* = \alpha_\tau(A^*)$; if we would consider the algebra of local observables instead of $\mathfrak{R}$, we would call α a *-automorphism) with

$$\alpha_{\tau_1} \cdot \alpha_{\tau_2} = \alpha_{\tau_1 + \tau_2}, \qquad \alpha_0 = \mathbf{1} \tag{2}$$

and

$$\alpha_\tau(A(x)) = (\alpha_\tau(A))(x), \qquad \forall\,\tau,\,\forall\,x \in \mathbb{R}^4, \tag{3}$$

where $A \to A(x) = U(x)\,AU^{-1}(x)$ is the translation by a four-vector x unitarily implemented on $\mathfrak{R}$ by one of the assumptions of quantum field theory.

Actually, (3) is already too restrictive. We shall later see that the scheme can easily be generalized to include some more general symmetry transformations, e.g. dilatations, homogeneous Lorentz transformations or conformal transformations.

As shown by the Wightman reconstruction theorem (Gelfand–Segal construction), the theory is, up to unitary equivalence, determined by the vacuum functional

$$A \to \varphi_0(A) = (\Omega\,|\,A\Omega), \qquad \forall\,A \in \mathfrak{R}\,.$$

Now, due to (3), one may construct a new vacuum functional

$$\varphi_\tau(A) = \varphi_0(\alpha_\tau(A)) = (\Omega\,|\,A_\tau\Omega)\,.$$

Then there are two cases—

i) $\varphi_\tau(A) = \varphi_0(A) \qquad \forall\,A \in \mathfrak{R},\ \forall\,\tau\,.$

This is the case of a *conserved symmetry*. Due to a simple theorem [4] it follows that there exists a unitary operator W_τ with

$$W_\tau\Omega = \Omega$$

$$\alpha_\tau(A) = W_\tau A W_\tau^{-1}$$

$$[W_\tau, U(x)] = 0$$

$$W_{\tau_1} \cdot W_{\tau_2} = W_{\tau_1 + \tau_2}\,.$$

For a one-parameter continuous group it follows furthermore by Stone's theorem that

$$W_\tau = \exp(iQ\tau)$$

with a self-adjoint operator Q, $Q\Omega = 0$.

ii) $\varphi_\tau(A) \neq \varphi_0(A)$ for some $A \in \mathfrak{R}\,.$

This is the case of a *spontaneously broken symmetry*. Then there is no such unitary operator W nor a Q as in case i).

I shall skip the question here, how one may hope to use this concept to describe broken symmetries as they are observed in nature.

The question I am going to discuss now is: Under what conditions may a symmetry transformation be spontaneously broken or must be conserved.

2. DISCRETE SYMMETRY TRANSFORMATIONS

There is not much known concerning discrete transformations in this context. However, there is the following simple example which shows that a discrete symmetry transformation may be spontaneously broken even if there is a gap above the vacuum in the energy momentum spectrum [5]:

Let $\phi(x) = \phi_m^2(x)$ be the Wick square of real free scalar field of mass $m > 0$. Define α on $\phi(x)$ by

$$\alpha\phi(x) = -\phi(x), \qquad \forall x.$$

There is a unique extension of α to $\Re$ by the requirement that α preserve the algebraic structure.* α clearly commutes with the 4 translations.

The symmetry transformation α is spontaneously broken, since the three-point function of ϕ firstly does not vanish and secondly changes its sign under α.

3. CONTINUOUS ONE-PARAMETER SYMMETRY TRANSFORMATIONS

Not much is known for the general case [6]. We therefore turn to the case most frequently discussed and which has its roots in Lagrangian quantum field theory. There $\gamma(A)$ is infinitesimally defined by

$$\frac{d}{d\tau}\alpha_\tau(A)\Big|_{\tau=0} = i \lim_{r\to\infty}\left[\int j_0(x)\,\vartheta_r(x)\,\eta(x^0)\,d^4x,\, A\right] \equiv$$

$$\equiv i \lim_{r\to\infty}[Q_r, A] \qquad \forall A \in \Re \qquad (4)$$

with

$$\vartheta_r(x) = \vartheta\left(\frac{|x|}{r}\right)$$

$$\vartheta(s)\in\mathscr{D}(\mathbb{R}^1), \qquad \vartheta=1 \quad \text{for} \quad 0\leqq s\leqq 1, \qquad \vartheta=0 \quad \text{for} \quad s\geqq 2$$

$$\eta(x^0)\in\mathscr{D}(\mathbb{R}^1) \quad \text{with} \quad \int_{-\infty}^{\infty}\eta(x^0)\,dx^0 = 1$$

and where $j_0(x)$ is from a hermitean local (in particular local with respect to the elements of $\Re$) current $j_\mu(x)$ which is conserved, $\partial^\mu j_\mu(x) = 0$. We assume in this section that the current is covariant under translations

$$j_\mu(x + a) = U(a)j_\mu(x)\,U^{-1}(a)$$

---

* As pointed out to the author by Prof. Araki, it is possible that in the example considered, α may not define an automorphism on the corresponding ring of bounded operators.

but not necessarily that it transforms like a covariant vector field under homogeneous Lorentz transformations.

It is well known and easy to see that $\partial^\mu j_\mu(x) = 0$, together with the covariance of j_μ under translations and the relative locality of $j_\mu(x)$ and A, entail that the infinitesimal transformation (4) commutes with the translations and is independent of the still free details of the functions ϑ and η. In fact, due to the relative locality, the limit on the r.h.s. of (4) is already reached for finite r.

It is not easy at all to go from (4) to the finite transformation α_τ. Under some assumptions [7] (to be then verified in specific models) this can be done and

$$\alpha_\tau(A) = \lim_{r \to \infty} \exp(iQ_r\tau)\, A \exp(-iQ_r\tau) = \exp(iQ_r\tau)\, A \exp(-iQ_r\tau) \quad \text{for} \quad r \geq r_0(A, \tau)$$

has the properties listed above.

It is easy to show now

Lemma 3.1:

Under the assumptions listed, α_τ is spontaneously broken if and only if

$$\lim_{r \to \infty} (\Omega | [Q_r, A]\, \Omega) \neq 0 \quad \text{for some} \quad A \in \mathfrak{R}.$$

Therefore one can use the latter condition as a criterion for deciding whether a symmetry is spontaneously broken.

Theorem 3.1 (Goldstone theorem) [8]:

Under the assumptions listed, one has

$$\lim_{r \to \infty} (\Omega | [Q_r, A]\, \Omega) = (\Omega | Q_r E_0 A\Omega) - (\Omega | A E_0 Q_r \Omega)$$

for $r \geq r_0(A)$ where $E_0 = \int_{p^2 = 0} dE(p)$, with $E(p)$ from $U(x) = \int e^{ipx} dE(p)$ is the projection on the mass zero states in $\mathcal{H}$.

Remark 1:

Theorem 1 stays true in case of an indefinite metric in the underlying Hilbert space if the Fourier transform of

$$(\Omega | j_\mu(x)\, A\Omega)$$

is a tempered measure on $\mathbb{R}^4$.

Remark 2:

The well-known verbal formulation of Theorem 1 is: There is no spontaneously broken symmetry unless there are zero mass states in $\mathcal{H}$. These zero mass states are in the sector

$$\overline{\{E_0 j_0(f)\, \Omega\}}_{f \in \mathfrak{D}(\mathbb{R}^4)}$$

of $\mathcal{H}$. Hence most of their quantum numbers are fixed by j_0.

Remark 3:

Theorem 1 can be extended to some symmetry transformations generated by translationally noncovariant currents, as we shall see later on.

4. QUANTUM NUMBERS OF GOLDSTONE PARTICLES

Let the assumptions be as above.

Concerning *internal quantum numbers*, there is no general restriction: Take the free field of zero mass $\phi_0(x)$ and define α_τ by $\alpha_\tau\phi(x) = \phi(x) + \tau$. Then $j_\mu = \partial_\mu\phi$. The Goldstone states in this trivial example [9] are generated by $\phi(x)$ from Ω and may apparently have any internal quantum number we want. However, there are general restrictions concerning the spin- or better *helicity-quantum numbers* of Goldstone states.

Everybody believes the following:

Theorem 4.1 :

Let $j_\mu(x)$ be a Lorentz covariant vector field, $\partial^\mu j_\mu(x) = 0$, and let the symmetry transformation be spontaneously broken. Then the Goldstone states are scalar. (The proof is like that for showing that in classical electrodynamics the infinite space integral over a charge density is a scalar.)

Assume now that $j_\mu(x)$ is a component of a Lorentz covariant Wightman field

$$j_\mu(x) = t_{\mu m}(x)$$

$\mu = 0, 1, 2, 3$, $\partial^\mu t_{\mu m}(x) = 0$, and m stands for a set of vector indices, such that under Lorentz transformations it changes into

$$U(\Lambda, a)\, t_{\mu m}(x)\, U^{-1}(\Lambda, a) = (\Lambda^{-1})_\mu^{\mu'}\, D(\Lambda^{-1})_m^{m'}\, t_{\mu' m'}(\Lambda x + a),$$

where $D_m^{m'}$ is some finite dimensional representation of the homogeneous Lorentz group.

One could suggest that it should be possible to get e.g. Goldstone particles of helicity $\pm\, 1$, in case $D(\Lambda) = \Lambda$, and similarly for higher dimensional representations. However, it turns out that this is impossible.

Theorem 4.2 :

Let $t_{\mu m}(x)$ be a Lorentz covariant Wightman field, i.e.

$$U(\Lambda, a)\, t_{\mu m}(x)\, U^{-1}(\Lambda, a) = D(\Lambda^{-1})_{\mu m}^{\mu' m'} t_{\mu' m'}(\Lambda x + a)$$

with $D_{\mu m}(\Lambda^{-1})$ a finite dimensional one-valued irreducible representation of the homogeneous Lorentz group, $\partial^\mu t_{\mu m}(x) = 0$.

Then

$$\lim_{r \to \infty} (\Omega | [t_{0m}(\vartheta_r \otimes \eta), A] \Omega) = 0, \qquad \forall A \in \mathfrak{R},$$

unless $t_{\mu m} = j_\mu(x)$ is a Lorentz covariant vector field.

In other words, there are no Goldstone states with integer helicity $\neq 0$.

Remark 1 :

The statement can be extended to cover reducible representations $D(\Lambda^{-1})$ as well. This will be demonstrated later on for a tensor of rank 2 (it furthermore stays true for half-integer helicities larger than 3/2, cf. ref. [10]).

*Remark 2** :

Definite metric of the underlying state space is an essential input for the preceding Theorem 3. To see this, consider the free electromagnetic potentials A_μ:

$$\square A_\mu(x) = 0$$

$$(\Omega | A_\mu(x) A_\nu(x) \Omega) = - i g_{\mu\nu} D(x - y)$$

$$(\Omega | A_\mu(x) \Omega) = 0.$$

Then

$$A_\mu(x) \to \alpha_\tau^{(\lambda)} A_\mu(x) = A_\mu(x) + \delta_{\mu\lambda}\tau, \qquad \lambda = 0, 1, 2, 3 \tag{5}$$

are four groups of symmetry transformations fulfilling all of our assumptions. There are four currents,

$$j_{\mu\lambda}(x) = \partial_\mu A_\lambda, \qquad \lambda = 0, \ldots, 3,$$

generating these transformations. Correspondingly there are four Goldstone particles, the four free photon states.

Obviously, the transformations (5) are subgroups of the gauge transformation

$$A_\mu(x) \to A_\mu(x) + \partial_\mu \Lambda(x), \qquad \square \Lambda(x) = 0$$

with $\Lambda(x) = x^\lambda \cdot \tau$. The argument found in the textbooks for showing that some gauge transformations change only the scalar and longitudinal photons does not apply to the transformations (5)!

5. SPECIAL CASE: TENSOR CURRENTS OF RANK 2

Instead of reproducing the full proof of Theorem 4.2, which can be found in ref. [10], I shall demonstrate how it works and can be extended to reducible representations by considering a special case.

---

* Compare also the article by Sen and Weil [11].

We need for this a lemma which is also useful for the proof of Theorem 3.1 above, as well as for other theorems to which we come later.

Lemma 5.1:

Let $j(x)$ be a Wightman field, $A \in \mathfrak{R}$ and $F_\Delta = \int_{p^2 \in \Delta} d^4 E(p)$, $E(p)$ from $U(x) = \int e^{ipx} d^4 E(p)$, Δ any Borel set of the positive real line. Then

$$(\Omega | j(x) F_\Delta A\Omega) - (\Omega | A F_\Delta j(x) \Omega)$$

vanishes for $x \notin C$ where C is some causal domain independent of Δ:

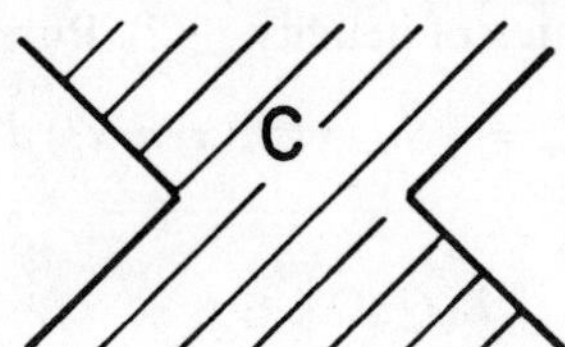

Proof:

Follows immediately from a Jost–Lehmann–Dyson representation given by Araki, Hepp and Ruelle [12].

i) Let $t_{\mu m}(x) = F_{\mu\nu}(x) = -F_{\nu\mu}(x)$ be an *antisymmetric tensor* of rank 2 ($E_0 F_{\mu\nu}$ generates from Ω states of helicity ± 1, as is well known from the free electromagnetic field) with $\partial^\mu F_{\mu\nu}(x) = 0$. Put

$$Q_{r\nu} = \int F_{0\nu}(x)\, \vartheta_r(x)\, \eta(x^0)\, d^4 x\,.$$

Proposition 5.1:

$$\lim_{r \to \infty} (\Omega | [Q_{r,\nu}, A]\, \Omega) = 0, \qquad \forall A \in \mathfrak{R}, \qquad \nu = 0, 1, 2, 3\,.$$

Proof:

Antisymmetry yields $F_{00} = 0$, hence $Q_{r0} = 0$.

From Theorem 3.1 we have

$$\lim_{r \to \infty} (\Omega | [Q_{ri}, A]\, \Omega) = \lim_{r \to \infty} \left[(\Omega | Q_{ri} E_0 A\Omega) - (\Omega | A E_0 Q_{ri} \Omega) \right]\,.$$

Consider therefore

$$h_i(x) = (\Omega | (\textstyle\int F_{0i}(x)\, \eta(x^0)\, dx^0)\, E_0 A\Omega) - (\Omega | A E_0 (\textstyle\int F_{0i}(x)\, \eta(x^0)\, dx^0)\, \Omega)\,.$$

Due to Lemma 5.1, this has compact support in x; according to a theorem of Borchers it is C^∞ in x. Hence the Fourier transform, $\tilde{h}_i(p)$, is analytic in p and of strong decrease. From $\partial^\mu F_{\mu\nu}(x) = 0$ it follows that $\partial^\nu F_{\mu\nu}(x) = 0$, i.e. $\partial^i F_{0i}(x) = 0$,

$$\sum_{i=1}^{3} p^i \tilde{h}_i(p) = 0\,, \qquad \forall p\,.$$

By taking the derivative with respect to p^j at $p = 0$, we get

$$\tilde{h}_j(p) = 0 \quad \text{for} \quad j = 1, 2, 3 \,.$$

Hence

$$\lim_{r \to \infty} (\Omega \,|\, [Q_{ri}, A] \,\Omega) = \lim_{r \to \infty} \int \tilde{\vartheta}_r(p) \, \tilde{h}_i(p) \, d^3p = \lim_{r \to \infty} \int r^3 \vartheta(rp) \, \tilde{h}_i(p) \, d^3p =$$

$$= \lim_{r \to \infty} \int \tilde{\vartheta}(q) \, \tilde{h}_i\!\left(\frac{q}{r}\right) d^3q = \int \tilde{\vartheta}(q) \, d^3q \, \tilde{h}_i(0) = 0 \,.$$

ii) Let $t_{\mu m}(x) = S_{\mu\nu}(x) = S_{\nu\mu}(x)$ be a *symmetric tensor* of rank 2, $\partial^\mu S_{\mu\nu}(x) = 0$ ($E_0 S_{\mu\nu}(x)$ generates from Ω states of helicity ± 2). Put

$$Q_{r\nu} = \int S_{0\nu}(x) \, \vartheta_r(x) \, \eta(x^0) \, d^4x$$

Lemma 5.2 :

$$\| E_0 Q_{r\nu} \Omega \| \leqq b < \infty \,.$$

Proof :

Consider the Källén-Lehmann representation of the two-point function for $S_{\mu\nu}(x) = S_{\nu\mu}(x)$ with $\partial^\mu S_{\mu\nu}(x) = 0$. It is

$$(\Omega \,|\, S_{\mu\nu}(x) \, S_{\kappa\sigma}(y) \, \Omega) = \int_{p^0 \geqq 0} d^4p \, e^{-ip(x-y)} \{ [g_{\mu\nu} g_{\kappa\sigma}(p^2)^2 -$$

$$- g_{\mu\nu} p_\kappa p_\sigma p^2 - g_{\kappa\sigma} p_\mu p_\nu p^2 + p_\mu p_\nu p_\kappa p_\sigma] \rho_1(p^2) + [g_{\mu\nu} g_{\nu\sigma}(p^2)^2 + g_{\nu\kappa} g_{\mu\sigma}(p^2)^2 -$$

$$- g_{\mu\kappa} p_\nu p_\sigma p^2 - g_{\nu\kappa} p_\mu p_\sigma p^2 - g_{\mu\sigma} p_\nu p_\kappa p^2 - g_{\nu\sigma} p_\mu p_\kappa p^2 + 2 p_\mu p_\nu p_\kappa p_\sigma] \rho_2(p^2) \} \,,$$

where ρ_1, ρ_2 are Lorentz-invariant measures ($g_{00} = 1, g_{11} = g_{22} = g_{33} = -1, g_{\mu\nu} = 0$ for $\mu \neq \nu$). Therefore

$$(\Omega \,|\, S_{00}(x) \, S_{00}(y) \, \Omega) = \int_{p^0 \geqq 0} e^{-ip(x-y)} |p|^4 \, [\rho_1(p^2) + 2\rho_2(p^2)] \, d^4p$$

$$(\Omega \,|\, S_{00}(x) \, E_0 S_{00}(y) \, \Omega) = a \int_{p^0 \geqq 0} e^{-ip(x-y)} |p|^4 \, \delta(p^2) \, d^4p \,, \qquad a \geqq 0$$

and

$$(\Omega \,|\, S_{0i}(x) \, S_{0i}(y) \, \Omega) = \int_{p^0 \geqq 0} e^{-ip(x-y)} \{ p_0^2 p_i^2 (\rho_1 + 2\rho_2) + (p^{02} - p^2)(p^2 - p_i^2)\rho_2 \} d^4p$$

$$(\Omega \,|\, S_{0i}(x) \, E_0 S_{0i}(y) \, \Omega) = a \int_{p^0 > 0} e^{-ip(x-y)} \, p_0^2 p_i^2 \, \delta(p^2) \, d^4p \,.$$

Hence

$$\| E_0 Q_{r0} \Omega \|^2 = a \int_{p^0 > 0} |\tilde{\vartheta}_r(p)|^2 \, |\tilde{\alpha}(p^0)|^2 \, |p|^4 \, \delta(p^2) \, d^4p =$$

$$= a \tfrac{1}{2} \int |\tilde{\vartheta}(q)|^2 \, |\tilde{\alpha}(|q|/r)|^2 \, |q|^3 \, d^3q$$

and

$$\lim_{r \to \infty} \| E_0 Q_{r0} \Omega \|^2 = a \tfrac{1}{2} \int |\tilde{\vartheta}(q)|^2 \, |q|^3 \, d^3q = b \,, \qquad 0 \leqq b < \infty \,.$$

In the same manner one gets

$$\lim_{r \to \infty} \|E_0 Q_{ri}\Omega\|^2 \le b.$$

Lemma 5.3:

If $\|E_0 Q_r\Omega\|^2 \le b < \infty$, then

$$\lim_{r \to \infty} (\Omega | [Q_r, A] \Omega) = 0, \qquad \forall A \in \mathfrak{R}.$$

Proof:

$\|E_0 Q_r\Omega\|^2 < \infty$ implies the existence of a weakly convergent subsequence, $E_0 Q_{r_n}\Omega \overset{w}{\to} \chi$. Hence

$$\lim_{n \to \infty} (\Omega | [Q_{r_n}, A] \Omega) = (\chi | A\Omega) - (A^+\Omega) \chi).$$

On the other hand, this equals, for all finite $x \in \mathbb{R}^4$,

$$\lim_{n \to \infty} (\Omega | [Q_{r_n}, U(x) A U^{-1}(x)] \Omega) = (\chi | U(x) A\Omega) - (U(x) A^+\Omega | \chi).$$

Choose $x = (0, \lambda e)$. According to the cluster theorem, $w\text{-}\lim_{\lambda \to \infty} U(x) = P_\Omega$, the projector on Ω. Hence we get

$$\lim_{n \to \infty} (\Omega | [Q_{r_n}, A] \Omega) = (\chi | \Omega) (\Omega | A\Omega) - (\Omega | A^+\Omega) (\Omega | \chi).$$

Now we may assume without loss of generality that $(\Omega | Q_r\Omega) = 0$ (otherwise replace Q_r by $Q_r - (\Omega | Q_r\Omega)$). Then $(\Omega | \chi) = (\chi | \Omega) = 0$ and Lemma 5.3 is proved.

By combining Lemmas 5.2 and 5.3 we get

Proposition 5.2:

For $S_{\mu\nu}(x) = S_{\nu\mu}(x)$ with $\partial^\mu S_{\mu\nu}(x) = 0$,

$$Q_{r\nu} = \int \vartheta_r(x) \eta(x^0) S_{0\nu}(x) d^4x;$$

it follows that

$$\lim_{r \to \infty} (\Omega | [Q_{r\nu}, A] \Omega) = 0, \qquad \forall A \in \mathfrak{R}.$$

iii) Consider now an *arbitrary* Lorentz covariant *tensor* $t_{\mu\nu}(x)$ of rank 2, with $\partial^\mu t_{\mu\nu}(x) = 0$.

Proposition 5.3:

$$\lim_{r \to \infty} (\Omega | [\int t_{0\nu}(x) \vartheta_r(x) \eta(x^0) d^4x, A] \Omega) = 0, \qquad \forall A \in \mathfrak{R}.$$

Proof:

Decompose into symmetric and antisymmetric parts

$$t_{\mu\nu} = S_{\mu\nu} + F_{\mu\nu}.$$

We could apply Propositions 5.1 and 5.2 if we knew $\partial^\mu S_{\mu\nu} = 0$ and $\partial^\mu F_{\mu\nu} = 0$ separately. But actually we need less, namely

$$\partial^\mu E_0 S_{\mu\nu}(x)\,\Omega = 0$$

$$\partial^\mu E_0 F_{\mu\nu}(x)\,\Omega = 0\,,$$

which can indeed be inferred:

$$E_0 S_{\mu\nu}(x)\,\Omega \quad \text{are states of helicity } \pm 2$$

$$E_0 F_{\mu\nu}(x)\,\Omega \quad \text{are states of helicity } \pm 1$$

Let $P_{\pm 1}$ be the projectors on the states with helicity ± 1. Then

$$\partial^\mu E_0 F_{\mu\nu}(x)\,\Omega = \partial^\mu E_0 (P_{+1} + P_{-1})\, t_{\mu\nu}(x)\,\Omega =$$

$$= E_0 (P_{+1} + P_{-1})\, \partial^\mu t_{\mu\nu}(x)\,\Omega = 0.$$

Likewise one gets $\partial^\mu E_0 S_{\mu\nu}(x)\,\Omega = 0$. Thus Proposition 5.3 is proved.

6. CONNECTION WITH THE GLOBAL INFINITESIMAL GENERATOR

There is another interesting result in this context, proved recently by Maison [13].

Theorem 6.1:

Let $j_\mu(x)$ be a translationally covariant Wightman field with $\partial^\mu j_\mu(x) = 0$, $Q_r = j_0(\vartheta_r \otimes \eta)$. Then

$$\lim_{r \to \infty} (\Omega | Q_r A\Omega) = \tfrac{1}{2} \lim_{r \to \infty} (\Omega | [Q_r, A]\, \Omega), \qquad \forall A \in \mathfrak{R}.$$

This was conjectured to be true by Swieca (1967) and proved already earlier for the case where a gap in the energy momentum spectrum exists above the vacuum. The proof makes use of the assumed property of $\vartheta_r(x)$ to depend only on $|x|$. The importance of this point had been stressed by Ferrari and Picasso [14].

Corollary:

Under the assumptions of Theorem 4, it follows that for a conserved symmetry transformation

$$\lim_{r \to \infty} (A\Omega | Q_r B\Omega) = (A\Omega | QB\Omega), \qquad \forall A, B \in \mathfrak{R}$$

if Q is the infinitesimal generator of the conserved symmetry transformation [15].

Remark 1:

Theorem 6.1 and hence the corollary stay true if one inserts for either A or B a quasilocal operator which is obtained by $\int C(x) f(x)\, d^4x$ with $C \in \mathfrak{R}$ and

$f \in \mathfrak{S}(\mathbb{R}^4)$. This is true because the smearing with $f(x)$ amounts to a multiplication in p-space with the function $\tilde{f}(p) \in \gamma(\mathbb{R}^4)$ [see note added in proof].

Remark 2:

In most cases of interest, one has $\|Q_r\Omega\| \to \infty$ for $r \to \infty$. In these cases apparently $Q_r\Omega \overset{w}{\to} 0$ does not hold. However, as shown in ref. [16], there are other cases.

Remark 3:

Theorem 6.1 stays true for the case of an indefinite metric in the underlying state space as long as the Fourier transform of $(\Omega \,|\, j_\mu(x)\, A\Omega)$ is a tempered measure on $\mathbb{R}^4$.

7. CURRENTS WHICH ARE NOT COVARIANT UNDER TRANSLATIONS

There are important symmetry transformations, like homogeneous Lorentz transformations, dilatations and conformal transformations which are not generated by translationally covariant currents.

i) *Dilatation*

$$x_\mu \to e^{-\lambda} x_\mu, \qquad -\infty < \lambda < \infty.$$

Classical field theory yields for this the conserved current [17]

$$J_\mu = x^\nu \Theta_{\mu\nu}(x), \tag{6}$$

where $\Theta_{\mu\nu}(x)$ is the suitably chosen traceless symmetric energy-momentum tensor density. Let us denote the corresponding finite transformation on the fields by α_λ:

$$x_\mu \to e^{-\lambda} x_\mu, \qquad \mu = 0, 1, 2, 3$$

$$A \to d_\lambda(A) \qquad \forall\, A \in \mathfrak{R} \tag{7}$$

(for free zero mass fields $d_\lambda(\phi(x)) = e^{-\lambda}\phi(e^{-\lambda}x)$).

Let us assume that we have on $\mathfrak{R}$ a map (7) and a generating current (6), where $\Theta_{\mu\nu}$ is a Lorentz covariant Wightman field with $\partial^\mu \Theta_{\mu\nu}(x) = 0$, $\Theta_{\mu\nu} = \Theta_{\nu\mu}$, $\Theta^\mu_\mu(x) = 0$. The noncovariance of (6) under translations corresponds to the fact that d_λ does not commute with the translations as required in (3), as it must be for dilatations which are required instead to have the property

$$d_\lambda(U(x)\, A U^{-1}(x)) = U(e^\lambda x)\, d_\lambda(A)\, U^{-1}(e^\lambda x). \tag{3'}$$

Hence $A \to (\Omega \,|\, d_\lambda(A)\,\Omega)$, $\forall\, A \in \mathfrak{R}$ is again a vacuum functional on $\mathfrak{R}$. We therefore can extend the notion of conserved or spontaneously broken symmetry to the dilatation, and Lemma 3.1 applies.

ii) *Conformal transformations*

$$x_\mu \to \frac{x_\mu + \alpha_\mu x^2}{1 + \alpha x + \alpha^2 x^2},$$

where α is any four vector. Classical field theory yields 4 conserved conformal currents

$$C_{\mu\nu}(x) = (2x_\nu x_\lambda - g_{\nu\lambda} x^2)\,\Theta_\mu^\lambda(x), \quad \partial^\mu C_{\mu\nu}(x) = 0$$

with $\Theta_{\mu\nu}(x)$ as above [17]. Here it is not clear whether a finite conformal transformation can be defined on $\mathfrak{R}$. In any case, one can investigate the infinitesimal transformations defined by $C_{\mu\nu}(x)$ and (4), assuming that $\Theta_{\mu\nu}(x)$ is a Lorentz covariant Wightman field with $\partial^\mu\Theta_{\mu\nu}(x) = 0$, $\Theta_{\mu\nu}(x) = \Theta_{\nu\mu}(x)$, $\Theta_\mu^\mu(x) = 0$.

iii) From $\partial^\mu J_\mu(x) = 0$, $\partial^\mu C_{\mu\nu}(x) = 0$ and relative locality it is easily seen that the infinitesimal transformations

$$\lim_{r\to\infty} \left[J_0(\vartheta_r \otimes \eta), A \right], \qquad \lim_{r\to\infty} \left[C_{0\nu}(\vartheta_r \otimes \eta), A \right]$$

on $\mathfrak{R}$ are independent of the still free details of ϑ_r and η as above.

It is also easily shown that

$$(\Omega | [J_0(\vartheta_r \otimes \eta), A(x)]\,\Omega) = (\Omega | [J_0(\vartheta_r \otimes \eta), A]\,\Omega), \qquad \forall\, x \in \mathbb{R}^4,$$

since the transformations implied by $\Theta_{\mu\nu}(x)$ must be conserved symmetries as seen above.

The analogous relation need not be true for conformal transformations, because dilatations may in fact be spontaneously broken.

Theorem 7.1 (extension of Theorem 3.1):

Let

$$D_r = \int x^\nu \Theta_{0\nu}(x)\,\vartheta_r(x)\,\eta(x^0)\,d^4x$$

or

$$K_{r\nu} = \int (2x_\nu x_\lambda - g_{\nu\lambda} x^2)\,\Theta_0^\lambda\,d^4x,$$

respectively, $\Theta_{\mu\nu}(x)$ be a Lorentz covariant Wightman field, $\partial^\mu\Theta_{\mu\nu}(x) = 0$, $\Theta_{\mu\nu} = \Theta_{\nu\mu}$, $\Theta_\mu^\mu = 0$. Then for sufficiently large r

$$(\Omega | [D_r, A]\,\Omega) = (\Omega | [D_r E_0 A - A E_0 D_r]\,\Omega)$$

or

$$(\Omega | [K_{r\nu}, A]\,\Omega) = (\Omega | [K_{r\nu} E_0 A - A E_0 K_{r\nu}]\,\Omega),$$

respectively, with E_0 from Theorem 3.1.

Proof:

We formulate the proof for dilatations. For conformal transformations it works in the same way.

Put $F_\varepsilon = F_{(\varepsilon,\infty)}$ for $\varepsilon > 0$, F_Δ from Lemma 5.1. By that lemma we know that

$$h_\varepsilon^{\mu\nu}(y) = (\Omega|\Theta_{\mu\nu}(x)\, F_\varepsilon A\Omega) - (\Omega|AF_\varepsilon\Theta_{\mu\nu}(x)\,\Omega)$$

vanishes for $y \notin$ causal domain independent of ε.

Consider

$$(\Omega|D_r F_\varepsilon A - AF_\varepsilon D_r\Omega) =$$

$$= \int \vartheta_r(x)\,\eta(x^0)\,x^\nu (\Omega|[\Theta_{\mu\nu}(x)\,F_\varepsilon A - AF_\varepsilon\Theta_{\mu\nu}(x)]\,\Omega)\,d^4x\big|_{\mu=0} = \tag{8}$$

$$= \int \vartheta_r(x)\,\eta(x^0)\,(\Omega|[J_\mu(x)\,F_\varepsilon A - AF_\varepsilon J_\mu(x)]\,\Omega)\,d^4x\big|_{\mu=0}.$$

Because of the lemma, the integrand vanishes outside of a causal domain. Since $\partial^\mu J_\mu(x) = 0$, the value of (8) does not depend on the details of $\eta(x^0)$ as long as $\int \eta(x^0)\,dx^0 = \tilde\eta(0) = 1$. On the other hand, the Fourier transform of the integrand in (8) vanishes for $p^2 < \varepsilon$, i.e. for $p^0 < \varepsilon$, hence (8) vanishes. Therefore

$$\lim_{\varepsilon\to 0}\lim_{r\to\infty} (\Omega|[D_r F_\varepsilon A - AF_\varepsilon D_r]\,\Omega) = 0.$$

Due to the lemma, the limit is reached already for finite r independently of ε and we therefore have

$$0 = \lim_{r\to\infty}\lim_{\varepsilon\to 0} (\Omega|[D_r F_\varepsilon A - AF_\varepsilon D_r]\,\Omega) =$$

$$= (\Omega|[D_r(1 - E_0)\,A - A(1 - E_0)\,D_r]\,\Omega)$$

for $r \geqq r_0(A)$. Q.E.D.

From this it does not follow that dilatations and conformal transformations may indeed be spontaneously broken. However, in ref. [18] an example is given (trivial as far as physics is concerned) showing that dilatations may be spontaneously broken. A similar example may be given for conformal transformations (taking the assertion of Lemma 3.1 as definition).

Theorem 7.2 (extension of Theorem 6.1):

Let $J_\mu, D_r, \Theta_{\mu\nu}$ be as above. Then

$$\lim_{r\to\infty} (\Omega|D_r A\Omega) = \lim_{r\to\infty} \tfrac{1}{2}(\Omega|[D_r, A]\,\Omega), \qquad \forall A \in \mathfrak{R}.$$

Proof:

The proof is completely analogous to that of Theorem 6.1 if, in addition, one takes into account the properties assumed for $\Theta_{\mu\nu}$.

Corollary 1:

Let $J_{\mu\rho}(x) = x^\nu \partial_\rho \Theta_{\mu\nu}(x)$, $\Theta_{\mu\nu}(x)$ as above. (Then $\partial^\mu J_{\mu\rho}(x) = 0$.) Then

$$\lim_{r\to\infty} (\Omega|J_{0\rho}\,(\vartheta_r \otimes \eta)\,A\Omega) = \tfrac{1}{2}\lim_{r\to\infty} (\Omega|[J_{0\rho}(\vartheta_r \otimes \eta), A]\,\Omega).$$

164   HELMUT REEH

Proof:

The expression considered differs from that considered in Theorem 7.2 by an additional ∂_ρ or by a factor p_ρ in p-space. Since such a factor can only improve the convergence for $r \to \infty$, the first equation of the statement follows. Concerning the second, observe that for sufficiently large r

$$\left[\int x^\nu (\partial_\rho \Theta_{0\nu}(x))\, \vartheta_r(x)\, \eta(x^0)\, d^4x,\, A \right] =$$

$$= - \left[\int \Theta_{0\rho}(x)\, \vartheta_r(x)\, \eta(x^0)\, d^4x,\, A \right] + \left[\int (\partial_\rho x^\nu \Theta_{0\nu}(x))\, \vartheta_r(x)\, \eta(x^0)\, d^4x,\, A \right] =$$

$$= - \left[\int \Theta_{0\rho}(x)\, \vartheta_r(x)\, \eta(x^0)\, d^4x,\, A \right], \tag{9}$$

since the second term on the right-hand side vanishes because of relative locality of $\Theta_{\mu\nu}(x)$ and A, and (for $\rho = 0$) since $\partial_\mu x^\nu \Theta_{\mu\nu}(x) = 0$. But the vacuum expectation value of the remaining term vanishes by Theorem 4.2.

Corollary 2:

Let D_r or $J_{0\rho}(\vartheta_r \otimes \eta)$ respectively belong to a conserved symmetry and let D and P respectively be the corresponding infinitesimal generators. Then

$$\text{and} \qquad \lim_{r \to \infty} (B\Omega | D_r A\Omega) = (B\Omega | DA\Omega)$$

and

$$\lim_{r \to \infty} (B\Omega | J_{0\rho}(\vartheta_r \otimes \eta)\, A\Omega) = (B\Omega | P_\rho A\Omega),$$

respectively, $\quad \forall\, A, B \in \mathfrak{R}$.

Proof:

Same as for the corollary of Theorem 6.1.

Remark:

As in case of Theorem 6.1; and for the same reason, Theorem 7.2, Corollary 1 and Corollary 2 stay true if A (or B) is replaced by $\int C(x) f(x)\, d^4x$, $C \in \mathfrak{R}$, $f \in \mathfrak{S}(\mathbb{R}^4)$.

We will now see that Theorem 7.2 cannot be extended to all conserved currents which are not translationally covariant, in particular not to conformal currents.

Example:

Let $C_{\mu\nu}(x) = (2x_\nu x_\lambda - g_{\nu\lambda} x^2)\, \Theta_\mu^\lambda$. Put $\Theta_\mu^\lambda = \partial_\mu \partial^\lambda \phi_0$, where ϕ_0 is the real free scalar field of zero mass. Then for $f \in \mathscr{D}(\mathbb{R}^4)$

$$(\Omega | C_{00}(\vartheta_r \otimes \eta)\, \phi(f)\, \Omega) \sim r \cdot \int (\Delta\, \tilde\vartheta(\boldsymbol{q}))\, |\boldsymbol{q}|\, d^3 \boldsymbol{q}\, \tilde\eta(0)\, \tilde f(0) =$$

$$= r \cdot \int \tilde\vartheta(\boldsymbol{q})\, \frac{2}{|\boldsymbol{q}|}\, d^3 \boldsymbol{q}\, \tilde\eta(0)\, \tilde f(0) = r \cdot c \cdot \int_0^\infty \vartheta(|\boldsymbol{x}|)\, d|\boldsymbol{x}| \cdot \tilde\eta(0) \cdot \tilde f(0)$$

for $r \to \infty$. This diverges for some $f(x)$ except for very special choices of $\vartheta(|\boldsymbol{x}|)$.

Proof:

i) $\int \tilde{\vartheta}_r(x)\,\eta(x^0)\,2x_0 x_\lambda \partial_0 \partial^\lambda\,(\Omega|\phi(x)\,\phi(f)\,\Omega)\,d^4x =$

$$= \int \vartheta_r(x)\,(\eta(x^0)\,2x_0)\,x_\lambda \partial_0 \partial^\lambda\,(\Omega|\phi(x)\,\phi(f)\,\Omega)\,d^4x =$$

$$= \tfrac{1}{2}\int \vartheta_r(x)\,\hat{\eta}(x^0)\,x_\nu \partial_0 \partial^\lambda\,(\Omega|[\phi(x),\,\phi(f)]\,\Omega)\,d^4x$$

for $r \to \infty$ by Theorem 7.1 with η replaced by $\hat{\eta} = 2x_0\eta(x^0)$ (or by explicit calculation), and this is certainly finite.

ii) $\lim\limits_{r \to \infty} \int \vartheta_r(x)\,\eta(x^0)\,x_0^2 \partial_0 \partial_0\,(\Omega|\phi(x)\,\phi(f)\,\Omega)\,d^4x = 0$ due to Theorems 3.1 and 4.2.

$$\int \vartheta_r(x)\,\eta(x^0)\,x_i x^i \partial_0 \partial_0\,(\Omega|\phi(x)\,\phi(f)\,\Omega)\,d^4x =$$

$$= \int \partial_i \partial^i \tilde{\vartheta}_r(p)\,\tilde{\eta}(p^0)\,p_0^2 \delta(p^2)\,\Theta(p^0)\tilde{f}(p)\,d^4p \sim$$

$$\sim \tilde{\eta}(0)\tilde{f}(0)\,r \cdot \int (\Delta\tilde{\vartheta}(q))|q|\,d^3q$$

$$\int (\Delta\vartheta(q))|q|\,d^3q = \int \vartheta(q)\,(2/|q|)\,d^3q = c \cdot \int \vartheta(x)\,|x|^{-2}\,d^3x\,.$$

(For the last equation see, e.g., [19].)

I conclude with an application of Corollary 2 proposed by Bose and McGlinn [20]: Assume that $\Theta_{\mu\nu}(x)$ is the energy-momentum density. Then we have for infinitesimal translations

$$i[P_\rho, A] = \lim_{r \to \infty}\,[\Theta_{0\nu}(\vartheta_r \otimes \eta),\,A]\,.$$

Consider

$$i[P_\rho, D_r] = \int x^\nu \partial_\rho \Theta_{0\nu}(x)\,\vartheta_r(x)\,\eta(x^0)\,d^4x\,.$$

Applying equation (9) we get

$$[i[P_\rho, D_r],\,A] = -[P_\rho, A]$$

and by Corollary 2

$$\lim_{r \to \infty}\,(B\Omega|i[P_\rho, D_r]\,A\Omega) = -(B\Omega|P_\rho A\Omega) \quad \forall\,A, B \in \mathfrak{R}\,. \tag{10}$$

Consider now the identity

$$[D_r, P_\rho P^\rho] = [D_r, P_\rho]\,P^\rho + P^\rho[D_r, P_\rho]\,.$$

Since $[P^\rho, A]$ is again a local operator, one gets by means of (10)

$$(B\Omega|[D_r, P_\rho P^\rho]\,A\Omega) = 2i(B\Omega|P_\rho P^\rho A\Omega)\,. \tag{11}$$

Assume now that (11) stays to be true for $A = \int C(x)f(x)\,d^4x$ and $B = \int D(x)\,g(x) \cdot dx$, $C, D \in \mathfrak{R}$ $f, g \in \mathfrak{S}(\mathbb{R}^4)$. (As said above, it can be proved that eqn. (11) stays true if one of A, B is replaced by such an operator.)

Assume furthermore that the one particle states φ_M possibly contained in the theory considered may be generated from Ω by such quasilocal operators. (Since we assume the existence of a conserved dilatation current, there is no gap in the energy-momentum spectrum above the vacuum. Hence the assumption is not trivial at all !) Let M denote the mass of those particles. Then from (11) it follows that

$$M^2 = 0$$

even for a theory with spontaneously broken dilatation.

REFERENCES

[1] See, e.g.:
C. Orzalesi, *Rev. Mod. Phys.*, **42**, 381 (1970);
J. A. Swieca, Cargese lecture, 1969;
H. Reeh, *Fortschr. d. Physik*, **16**, 687 (1968);
D. Kastler, Rochester Conference, 1967:
D. W. Robinson, Istanbul lecture, 1966.

[2] R. F. Streater and A. S. Wightman, "PCT. Spin and Statistics and All That". Benjamin, New York (1964).

[3] See, e.g.:
R. Haag and D. Kastler, *J. Math. Phys.*, **5**, 848 (1964);
H. Araki, *Progr. Theor. Phys.*, **32**, 844 (1964);
as well as the work of I. Segal.

[4] For algebras of bounded operators this can be found in Naimark's book on "Normed Rings" as well as in Dixmier's books.

[5] M. Rinke (private communication).

[6] H. Reeh, *Nuovo Cimento*, **51A**, 638 (1967).

[7] D. Maison, *Commun. Math. Phys.*, **14**, 56 (1969); "Some Remarks on Symmetries Generated by Local Currents" (preprint, Munich, 1970).
In case of conserved internal symmetries, there are other possibilities to arrive at α_τ by showing at first that there exists a self-adjoint global generator Q:
J. A. Swieca, Cargese lecture, 1969;
K. Kraus and L. J. Landau, "Conserved Currents and Symmetry Transformations in Local Scattering Theory" (preprint, 1971).

[8] In this general form at first proved by H. Ezawa and J. A. Swieca: *Commun. Math. Phys.*, **5**, 330 (1967).

[9] The triviality of this transformation can also be seen as follows: Denote by $\mathscr{A}(O)$ the algebra of bounded operators generated by $e^{i\phi_0(f)}$, f real, $f \in \mathscr{D}(\mathbb{R}^4)$, $\mathrm{supp} f \subset O$, O a finite open subregion of $\mathbb{R}^4$, and $\mathscr{A} = \vee_{O \in \mathbb{R}^4} \mathscr{A}(O)$ the algebraic closure of all $\mathscr{A}(O)$. Then by $\phi_0(f) \to \phi_0(f) + \tau \int f(x)\,d^4x$ a *-automorphism is defined on $\mathscr{A}$. Denote by $\mathscr{B}(O)$ the subalgebra of $\mathscr{A}(O)$ such that $\int f(x)\,d^4x = = 0$ and $\mathscr{B} = \vee_{O \in \mathbb{R}^4} \mathscr{B}(O)$. Then: i) By standard arguments $\mathscr{B}$ is irreducible. Therefore it describes the same physics as $\mathscr{A}$ does. ii) $\mathscr{B}$ is invariant under α_τ.

[10] D. Maison and H. Reeh, *Commun. Math. Phys.* (to be published).

[11] R. N. Sen and C. Weil, *Nuovo Cimento,* **6A**, 581 (1971).

[12] H. Araki, K. Hepp, and D. Ruelle, *Helv. phys. acta,* **35**, 164 (1962).

[13] D. Maison (to be published).

[14] R. Ferrari and L. E. Picasso (private communication).

[15] The simple proof of the corollary can be found, e.g., in H. Reeh, see ref. [1] above.

[16] D. Maison and H. Reeh, *Nuovo Cimento,* **1A**, 78 (1971).

[17] See, e.g.:

C. G. Callan, S. Coleman, and R. Jackiw, *Annals of Phys.,* **59**, 47 (1970);

D. J. Gross and J. Wess (preprint, Ref. TH. 1076 CERN).

[18] H. Reeh, *Nuovo Cimento* (in press).

[19] I. M. Gelfand and G. E. Schilow, "Verallgemeinerte Funktionen", Vol. I, page 187. Berlin (1960).

[20] S. K. Bose and W. D. McGlinn, "Space Time Symmetries and the Spontaneous Breakdown of Dilation Invariance" (preprint, Notre Dame, 1970).

Note added in proof: Theorem 6.1 and the Corollary have been extended to arbitrary quasilocal operators by A. H. Völkel, "On Self-Adjoint Extensions of Global Charge Operators", Preprint, Rio de Janeiro, April 1972.

The Galilei Group and Landau Excitations

R. N. SEN

Department of Physics, Technion—Israel Institute of Technology, Haifa (Israel)

Abstract

The behavior of low-lying excitations in superfluid helium under Galilei transformations is investigated. It appears that the ten-parameter Galilei group acts on them in part linearly, and in part nonlinearly. The latter appears to give rise to the possibility of distinguishing between different inertial frames. In one dimension, the structure is that of a vector bundle over the base space of the boosts. The treatment is elementary and most mathematical ideas are developed *ab initio*.

I. INTRODUCTION

The present article is based on two observations. The first [1] is that the Galilei transformation properties of the energy and momentum of low-lying ("Landau") excitations in superfluid Helium

$$\left. \begin{array}{ll} \boldsymbol{p} \to \boldsymbol{p}, & E \to E + \boldsymbol{p} \cdot \boldsymbol{v} \\ \boldsymbol{x} \to \boldsymbol{x} + \boldsymbol{v}t, & t \to t \end{array} \right\} \tag{1.1}$$

under

are the same as those of $\boldsymbol{p}$ and E in the *true* unitary representations of the Galilei group [2]. This suggests that the states of the Landau excitations span a Hilbert space which carries a true representation of the Galilei group. If so, this representation is likely to be reducible because, as Inönü and Wigner [3] have pointed out, within *irreducible* representations there exist neither localizable states nor states with definite velocity. Therefore the physical interpretation of these representations is problematical. However, this problem disappears if irreducibility is dropped; as shown by Wightman [4], localizability is retrieved in *reducible* representations which contain all irreducibles of the same helicity with the same multiplicity.

This sets the stage for the second observation. In the representations involved, there is no relation between E and $\boldsymbol{p}$—in fact, all real values of E occur for a given

p (as is evident from eq. (1.1)). However, it is an empirical fact that there exists a dispersion law

$$E = E(p), p = |p| \tag{1.2}$$

for the Landau excitations—the famous phonon–roton curve of Landau [5]. That is, E is determined *uniquely* by p in the laboratory frame, and therefore, from (1.1), in any inertial frame. In other words, an arbitrary superposition of states, which is permissible in any linear representation, is *not* observed in any inertial frame. The conclusion is that the group elements which connect different inertial frames—the boosts—act *nonlinearly* on the set of states of the Landau excitations. The group operations which take place *within* an inertial frame—rotations, space and time translations—act linearly, and localizable states exist within each linear manifold [1].

In this article we shall make a preliminary analysis of the geometrical—i.e. nonlinear—elements in the structure of the Landau excitations and their relationship to the linear elements. Our building blocks will be certain true representations of the Galilei group. Since this article is addressed to a mixed audience, the only mathematics assumed will be "a rudimentary knowledge of the Latin and Greek alphabets". In section II we define the Galilei group $\mathscr{G}$ and establish the basic notations. In section III we classify the irreducible unitary representations of $\mathscr{G}$. This section contains an elementary account of induced representations and the representations of the Euclidean group in two and three dimensions. The results for $\mathscr{G}$ are assembled in section III d. In section IV we give a physical interpretation of the representations in which we are interested. In section V we analyze, albeit incompletely, the structure of the Landau excitations, and finally, in section VI, we make a general comment.

II. THE GALILEI GROUP AND ITS LIE ALGEBRA

The inhomogeneous Galilei group $\mathscr{G}$ is a ten-parameter group of linear transformations on x and t of the following form [3]:

$$\left. \begin{array}{l} t' = t + b \\ x' = Rx + vt + a \end{array} \right\} \tag{2.1}$$

where R is a 3×3 orthogonal matrix, v and a are 3-vectors and b is a real number. We will write the element g of $\mathscr{G}$ which effects the transformation (2.1) as

$$g = (b, a, v, R). \tag{2.2}$$

From (2.1) we can easily ascertain that the group multiplication law is

$$(b', a', v', R')(b, a, v, R) = (b' + b, a' + R'a + bv', v' + R'v, R'R). \tag{2.3}$$

The identity element is

$$\mathbb{1} = (0, \mathbf{0}, \mathbf{0}, 1)$$

and the inverse of g is

$$g^{-1} = (b, \mathbf{a}, \mathbf{v}, R)^{-1} = (-b, -R^{-1}(\mathbf{a} - b\mathbf{v}), -R^{-1}\mathbf{v}, R^{-1}).$$

It is evident that the Galilei group contains the following subgroups, which we shall call *elementary:* (a) time translations; (b) space translations; (c) *pure* Galilei transformations or *boosts* $\mathbf{x}' = \mathbf{x} + \mathbf{v}t$, $t' = t$, sometimes called "accelerations"; and (d) 3-dimensional rotations. We will reserve the letters $b, \mathbf{a}, \mathbf{v}$ and R for these elementary operations. The corresponding group element will be denoted by enclosing the letter in heavy brackets, thus $(b), (\mathbf{a}), (\mathbf{v})$ and (R). The elementary subgroups themselves will be denoted by $\mathscr{T}, \mathscr{S}, \mathscr{V}$ and $\mathscr{R}$ respectively.

The same heavy brackets enclosing two or more elementary operations will denote elements of the "composite" subgroups. The more important of these will be given special names. Thus, $(b, \mathbf{a})$ is an element of the Abelian subgroup $\mathfrak{T}_4 = \mathscr{S} \times \mathscr{T}$ ($\times$ denotes direct product) of space *and* time translations. $(\mathbf{v}, R)$ is an element of the subgroup $\mathfrak{G}_3$, the *homogeneous* Galilei group in three dimensions. $(\mathbf{a}, R)$ is an element of the three-dimensional *Euclidean* group $\mathfrak{E}_3$, which is also a subgroup of $\mathscr{G}$. It is easily verified that $\mathfrak{E}_3$ and $\mathfrak{G}_3$ are isomorphic. In fact, $\mathfrak{E}_3 = \mathscr{S} \wedge \mathscr{R}$ and $\mathfrak{G}_3 = \mathscr{V} \wedge \mathscr{R}$,* where the "$\wedge$" denotes the semidirect product.** Later we shall have occasion to refer to the homogeneous Galilei group and the Euclidean group in *two* dimensions. These will, naturally, be denoted by $\mathfrak{G}_2$ and $\mathfrak{E}_2$, respectively.

It is easy to check that the Galilei group can be expressed in the following ways in terms of the elementary subgroups:

$$\mathscr{G} = (\mathscr{S} \times \mathscr{T}) \wedge (\mathscr{V} \wedge \mathscr{R}) = \mathfrak{T}_4 \wedge \mathfrak{G}_3 = \mathfrak{T}_4 \wedge (\mathfrak{B}_3 \wedge \mathfrak{R}_3)$$

$$= (\mathscr{S} \times \mathscr{V}) \wedge (\mathscr{T} \times \mathscr{R}).$$

The maximal Abelian invariant subgroups of $\mathscr{G}$ are a) the four-parameter group $\mathfrak{T}_4$, and b) the six-parameter group $\mathscr{S} \times \mathscr{V}$.

The Lie algebra of $\mathscr{G}$ can be determined from (2.3) in standard fashion. Let H be the generator of time translations $((b) = \exp i(bH))$, P_i those of space translation, K_i those of the boosts and J_i those of the rotations. Here $i = 1, 2, 3$ is a vectorial

---

* We will sometimes write $\mathfrak{E}_3 = \mathfrak{T}_3 \wedge \mathfrak{R}_3$ and $\mathfrak{G}_3 = \mathfrak{B}_3 \wedge \mathfrak{R}_3$, in an obvious notation.

** Unlike the direct product, the semidirect product cannot be defined for arbitrary pairs of groups. For $\mathfrak{A} \wedge \mathfrak{B}$ to be definable, $\mathfrak{B}$ must be a subgroup of the group of automorphisms of $\mathfrak{A}$, so that if $b \in \mathfrak{B}$, $a \in \mathfrak{A}$, the quantity $b(a)$ is again an element of $\mathfrak{A}$, obtained by applying b to a. Then the semidirect product $\mathfrak{A} \wedge \mathfrak{B}$ consists of the ordered pairs (a, b), with the multiplication law $(a, b)(a', b') = (a \cdot b(a'), bb')$, where $b(a')$ is an element of $\mathfrak{A}$ and $\cdot$ denotes the multiplication in $\mathfrak{A}$. If $\mathfrak{A}$ is Abelian, the $\cdot$ is generally replaced by $+$, thus: $(a, b)(a', b') = (a + b(a'), bb')$.

subscript. Then the nonvanishing Lie brackets are:

$$\left.\begin{aligned}
\left[J_i, J_j\right] &= i\varepsilon_{ijk}J_k \\
\left[J_i, K_j\right] &= i\varepsilon_{ijk}K_k \\
\left[J_i, P_j\right] &= i\varepsilon_{ijk}P_k \\
\\
\left[K_i, H\right] &= iP_i.
\end{aligned}\right\} \tag{2.4}$$

and

From these commutation relations it may be verified that

$$P^2 \tag{2.5a}$$

and

$$N^2 \equiv (K \times P)^2 = K^2P^2 - (K \cdot P)^2, \tag{2.5b}$$

where the cross denotes the vector product and the dot the scalar product, commute with all generators of the Galilei group. They are therefore constant within irreducible representations. We shall be interested in those irreducible representations in which $N^2 = 0$. In these cases there exists an additional operator which is constant within an irreducible representation. This is the operator

$$\Sigma = J \cdot P \tag{2.6}$$

which commutes with H, P and J and satisfies, in addition,

$$[\Sigma, K] = iK \times P = iN.$$

Thus Σ also commutes with all operators on the eigenspace of N with eigenvalue zero.

III. IRREDUCIBLE UNITARY REPRESENTATIONS OF $\mathscr{G}$

In this section we will determine the irreducible unitary representations (hereafter abbreviated IURs) of the Galilei group by the method of Wigner [6] and Mackey [7]. In this method representations of the entire (noncompact) group are "induced" from those of a maximal Abelian invariant subgroup via the "little groups". Since we shall not assume that this method is well known to all the readers, the computations will be spelled out in some detail.

Consider a group $\mathfrak{K}$ which is a semidirect product, $\mathfrak{K} = \mathfrak{A} \wedge \mathfrak{C}$, where $\mathfrak{A}$ is Abelian and $\mathfrak{C}$ is locally compact. Upon restriction to $\mathfrak{A}$, an IUR of $\mathfrak{K}$ will split, in an essentially unique manner, into a *continuous* direct sum or *direct integral* of IURs of $\mathfrak{A}$. This suggests that one might be able to reverse the argument and *induce* IURs of $\mathfrak{K}$ from those of $\mathfrak{A}$. Since $\mathfrak{A}$ is Abelian, the latter are well known.

Obviously, in order to proceed further, one should investigate the "action" of the elements of $\mathfrak{C}$ on the "space of irreducible representations" of $\mathfrak{A}$ [8]. Under the action of $\mathfrak{C}$ this space breaks up into disjoint *orbits*. We remind the reader of the definitions. Let M be a topological space (i.e. a set of points with enough continuity properties) on which a group $\mathfrak{H}$ acts as a transformation group, i.e. for each $p \in M$ and $h \in \mathfrak{H}$ there exists a unique point hp in M; it is the *transform* of p by h. For fixed p, the set of points hp, $h \in \mathfrak{H}$, is a subset of M which is called either (a) the orbit of $\mathfrak{H}$ *at* p, or (b) the orbit of p in M under $\mathfrak{H}$. Clearly two orbits are either disjoint or identical. We shall denote the orbit of $\mathfrak{H}$ at p by $\mathfrak{H}p$.

For a given p, there may be elements $h \in \mathfrak{H}$ such that $hp = p$. The set of such elements forms a group (the proof is trivial). This group is called the *little group* or *isotropy group* of $\mathfrak{H}$ at p. We shall denote it by $\mathfrak{H}^p$.

Let p and p' belong to the same orbit. Then there exists at least one element $g \in \mathfrak{H}$ such that $gp = p'$. If $h \in \mathfrak{H}^p$, then $hp = p$, and $ghp = gp = p'$, or $(ghg^{-1}) gp = p'$, or $(ghg^{-1}) p' = p'$. Hence $ghg^{-1} \in \mathfrak{H}^{p'} = g\mathfrak{H}^p g^{-1}$. $\mathfrak{H}^p$ and $\mathfrak{H}^{p'}$ are conjugate subgroups, and therefore isomorphic. *Little groups belonging to different points on the same orbit are isomorphic.*

Given two points p and p' on an orbit, the equation $gp = p'$ does *not* have a unique solution g in $\mathfrak{H}$. If g is any solution, so is $g^{p'} g g^p$, where $g^{p'} \in \mathfrak{H}^{p'}$ and $g^p \in \mathfrak{H}^p$. We shall overcome this lack of uniqueness by imposing further conditions as follows. Choose a fixed point p_0 on the orbit $\mathfrak{H}p$ under consideration. Let $g(p)$ be a group element which maps p_0 onto p, i.e.

$$g(p)\, p_0 = p. \tag{3.1}$$

We now require $g(p)$ to be a *continuous* function of p on the orbit, with $g(p_0) = \mathbb{1}$, the identity in $\mathfrak{H}$. These requirements make $g(p)$ unique and allow (3.1) to be inverted:

$$g^{-1}(p)\, p = p_0, \tag{3.1'}$$

a relation which holds for every $p \in \mathfrak{H}p$.

Let k be an arbitrary element of $\mathfrak{H}$. Then kp is a unique point in $\mathfrak{H}p$. We define an element $u(\tilde{k}, p_0) \in \mathfrak{H}$ by

$$k = g(kp)\, u(\tilde{k}, p_0)\, g^{-1}(p). \tag{3.2}$$

(The reason for the tilde symbol on the k in $u(\tilde{k}, p_0)$ will become apparent in a moment.) Now let both sides of (3.2) act on p:

$$kp = g(kp)\, u(\tilde{k}, p_0)\, g^{-1}(p)\, p$$

or

$$g^{-1}(kp) \cdot kp = u(\tilde{k}, p_0)\, g^{-1}(p)\, p.$$

Using (3.1') on both sides, this becomes

$$p_0 = u(\tilde{k}, p_0)\, p_0,$$

i.e. $u(\tilde{k}, p_0)$ *belongs to the little group* $\mathfrak{H}^{p_0}$ *of* p_0. Although for any k the element $u(\tilde{k}, p_0)$ is uniquely defined, the converse is generally false; distinct elements k may be mapped to the same $u(\tilde{k}, p_0)$. The $\tilde{k}$ in $u(\tilde{k}, p_0)$ denotes the equivalence class of all k's in $\mathfrak{H}$ which lead to the same $u(\tilde{k}, p_0)$.

Equation (3.2) factorizes an arbitrary element k of $\mathfrak{H}$ *uniquely* (for given p_0 and p) into elements $g(kp)$ and $g^{-1}(p)$ which acts nontrivially on the orbit $\mathfrak{H}p_0$, and an element $u(\tilde{k}, p_0)$ which belongs to $\mathfrak{H}^{p_0}$. The fact that there exists such a factorization for any pair of distinct points p and p_0 on $\mathfrak{H}p_0$ is of crucial importance for the inducing construction.

First of all, let us impose an additional structure on the set of points $\{p\}$ representing the orbit: that of a *linear vector space*. This is done by introducing the vectors $|p\rangle$ with the scalar product

$$\langle p|q\rangle = \delta(p - q),$$

where $p, q \in \mathfrak{H}p_0$ and the δ-function has the dimensionality of the orbit. Such a vector space is sometimes called a Dirac space [9]. The group elements $g(p)$ and $g^{-1}(p)$ act in this space according to

$$g(p)|p_0\rangle = |p\rangle, \quad g^{-1}(p)|p\rangle = |p_0\rangle, \tag{3.3}$$

which is just the transcription of (3.1) and (3.1'). Note that here we are using the same symbols for the (abstract) group elements and their operator representatives — and that the operators $g(p)$ are automatically unitary.

Return now to the little group $\mathfrak{H}^{p_0}$. If $\mathfrak{H}^{p_0}$ is either finite or Abelian or compact, its IURs are known. We assume, for the sake of definiteness, that $\mathfrak{H}^{p_0}$ is compact, in which case it has a countable number of (finite-dimensional) IURs, and the states within an IUR can be labeled by a discrete index. Choose one particular IUR:

$$u(\tilde{k}, p_0)|\xi\rangle = \sum_{\eta} \Delta_{\xi\eta}(\tilde{k})|\eta\rangle. \tag{3.4}$$

Here the elements $u(\tilde{k}, p_0)$ of $\mathfrak{H}^{p_0}$ are parametrized by $\tilde{k}$, $|\xi\rangle$ are the states which span the IUR under consideration, and $\Delta_{\xi\eta}(\tilde{k})$ are the matrix elements of the group operators u. The states are orthonormalized, as usual:

$$\langle \xi|\eta\rangle = \delta_{\xi\eta}.$$

Let us now construct the product space consisting of the vectors

$$|p, \xi\rangle \equiv |p\rangle \otimes |\xi\rangle$$

and investigate the action of elements of $\mathfrak{H}$ on this space, using (3.2):

$$k|p, \xi\rangle = g(kp)\, u(\tilde{k}, p_0)\, g^{-1}(p)|p, \xi\rangle.$$

First, from (3.3), $g^{-1}(p)|p, \xi\rangle = |p_0, \xi\rangle$, so that

$$k|p, \xi\rangle = g(kp)\, u(\tilde{k}, p_0)|p_0, \xi\rangle.$$

Then, from (3.4),

$$k|p, \xi\rangle = g(kp) \sum_{\eta} \Delta_{\xi\eta}(\tilde{k})|p_0, \eta\rangle$$

$$= \sum_{\eta} \Delta_{\xi\eta}(\tilde{k}) \, g(kp)|p_0, \eta\rangle.$$

Finally, using (3.3) again,

$$g(kp)|p_0, \eta\rangle = g(kp) \, g^{-1}(p) \, |p, \eta\rangle$$

$$= g(kp) \, g^{-1}(kp)|kp, \eta\rangle$$

$$= |kp, \eta\rangle,$$

so that

$$k|p, \xi\rangle = \sum_{\eta} \Delta_{\xi\eta}(\tilde{k})|kp, \eta\rangle. \tag{3.5}$$

We have thus obtained a representation of the group $\mathfrak{H}$ which is both irreducible and unitary. We have constructed this representation out of a space M on which the group acts. These IURs of $\mathfrak{H}$ are labeled by two quantities: (a) an orbit of $\mathfrak{H}$ in M, and (b) an IUR of the corresponding little group.

In order to construct representations of the semidirect product mentioned at the beginning, we take M to be the "space of irreducible representations" of $\mathfrak{A}$, and $\mathfrak{C}$ to be the group $\mathfrak{H}$. It can be proved that this method yields all irreducible representations for the Galilei group (as well as for the Poincaré group). In practical terms, we have to investigate the action of $\mathfrak{C}$ on the space of irreducible representations of $\mathfrak{A}$, and calculate the little groups and their IURs.

In the following subsections this procedure is carried out explicitly for $\mathcal{G}$.

a) Calculation of the Little Groups

For the Abelian invariant subgroup we choose $\mathfrak{T}_4$ rather than the six-parameter subgroup $\mathcal{S} \times \mathcal{V}$. This choice is more instructive and the results are easier to interpret physically [10].

The generic element of $\mathfrak{T}_4$ is (b, a). Consider the following mapping from $\mathfrak{T}_4$ to the complex numbers of modulus unity:

$$(b, a) \xrightarrow{[E, p]} \exp i(p \cdot a - Eb), \tag{3.6}$$

where $p \cdot a = p_1 a_1 + p_2 a_2 + p_3 a_3$ and E, p_i are reals. For fixed E and p the above formula, as is well known, gives a continuous IUR—necessarily one-dimensional —of $\mathfrak{T}_4$. The quantity on the right-hand side of (3.6) is called a *character* of the group $\mathfrak{T}_4$ and is sometimes written as $\chi^{[E, p]}(b, a)$. It is evident that the set of all characters of $\mathfrak{T}_4$ forms an Abelian group under ordinary multiplication:

$$\chi^{[E,\,p]}(b,\,a) \cdot \chi^{[E',\,p']}(b,\,a) = \chi^{[E+E',\,p+p']}(b,\,a).$$

This group is called the *dual* of the original group $\mathfrak{T}_4$, and is denoted by $\hat{\mathfrak{T}}_4$. The use of the term "dual" is justified by the celebrated *Pontrjagin duality theorem* [11], which asserts that the duality is a reciprocal relationship: if $\mathfrak{R}$ is any Abelian group and $\hat{\mathfrak{R}}$ its dual, then the dual of $\hat{\mathfrak{R}}$ is again $\mathfrak{R}$:

$$\hat{\hat{\mathfrak{R}}} = \mathfrak{R}.$$

Since $\mathscr{G} = \mathfrak{T}_4 \wedge \mathfrak{G}_3$, we have to ascertain the action of $\mathfrak{G}_3$ on $\hat{\mathfrak{T}}_4$, which we earlier called the "space of irreducible representations of $\mathfrak{T}_4$". An element of $\mathfrak{G}_3$ in $\mathscr{G}$ is of the form $(0,\,0,\,v,\,R)$. An element of $\mathfrak{T}_4$ in $\mathscr{G}$ is of the form $(b,\,a,\,0,\,1)$. From the multiplication law (2.3) we obtain

$$(0,\,0,\,v,\,R)(b,\,a,\,0,\,1) = (b,\,Ra + bv,\,v,\,R).$$

Hence the action of $\mathfrak{G}_3$ on $\mathfrak{T}_4$ defines itself naturally to be

$$(v,\,R)\,(b,\,a) = (b',\,a'), \tag{3.7}$$

where

$$\left.\begin{array}{l} b' = b \\ a' = Ra + bv. \end{array}\right\} \tag{3.8}$$

The action of $\mathfrak{G}_3$ on $\hat{\mathfrak{T}}_4$ now follows from a characteristic duality requirement, which is that the elements of $\hat{\mathfrak{T}}_4$ and $\mathfrak{T}_4$ transform cogrediently to each other under $\mathfrak{G}_3$—i.e. the exponents

$$\boldsymbol{p} \cdot \boldsymbol{a} - Eb$$

occurring in the characters remain invariant. Denoting the element $(E,\,\boldsymbol{p})$ of $\hat{\mathfrak{T}}_4$ by $[E,\,\boldsymbol{p}]$, we obtain

$$(v,\,R) \cdot [E,\,\boldsymbol{p}] = [E',\,\boldsymbol{p}'], \tag{3.9}$$

where

$$\left.\begin{array}{l} E' = E + \boldsymbol{v} \cdot R\boldsymbol{p} \\ \boldsymbol{p}' = R\boldsymbol{p}. \end{array}\right\} \tag{3.10}$$

The next step is to investigate how the space $\hat{\mathfrak{T}}_4$ splits into orbits under the action of $\mathfrak{G}_3$. From (3.9) and (3.10) we see that the orbits of $\mathfrak{G}_3$ in $\hat{\mathfrak{T}}_4$ are the cylinders

$$\boldsymbol{p}^2 = \text{const.}, \quad -\infty < E < \infty,$$

for $\boldsymbol{p} \neq 0$. The little groups of all these orbits are isomorphic. For $\boldsymbol{p} = 0$, the orbits are single points on the E-axis and their little groups are also isomorphic. Thus there are two different little groups, depending on whether or not $\boldsymbol{p}$ vanishes.

Case 1: $p \neq 0$.

To calculate the little group, take as "fiducial vector" the vector $(0, p_0)$, i.e. $E = 0$, $p = p_0 \neq 0$. The little group is that subgroup of $\mathfrak{G}_3$ which leaves this vector invariant. From (3.9) and (3.10) it is at once evident that this is *the homogeneous Galilei group* $\mathfrak{G}_2$ *in the plane perpendicular to* p_0.

Case 2: $p = 0$.

The fiducial vector is $(0, \mathbf{0})$. Every homogeneous Galilei transformation leaves this vector invariant, and hence *the little group is* $\mathfrak{G}_3$ *itself*.

What remains to be done is (a) to calculate the IURs of the little groups $\mathfrak{G}_2$ and $\mathfrak{G}_3$, and (b) to assemble, from these and the orbits, the IURs of $\mathscr{G}$.

b) Irreducible Unitary Representations of $\mathfrak{E}_2(\mathfrak{G}_2)$

The homogeneous Galilei group $\mathfrak{G}_2$ is isomorphic to the two-dimensional Euclidean group $\mathfrak{E}_2$. The elements of the latter are $(\boldsymbol{\alpha}, \rho)$, where $\boldsymbol{\alpha}$ is a 2-vector and ρ a 2×2 orthogonal matrix. The multiplication law is the familiar one:

$$(\boldsymbol{\alpha}_1, \rho_1)(\boldsymbol{\alpha}_2, \rho_2) = (\boldsymbol{\alpha}_1 + \rho_1 \boldsymbol{\alpha}_2, \rho_1 \rho_2); \tag{3.11}$$

that is, $\mathfrak{E}_2 = \mathfrak{T}_2 \wedge \mathfrak{R}_2$, where $\mathfrak{T}_2$ and $\mathfrak{R}_2$ are respectively the groups of translations and rotations in two dimensions. The characters of $\mathfrak{T}_2$ are $\exp i(\boldsymbol{\eta} \cdot \boldsymbol{\alpha})$, where $\boldsymbol{\eta}$ is a point in $\hat{\mathfrak{T}}_2$. The orbits of $\mathfrak{R}_2$ in $\hat{\mathfrak{T}}_2$ are the circles $\eta^2 = $ const., and there are two classes of them: a) $\eta^2 > 0$, and b) $\boldsymbol{\eta} = 0$. In case a), the little group is that subgroup of $\mathfrak{R}_2$ which leaves a nonzero vector invariant. This subgroup contains only the identity element—the little group is trivial. In case b) the little group consists of that subgroup of $\mathfrak{R}_2$ which leaves the null vector invariant, i.e. $\mathfrak{R}_2$ itself. Correspondingly, there are two classes of representations of $\mathfrak{E}_2$.

Let us first consider case a), for which the little group is trivial. Take a set of δ-normalized vectors $|\boldsymbol{\eta}\rangle$:

$$\langle \boldsymbol{\eta} | \boldsymbol{\eta}' \rangle = \delta(\boldsymbol{\eta} - \boldsymbol{\eta}'). \tag{3.12}$$

We denote the representative of the abstract group element by the letter U. Thus $U(\boldsymbol{\alpha}, 1)$ are the translation operators and $U(\mathbf{0}, \rho)$ the rotation operators. We have

$$U(\boldsymbol{\alpha}, 1)|\boldsymbol{\eta}\rangle = e^{i\boldsymbol{\eta} \cdot \boldsymbol{\alpha}}|\boldsymbol{\eta}\rangle,$$

$$U(\mathbf{0}, \rho)|\boldsymbol{\eta}\rangle = |\rho\boldsymbol{\eta}\rangle,$$

so that

$$U(\boldsymbol{\alpha}, \rho)|\boldsymbol{\eta}\rangle = e^{i\boldsymbol{\eta} \cdot \boldsymbol{\alpha}}|\rho\boldsymbol{\eta}\rangle. \tag{3.13}$$

In the Dirac space with scalar product (3.12), $U(\boldsymbol{\alpha}, \rho)$ as defined above gives a unitary representation of $\mathfrak{E}_2$. The irreducible representations are labeled by a positive number $\mathcal{O}$:

$$\mathcal{O} = |\boldsymbol{\eta}|.$$

To construct a unitary representation on an ordinary Hilbert space we consider, instead of the Dirac space of vectors $|\boldsymbol{\eta}\rangle$, the Hilbert space of square-integrable complex functions $L_2(\mathcal{O})$ on the circle $|\boldsymbol{\eta}| = \mathcal{O}$. Let $\psi(\boldsymbol{\eta}) \in L_2(\mathcal{O})$. The argument $\boldsymbol{\eta}$ of the vector $\psi \in L_2(\mathcal{O})$ is a point on the circle $\mathcal{O}$. The scalar product is

$$(\phi, \psi) = \int \phi^*(\boldsymbol{\eta}) \psi(\boldsymbol{\eta}) \, \delta(\mathcal{O}^2 - \boldsymbol{\eta}^2) \, d^2\eta. \tag{3.14}$$

The group element $(\boldsymbol{\alpha}, \rho)$ of $\mathfrak{E}_2$ is represented by the operator $U(\boldsymbol{\alpha}, \rho)$:

$$U(\boldsymbol{\alpha}, \rho) \psi(\boldsymbol{\eta}) = e^{i\boldsymbol{\eta} \cdot \boldsymbol{\alpha}} \psi(\rho\boldsymbol{\eta}). \tag{3.15}$$

The unitarity of U is obvious.

In case b), the little group is $\mathfrak{R}_2$ itself. Its irreducible representations are labeled by an integer n:

$$\rho(\theta) \to e^{in\theta}, \tag{3.16}$$

where θ is the angle of rotation which labels the element ρ of $\mathfrak{R}_2$.

We see from (3.16) that the dual of $\mathfrak{R}_2$ is *discrete*: it is the additive group of all integers. This is an example of yet another theorem of Pontrjagin, which states that the dual of a *compact* commutative group is discrete, and *vice versa* [11].

Returning to the representations of $\mathfrak{E}_2$, the fact that $\boldsymbol{\eta} = 0$ means that the translations are represented by the identity. Thus the representations are

$$U(\boldsymbol{\alpha}, \rho(\theta)) = e^{in\theta}.$$

They are actually representations of the factor group $\mathfrak{E}_2/\mathfrak{T}_2 = \mathfrak{R}_2$.

c) Irredicible Unitary Representations of $\mathfrak{E}_3(\mathfrak{G}_3)$

Since $\mathfrak{E}_3 = \mathfrak{T}_3 \wedge \mathfrak{R}_3$ and the characters of $\mathfrak{T}_3$ are $\exp i(s \cdot q)$, where s is a point in $\hat{\mathfrak{T}}_3$, it follows that the orbits are a) the spheres $s^2 = \text{const.} > 0$, and b) the point $s = 0$. In case a) the little group is that subgroup of $\mathfrak{R}_3$ which leaves a non-zero vector s_0 invariant. This is the group of two-dimensional rotations $\mathfrak{R}_2$ in the plane perpendicular to s_0. In case b), the little group is $\mathfrak{R}_3$ itself.

In case a), an irreducible representation is characterized by a positive number $s^2 > 0$ and an integer n. The states are labeled by $|s, n\rangle$, where n characterizes an IUR of the little group $\mathfrak{R}_2$ of rotations which leave the fiducial vector s_0 invariant. According to (3.2), an arbitrary rotation R can be decomposed uniquely relative to s and s_0 as

$$R = r(Rs) \, u(\tilde{R}, s_0) \, r^{-1}(s),$$

where $r(s)$ is the rotation which takes s_0 to s, i.e.

$$r(s)\,s_0 = s,$$

with $r(s_0) = 1$ and continuity of $r(s)$ in s being assumed. Thus $r(s)$ is the positive rotation by the angle between s and s_0 around the axis which is perpendicular to the plane of s and s_0. Then $u(\tilde{R}, s_0)$ is a rotation by an angle θ around the axis s_0, and θ can be calculated explicitly from

$$u(\tilde{R}, s_0) \equiv u(\theta, s_0) = r^{-1}(Rs)\, Rr(s).$$

Hence

$$U(0, R)\,|s, n\rangle = e^{in\theta}\,|Rs, n\rangle$$

and

$$U(a, R)\,|s, n\rangle = e^{ia\cdot s}e^{in\theta}\,|Rs, n\rangle.$$

In case b) the elements of $\mathfrak{T}_3$ are trivially represented, so that the representations are just those of $\mathfrak{R}_3$:

$$U(a, R)\,|j; m\rangle = \sum_{m'} D^j_{mm'}(R)\,|j; m'\rangle.$$

Notice that half-integral spins would arise upon passage to the covering group of $\mathfrak{R}_3$ — i.e. the covering group of $\mathscr{G}$.

d) Irreducible Unitary Representations of $\mathscr{G}$

We have only to assemble the results which we have obtained so far. There are four classes of IURs. We remind the reader that $p \in \hat{\mathfrak{T}}_3 (= \mathscr{P})$ and $\mathfrak{H}^x$ stands for the little group of $\mathfrak{H}$ at x.

i) $p \neq 0, \mathfrak{G}_3^p = \mathfrak{G}_2 = \mathfrak{B}_2 \wedge \mathfrak{R}_2$ in the plane perpendicular to p. Let $\eta \in \hat{\mathfrak{B}}_2$. Then there are two classes of representations:

Class I: $\eta \neq 0, \mathfrak{R}_2^\eta = 1$.

Class II: $\eta = 0, \mathfrak{R}_2^\eta = \mathfrak{R}_2$.

ii) $p = 0, \mathfrak{G}_3^p = \mathfrak{G}_3 = \mathfrak{B}_3 \wedge \mathfrak{R}_3$. Let $q \in \hat{\mathfrak{B}}_3$. Then there are two classes of representations:

Class III: $q \neq 0, \mathfrak{R}_3^q = \mathfrak{R}_2$ in the plane perpendicular to q.

Class IV: $q = 0, \mathfrak{R}_3^q = \mathfrak{R}_3$.

Writing down the states and the group operators within these irreducible representations is a simple matter which involves, at most, the use of equation (3.2). For classes I, III and IV we leave this task to the reader. For class II we have, explicitly,

$$U(b, a, v, R)\,|E, p, n\rangle = \exp i(Eb - p \cdot a)\exp in\theta(R, p)\,|E + v \cdot Rp, Rp, n\rangle,$$

where $\theta(R, p) = \cos^{-1}(q \cdot Rq)$, q being any unit vector perpendicular to p. That is, θ is the angle by which R effects a rotation *around* p.

It is evident that these representations are characterized by a positive number p^2 and an integer n. We invite the reader to determine the matrix elements of the operators N and Σ, defined by equations (2.5b) and (2.6), in these representations. He will then find that we are more-or-less justified in calling the class II, $n = 0$ IURs the *zero-helicity representations*.

The class II, $n = 0$ representation characterized by

$$P^2 = p^2$$

will be called the *P*-representation for brevity.

IV. PHYSICAL INTERPRETATION OF *P*-REPRESENTATIONS

Henceforth we shall be concerned exclusively with the *P*-representations of $\mathscr{G}$. Specializing the first formula on this page to $n = 0$, we obtain

$$U(b, a, v, R)\,|E, p\rangle = \exp i(Eb - p \cdot a)\,|E + v \cdot Rp, Rp\rangle \qquad (4.1)$$

with

$$\langle E, p\,|E', p'\rangle = \delta(E - E')\,\delta(p - p'). \qquad (4.2)$$

The quantity p^2 being constant within an IUR, the set of p-values reached in it is $\{Rp_0\}$, where p_0 is a fixed vector. This set is far too small to permit any kind of localization in the space dual to that of the p's, and, moreover, the identification of p with momentum becomes questionable. We shall consider here the question of physical interpretation of p independently of that of localizability.

In the normalization (4.2), a state within an IUR is labeled by a point in $\hat{\mathfrak{T}}_4$. The observation upon which the physical interpretation of p hinges is that a character of $\mathfrak{T}_4$,

$$\chi^{[E,p]}(t, x) = \exp i(p \cdot x - Et), \qquad (4.3)$$

can be factorized (albeit in a nonunique manner) into plane Schrödinger waves of positive and negative mass:

$$\exp i(p \cdot x - Et) = \exp i(k \cdot x - \omega t)\exp i(-k' \cdot x + \omega' t), \qquad (4.4)$$

where $\omega = k^2/2\mu$, $\omega' = k'^2/2\mu$ and

$$p = k - k', \qquad E = \omega - \omega' = p \cdot q, \qquad (4.5)$$

with

$$q = \frac{1}{2\mu}(k + k').$$

The two factors in the right-hand side of (4.4) are respectively solutions of

$$i\frac{\partial\psi}{\partial t} = \mp\frac{1}{2\mu}\nabla^2\psi.$$

Since under a Galilei boost $x \to x + vt$ the momentum of a particle of mass μ transforms as

$$k \to k + \mu v, \tag{4.6}$$

it follows from (4.5) that

$$\left.\begin{array}{c} p \to p \\ E \to E + p\cdot v \end{array}\right\} \tag{4.7}$$

and

under the same boost. ((4.7) is just (3.10) with $R = 1$.) Hence it follows that $p\cdot x - Et$ is invariant under homogeneous Galilei transformations, *where p and E are now given by* (4.5). That is, they have the meaning of momentum and energy differences between two states of a massive particle.

The factorization (4.4) permits the "process" shown in Figure 1. One may regard it as the emission (or absorption) of energy and momentum without charge of mass by a massive particle.

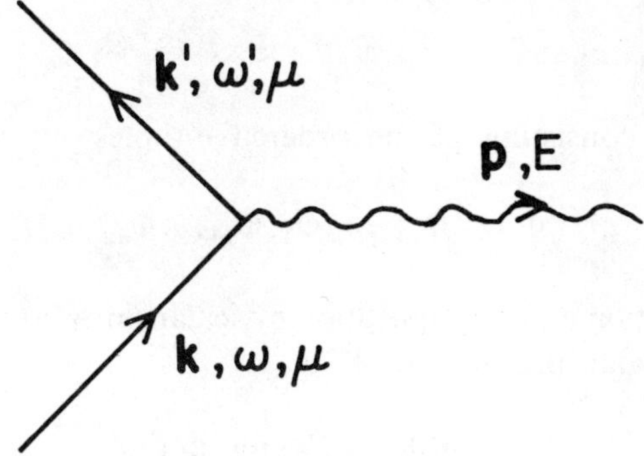

Figure 1

Figure 1 is consistent with Galilei invariance, because representations of $\mathcal{G}$ corresponding to the wavy line exist. This has no counterpart in relativistic mechanics, and therefore the process shown in Fig. 1 is kinematically forbidden* in relativistic mechanics.

---

* It appears possible, however, to generate zero-energy representations (the little group is the entire homogeneous Lorentz group) of the Poincaré group by an analogue of Figure 1. Take a spinning particle, set $k = k'$ and $\omega = \omega'$, and consider *different* spin projections in the initial and final states. The wavy line carries only the "spin defect" [12].

V. MATHEMATICAL STRUCTURE OF LANDAU EXCITATIONS

In this section we shall attempt to formulate the problems involved in the mathematical structure of Landau excitations in superfluid Helium. In order to do this, we shall first recapitulate the notion of a direct integral of Hilbert spaces. Then we shall construct a direct integral of the P-representations. Next, we shall choose a particular subspace $\mathcal{H}_0$ of the direct integral space $\mathcal{H}$ and construct certain subspaces $\mathcal{H}_v$ by applying the boosts (v) to $\mathcal{H}_0$. We will then look into the structure of the union of all the $\mathcal{H}_v$:

$$S = \bigcup_v \mathcal{H}_v , \tag{5.1}$$

in which the boosts act *nonlinearly*. It will turn out that in *one dimension* the mathematical structure is that which is called a vector bundle, based on the boosts.

a) Direct Sums and Integrals

We recall the definition of the direct sum of a finite number of Hilbert spaces $\mathcal{H}_i, i = 1, 2, \ldots n$. Let Φ_i, Φ_i' be vectors and (Φ_i, Φ_i') the scalar product in $\mathcal{H}_i$. Then the *direct sum*

$$\mathcal{H} = \mathcal{H}_1 \oplus \mathcal{H}_2 \oplus \ldots \oplus \mathcal{H}_n \tag{5.2}$$

is the vector space consisting of the ordered n-tuples

$$\Phi = \{\Phi_i\} = \{\Phi_1, \Phi_2, \ldots \Phi_n\}, \tag{5.3}$$

where $\Phi_j \in \mathcal{H}_j$. Addition and multiplication by scalars in $\mathcal{H}$ are defined in an obvious manner, and the scalar product in $\mathcal{H}$ is

$$(\Phi, \Phi') = \sum_{i=1}^{n} (\Phi_i, \Phi_i'). \tag{5.4}$$

If we now wish to let $n \to \infty$, the straightforward generalization of (5.3) may lead to quantities with infinite norm. We therefore agree to consider (for $n = \infty$) as vectors in $\mathcal{H}$ only those sequences $\Phi = \{\Phi_1, \Phi_2, \ldots\}$ which satisfy the convergence condition

$$\|\Phi\|^2 = \sum_{i=1}^{\infty} \|\Phi_i\|^2 < \infty. \tag{5.5}$$

The generalization from direct sum to direct integral is intuitively obvious. Let $\{\mathcal{H}(t)\}$ be a family of Hilbert spaces parametrized by a continuous parameter t in the interval (a, b), and let $\mu(t)$ be a measure on this interval. Let $\phi(t), \psi(t)$ be vectors in $\mathcal{H}(t)$ and $(\phi(t), \psi(t))$ the scalar product in it. Then the *direct integral*

$$\mathcal{H} = \int^{\oplus} \mathcal{H}(t)\,d\mu(t) \tag{5.6}$$

is the space of (normalizable) vectors Φ which have components $\phi(t)$ in $\mathcal{H}(t)$, and the scalar product in $\mathcal{H}$ is

$$(\phi, \psi) = \int_b^a (\phi(t), \psi(t))\,d\mu(t). \tag{5.7}$$

We are interested in the direct integral of the P-representations. Our measure $\mu(P)$ will be the ordinary Lebesgue measure dP on the interval $0 < P < \infty$. This choice is dictated by the requirement of localizability [13].

b) Construction of the Spaces $\mathcal{H}_v$

We denote by $\mathcal{H}$ the direct integral

$$\mathcal{H} = \int^{\oplus} \mathcal{H}(P)\,dP$$

of the Hilbert spaces $\mathcal{H}(P)$ which carry the P-representations of $\mathcal{G}$, with the measure $d\mu(P) = dP$. In terms of p, $dP = p^2\,dp$, where $p = |\mathbf{p}|$. Hence if $\psi(\mathbf{p}, E)$ denotes a state in $\mathcal{H}$

$$\|\psi\| = \int p^2\,dp \left[\int_P d\Omega\,dE\,|\psi(\mathbf{p}, E)|^2 \right]$$

$$= \int |\psi(\mathbf{p}, E)|^2\,d^3p\,dE,$$

where the quantity in square brackets in the first line is the norm in the P-representation, the $d\Omega$ integration extending over the *surface* of the sphere $p^2 = P^2$.

The space $\mathcal{H}_0$ is constructed as follows: Take a real-valued function

$$E = E(p) \tag{5.8}$$

of p which is sufficiently smooth. Choose from the P-representation the state with a given $\mathbf{p}$ and the value of E assigned to it by (5.8). The linear superposition of all such states is a subspace of $\mathcal{H}$. Apply the rotation operators $(\mathbf{R})$ to this subspace to obtain a larger subspace, which is invariant under rotations. This is the subspace $\mathcal{H}_0$. It contains all $\mathbf{p}$-values, but E is defined uniquely for each $\mathbf{p}$ by (5.8). $\mathcal{H}_0$ is rotation-invariant by construction, and is trivially invariant under space and time translations.

The space $\mathcal{H}_v$ is defined as follows: It is the set of all vectors of the form $(\mathbf{v})\,\phi$, where $\phi \in \mathcal{H}_0$. $\mathcal{H}_v$ is evidently a linear vector space which is invariant under space and time translations, but not under rotations:

$$\text{(R)} \quad \mathscr{H}_v = \mathscr{H}_{Rv},$$
(5.9)

which is what one would expect.

c) The Problem of Landau Excitations

Consider, first of all, the one-dimensional case. Then the rotations do not exist, and a relation $E = E(p)$ implies that $E + pv$ is a monotonically increasing function of v for any p. Hence the subspaces $\mathscr{H}_v$ have no vectors in common; the boosts v partition the total space $\mathscr{H}$ into mutually disjoint subspaces $\mathscr{H}_v$. Furthermore, a boost u maps the subspace $\mathscr{H}_v$ one-to-one onto the subspace $\mathscr{H}_{u+v}$, and each subspace $\mathscr{H}_v$ is invariant under space and time translations. The "space of boosts" v is the real line $-\infty < v < \infty$, with the topology of the real line.

Such a structure is called a *vector bundle*. Abstractly, a vector bundle is a triplet consisting of: (a) a *base* space B, here the space of boosts, (b) a *total* space F, here $\mathscr{H}$ and (c) a many-one mapping π from the total space F onto the base space B, such that the set of points $\pi^{-1}(u)$ of F which are mapped onto the *same* point u of B is a vector space. No meaning is ascribed, in general, to $\pi^{-1}(u) + \pi^{-1}(v)$, where u and v are different. $\pi^{-1}(u)$ is called the "fiber" at u [14].

A group action on a vector bundle is clearly linear as long as it stays within a fiber; clearly nonlinear as soon as it maps one fiber onto another. Hence the vector bundle concept is the "canonical" one for describing a mixed—linear and non-linear—action of groups.

In three dimensions, the quantity $E(p) + \boldsymbol{p} \cdot \boldsymbol{v}$ does *not* define a $\boldsymbol{v}$ uniquely for given E and $\boldsymbol{p}$, and therefore the spaces $\mathscr{H}_v$ constructed earlier have elements in common. Therefore the simplicity of the one-dimensional structure is lost.

VI. CONCLUDING REMARK

It appears that the fiber bundle concept is a useful one for describing mixed linear and nonlinear group action. The list of situations where the need for such a mixed description arises is endless. Finally, I think that some of the rather confusing ideas about "broken symmetries" can be understood much more clearly in terms of non-linear group action on a topological space.

NOTES AND REFERENCES

[1] This observation was first made by D. Zahavi. See R. N. Sen and D. Zahavi, *Physica,* **59**, 379 (1972).

[2] Compare formulas (8.5) and (8.6), pp. 71–72, in "An Introduction to Liquid Helium", by J. Wilks, Clarendon Press, Oxford (1970), with equation (IV.1) *et seq.,* p. 783, in J. M. Levy–Leblond, *J. Math. Phys.,* **4**, 776 (1963).

[3] E. Inönü and E. P. Wigner, *Nuovo cimento,* **IX**, 706 (1952).

[4] A. S. Wightman, *Rev. Mod. Phys.,* **34**, 845 (1962).

[5] L. Landau, *Journal of Physics,* **XI**, 91 (1947); reprinted in "An Introduction to the Theory of Superfluidity", by I. M. Khalatnikov (Translated from the Russian by P. C. Hohenberg), W. A. Benjamin, New York–Amsterdam (1965).
The Landau form has been confirmed directly by neutron scattering experiments at $1.1°K$ and above.
For the most recent neutron diffraction data, see R. A. Cowley and A. D. B. Woods, *Can. J. Phys.,* **49**, 177 (1971).

[6] Wigner's fundamental paper appeared in *Ann. Math.,* **40**, 149 (1939). It is reprinted in "Symmetry Groups in Nuclear and Particle Physics", edited by F. J. Dyson, W. A. Benjamin, New York–Amsterdam (1966).

[7] For the physicist, we recommend "Induced Representations of Groups and Quantum Mechanics", by G. W. Mackey, Benjamin–Boringhieri, New York–Amsterdam–Torino (1968), where further references are given. The mathematically inclined reader may consult the mimeographed notes of Mackey's 1971 lectures on "Infinite Dimensional Group Representations and their Applications", Forschungsinstitut für Mathematik, ETH, Zürich (Spring, 1971).

[8] Note that, in the first instance, we are using only the continuity properties of this space. See the contribution by L. Michel in this volume.

[9] I believe that this terminology, as well as the relevant work in this direction, is due to A. Grossmann. See the footnote on p. 42 of P. Moussa and R. Stora, "Some Remarks on the Product of Irreducible Representations of the Inhomogeneous Lorentz Group", in "Lectures in Theoretical Physics", Vol. VII A: "Lorentz Group", edited by W. E. Brittin and A. O. Barut, The University of Colorado Press, Boulder (1965).

[10] This is the method followed by Levy–Leblond, ref. [2]. Inönü and Wigner (ref. [3]) use the six-parameter group.

[11] The results of L. Pontrjagin quoted in this section may be found in Chapter V of his book, "Topological Groups", Translated from the Russian by Emma Lehmer, Princeton University Press (1939).

[12] This term, as far as I know, is due to H. Bacry.

[13] The handwaving discussion given in ref. [1] can be misleading; if one takes it too literally, one would be able to construct states which are localizable in time! The reader is advised to consult Wightman, ref. [4].

[14] For a first encounter with vector bundles, consult Chapter 9 of "Lie Groups for Physicists", by Robert Hermann, W. A. Benjamin, New York–Amsterdam (1966). The geometrical background may be found in "Introduction to Differentiable Manifolds", by L. Auslander and R. MacKenzie, McGraw–Hill, New York (1963).

Note added in proof: The space $\mathcal{H}_0$ defined above consists of vectors which are δ-normalized with respect to energy. To obtain proper vectors in $\mathcal{H}$ one has to introduce a finite energy width for the Landau excitations. Such widths are traditionally interpreted as instability.

On the Ising Model with Long Range Interaction. III: A Rigorous Lower Bound of the Free Energy*

A. J. F. SIEGERT

Physics Department, Northwestern University, Evanston (Illinois)

Abstract

A rigorous lower bound of the free energy of a wide class of Ising models is obtained, to supplement the known upper bound. The bounds are evaluated for the case of weak long-range interaction.

I. INTRODUCTION

In the Weiss theory of ferromagnetism the interaction between the elementary magnets is replaced by a properly chosen spatially homogeneous field. It is known [1] that the results of the Weiss theory applied to Ising models become exact in the limit of infinitely weak ferromagnetic interaction of infinite range, provided that the thermodynamic limit is taken first, and the reciprocal range and the strength of the interaction both approach zero so as to keep the sum of the interaction potentials acting on any one elementary magnet finite.

Even before this result had been rigorously proved, Brout [2] started a program of finding correction terms to the Weiss theory for small values of a parameter γ, defined so that the interaction energy of a pair of spins is proportional to γ, and the range of interaction is $\gamma^{-1/D}$, where D is the dimensionality of the lattice. Originally it was hoped that a power series in γ would yield information about the phase transition of the model, since it starts at $\gamma = 0$ with the Weiss theory, which shows a phase transition. However, by pointing out a discrepancy in his first-order results, Brout showed that this hope is not realized, and proposed a modification of this expansion. Since then, several authors have worked on this problem [3].

* Program supported by National Science Foundation Grant and Office of Naval Research Contract.

The complete expansion of the free energy in a series, in which the dominant order in γ of each term is easily obtained, was given in part I of the present paper [4] for any fixed temperature $T > T_W$, where T_W is the Curie temperature of the Weiss theory. Many terms in this series become infinite when T approaches T_W from above. This expansion was obtained by a variant of the method of random fields. A further modification of this expansion avoids the obvious failure of the expansion in a "critical region" of temperatures near T_W and reproduces as a first approximation the results of prior authors. In part II [5] it was shown that these results can be obtained by a resummation of the most divergent terms of the original expansion, if a certain integral diverges no worse than logarithmically at T_W. It was also shown that the terms not included in the resummed part are, in the critical region, individually of higher order than the resummed part.

Convergence properties of these expansions are known only for the one-dimensional model with exponential interaction, which has been solved exactly [3]. In section II of the present paper we obtain a rigorous lower bound of the free energy for a wide class of models to supplement the known [6] upper bound. The evaluation of the bounds was carried out for a somewhat more restricted class, defined in section III and referred to as "typical" models.

For any fixed temperature $T > T_W$, the bounds agree through first order in γ with Brout's [2] first-order term in the unmodified γ-expansion, and the difference between the bounds is of order γ^2. In the critical region, i.e. for $|T - T_W|$ of order $\gamma^{2/3}$, $\gamma \log(1/\gamma)$, and γ for the typical one-, two- and three-dimensional models, respectively, the difference between the bounds approaches zero as the square of the critical region. This difference is, however, not sufficiently small to justify rigorously the result of the modified γ-expansion.

The proper treatment of the thermodynamic limit shows that the bounds for the typical one- and two-dimensional models are analytic functions of the temperature, while the bounds for the typical three-dimensional model break into two analytic parts at temperatures below T_W in the critical region. Both bounds have, however, continuous second derivatives, and the temperatures at which the singularities occur in the two bounds differ by $O(\gamma)$.

II. DERIVATION OF THE LOWER AND UPPER BOUND

As in parts I and II of this paper, we consider an Ising model of n spins located at the points of a lattice in D dimensions. The energy of a pair of spins located at sites k and l is assumed to be $-J\gamma\rho_{kl}(\gamma)\mu_k\mu_l$, where J is a coupling parameter, assumed to be positive, γ the reciprocal of an effective number of neighbors, and μ_k, μ_l are spin variables at sites k, l and assume the values ± 1. With

$$v = J/kT \tag{2.1}$$

the partition function is

$$Q_n = \sum_{\{\mu\}} \exp\left(\tfrac{1}{2}v\gamma \sideset{}{'}\sum_{\substack{k,l \\ k \neq l}} \rho_{kl}\mu_k\mu_l\right).$$ (2.2)

To represent a purely ferromagnetic interaction it is assumed that the matrix elements ρ_{kl} are positive. It is assumed that ρ_{kl} depend only on the vectorial distance $k - l$ between lattice sites k and l, and ρ_{k-l} is used interchangeably with ρ_{kl}. It is assumed that

$$\rho_k = \rho_{-k}$$ (2.3)

and that the matrix with elements ρ_{kl} is positive definite. The diagonal elements $\rho_{kk} = \rho_0$ do not occur in the physical problem, and are specified to be equal to unity for convenience. The lattice is assumed to be toroidally connected to itself, and corresponding periodicity conditions are imposed on ρ_k. In going to the limit of an infinite lattice, it is assumed that

$$\lim_{n \to \infty} \sum_k \gamma\rho_k(\gamma) = C(\gamma)$$ (2.4)

exists, and that

$$\lim_{\gamma \to 0} C(\gamma) = C$$ (2.5)

exists.

As shown in part I, we can then write the partition function in the form

$$Q_n = 2^n e^{-nv\gamma/2}(2\pi)^{-n/2}[\det \rho]^{-1/2} \int_{-\infty}^{\infty} \cdots \int_{-\infty}^{\infty} \exp\left[-\tfrac{1}{2}\sum_{kl} x_k(\rho^{-1})_{kl}x_l\right] \times$$

$$\times \prod_k \cosh(\sqrt{v\gamma}x_k)\,dx_k.$$ (2.6)

Adding and subtracting $(v\gamma/2\alpha_1)\sum x_k^2$ in the exponent, one has

$$Q_n = 2^n e^{-nv\gamma/2}(2\pi)^{-n/2}[\det \rho]^{-1/2} \times$$

$$\times \int_{-\infty}^{\infty} \cdots \int_{-\infty}^{\infty} \exp\left[-\tfrac{1}{2}\sum x_k(\rho^{-1} - (v\gamma/\alpha_1)I)_{kl}\,x_l\right] \prod_k \cosh(\sqrt{v\gamma}x_k) \times$$

$$\times \exp\left[-(v\gamma/2\alpha_1)x_k^2\right]dx_k,$$ (2.7)

where I denotes the unit matrix.

Choosing a value of α_1 such that $(\rho^{-1} - (v\gamma/\alpha_1)I)$ is a positive definite matrix, one obtains the upper bound

$$Q_n \leqq 2^n e^{-nv\gamma/2}[\det \rho]^{-1/2}[\det(\rho^{-1} - (v\gamma/\alpha_1)I)]^{-1/2} \times$$

$$\times \max_x\left[\cosh(\sqrt{v\gamma}x)\exp(-(v\gamma/2\alpha_1)x^2)\right]^n$$

or

$$\frac{1}{n}\log Q_n \leqq \log 2 - \frac{v\gamma}{2} - \frac{1}{2n}\log \det\left(I - \frac{v\gamma}{\alpha_1}\rho\right) +$$
$$+ \max_y \left[\log \cosh y - y^2/2\alpha_1\right]. \qquad (2.9)$$

The largest value of $\log \cosh y - y^2/2\alpha_1$ occurs for $|y| = y(\alpha_1)$ where $y(\alpha_1)$ is the largest root of the equation

$$\tanh y = y/\alpha_1, \qquad (2.10)$$

so that

$$\frac{1}{n}\log Q_n \leqq \log 2 - \frac{v\gamma}{2} - \frac{1}{2n}\log \det\left(I - \frac{v\gamma}{\alpha_1}\rho\right) +$$
$$+ \log \cosh y(\alpha_1) - \tfrac{1}{2}y(\alpha_1)\tanh y(\alpha_1). \qquad (2.11)$$

The free energy ψ per spin is defined by

$$\psi = -\frac{kT}{n}\log Q_n \qquad (2.12)$$

and one has thus a lower bound for the free energy.

An upper bound for the free energy, i.e. a lower bound for Q_n is known [6]. It is easily obtained from the Gibbs inequality [7] for any α for which $I - (v\gamma/\alpha)\rho$ is positive definite, and is given by

$$\frac{1}{n}\log Q_n \geqq \log 2 - \frac{v\gamma}{2} - \frac{1}{2n}\log \det\left(I - \frac{v\gamma}{\alpha}\rho\right) +$$
$$+ (2\pi v\gamma\tilde{\rho}_0)^{-1/2} \int_{-\infty}^{\infty} \exp\left(-y^2/2v\gamma\tilde{\rho}_0\right)\left[\log \cosh y - \frac{y^2}{2\alpha}\right] dy, \qquad (2.13)$$

where

$$\tilde{\rho}_0 \equiv \tilde{\rho}_{kk}(v\gamma/\alpha) \qquad (2.14)$$

and the matrix $\tilde{\rho}(t)$ is defined as

$$\tilde{\rho}(t) = \rho/(1 - t\rho). \qquad (2.15)$$

From the assumed properties of ρ it follows that $\tilde{\rho}_{kl}$ depends only on the vectorial difference $k - l$.

For the purpose of the present paper it is sufficient to use the weaker inequality

$$\frac{1}{n}\log Q_n \geqq \log 2 - \frac{v\gamma}{2} - \frac{1}{2n}\log \det\left(I - \frac{v\gamma}{\alpha}\rho\right) +$$
$$+ \frac{v\gamma}{2}\tilde{\rho}_0\left(\frac{v\gamma}{\alpha}\right)\left(1 - \frac{1}{\alpha}\right) - \frac{1}{4}\left[v\gamma\tilde{\rho}_0\left(\frac{v\gamma}{\alpha}\right)\right]^2. \qquad (2.16)$$

III. OPTIMAL CHOICE OF PARAMETERS AND THE THERMODYNAMIC LIMIT

The lowest upper bound and the largest lower bound are obtained if $\alpha_1^{-1} = \tanh y/y$ is determined by

$$y^2 = v\gamma\tilde{\rho}_0(v\gamma \tanh y/y) \tag{3.1}$$

and α by

$$1 - \alpha^{-1} = v\gamma\tilde{\rho}_0\left(\frac{v\gamma}{\alpha}\right). \tag{3.2}$$

We are interested in the infinite model with interaction matrix $\rho_{kl}(\infty)$ and choose a sequence of finite models with periodic interaction as specified in reference [4, sec. 2] to approach the limit. We assume symmetry and homogeneity for $\rho_{kl}(\infty)$, so that $\rho_{k-l}(\infty)$ can be used interchangeably with $\rho_{kl}(\infty)$. We define $g(\omega)$ by

$$g(\omega) = \sum_k e^{ik\cdot\omega}\rho_k(\infty), \tag{3.3}$$

where ω is a vector in D dimensions, and the sum extends over all points of the infinite lattice. In order to simplify the discussion we will restrict the further exposition to "typical" models, for which

$$g^{-1}(\omega) = g^{-1}(0)\left[1 + a\gamma^{-2/D}\omega^2 + O(\omega^4)\right] \tag{3.4}$$

for small ω.* We further restrict the discussion to the one-dimensional lattice of equidistant points and the square and cubic lattice, with the lattice constant chosen as unit of length, and we let n be m^D, where m is an integer. The eigenvalues g_τ of ρ for finite n are then given by

$$g_\tau = g\left(\frac{2\pi\tau}{m}\right)\Big/ \sum_s \rho_{sm}, \tag{3.5}$$

where τ is a vector whose components assume integer values from zero to $m - 1$, and the sum extends over all vectors s whose components are positive or negative integers or zero.

From eqn. (2.15) one obtains

$$\tilde{\rho}_0(t) = \frac{1}{n}\sum_\tau \frac{g_\tau}{1 - tg_\tau} \tag{3.6}$$

and the further treatment follows closely the theory of the system of noninteracting Bosons [8].

---

* This excludes interactions for which $\sum k^2\rho_k(\infty)$ diverges, but this is less restrictive than it appears, since some "atypical" models will be seen to be essentially similar to "typical" models of higher dimensionality.

For any fixed $t < g(0)^{-1}$, we have

$$\lim_{n \to \infty} \tilde{\rho}_0(t) = (2\pi)^{-D} \int_{-\pi}^{\pi} \cdots \int_{-\pi}^{\pi} \frac{g(\omega)\, d^D\omega}{1 - tg(\omega)} \equiv R(t). \qquad (3.7)$$

For the "typical" one- and two-dimensional models $R(t) \to \infty$, when $t \to g(0)^{-1}$ from below. The limit $n \to \infty$ is then taken by replacing $\tilde{\rho}_0$ by R in eqns. (3.1) and (3.2), and therefore also in the inequalities (2.11) and (2.16).

For the "typical" three-dimensional model, however, $R(t)$ approaches a finite value $R(g(0)^{-1})$ when $t \to g(0)^{-1}$, and the inverse function $t(R)$ becomes a constant for $n \to \infty$ when $R \geqq R(g(0)^{-1})$. There is thus a value v'_c of v, such that for $v > v'_c$, $v\gamma/\alpha = g(0)^{-1}$, or

$$\lim_{n \to \infty} \alpha = v\gamma g(0) = v/v_W, \qquad (3.8)$$

since

$$v_W\gamma g(0) = 1, \qquad (3.9)$$

where v_W is defined by the Curie temperature T_W of the Weiss theory through eqn. (2.1). The value of v'_c is determined by

$$1 - \frac{v_W}{v'_c} = v'_c\gamma R(\gamma v_W) \qquad (3.10)$$

or to first order in γ by

$$v'_c \simeq v_W\left[1 + v_W\gamma R(\gamma v_W)\right]. \qquad (3.11)$$

The corresponding value v''_c of v for the other bound is obtained by the same argument, and one has in the limit $n \to \infty$ for $v \geqq v''_c$

$$\frac{\tanh y}{y} = \frac{v_W}{v}, \qquad (3.12)$$

with v''_c determined by

$$y^2\big|_{v=v''_c} = v''_c\gamma R(\gamma v_W) \qquad (3.13)$$

or to order γ by

$$v''_c \simeq v_W\left[1 + \tfrac{1}{3}v_W\gamma R(\gamma v_W)\right]. \qquad (3.14)$$

The third terms in the inequalities (2.11) and (2.16) can be written in the form

$$-\frac{1}{2n}\log \det(1 - \lambda\rho) = \frac{1}{2}\int_0^\lambda \tilde{\rho}_0(t)\, dt, \qquad (3.15)$$

with $\lambda = v\gamma \tanh y/y$ and $\lambda = v\gamma/\alpha$ respectively. In these terms $\tilde{\rho}_0(t)$ is replaced by $R(t)$ in the limit $n \to \infty$.

IV. SUMMARY AND DISCUSSION

Collecting these results, one has for the typical one- and two-dimensional models

$$\lim_{n \to \infty} \frac{1}{n} \log Q_n \geqq \log 2 - \frac{v\gamma}{2} + \frac{1}{2} \int_0^{v\gamma/\alpha} R(t)\, dt + \frac{v^2\gamma^2}{4} R^2\left(\frac{v\gamma}{\alpha}\right) \qquad (4.1)$$

$$\lim_{n \to \infty} \frac{1}{n} \log Q_n \leqq \log 2 - \frac{v\gamma}{2} + \frac{1}{2} \int_0^{v\gamma \tanh y / y} R(t)\, dt + \log \cosh y - \frac{1}{2} y \tanh y, \quad (4.2)$$

with

$$1 - \alpha^{-1} = v\gamma R\left(\frac{v\gamma}{\alpha}\right) \qquad (4.3)$$

and

$$y^2 = v\gamma R\left(\frac{v\gamma \tanh y}{y}\right), \qquad (4.4)$$

where R is defined by eqn. (3.7).

For the typical three-dimensional case, these results are valid for $v \leqq v_c'$ for the lower bound and for $v \leqq v_c''$ for the upper bound, where v_c' and v_c'' are determined by eqns. (3.10–14), and $v_c' \geqq v_c'' \geqq v_W$. For $v \geqq v_c'$

$$\lim_{n \to \infty} \frac{1}{n} \log Q_n \geqq \log 2 - \frac{v\gamma}{2} + \frac{1}{2} \int_0^{\gamma v_W} R(t)\, dt + \frac{1}{4}\left(1 - \frac{v_W}{v}\right)^2, \qquad (4.5)$$

and for $v \geqq v_c''$

$$\lim_{n \to \infty} \frac{1}{n} \log Q_n \leqq \log 2 - \frac{v\gamma}{2} + \frac{1}{2} \int_0^{\gamma v_W} R(t)\, dt + \log \cosh y - \frac{1}{2} y \tanh y, \quad (4.6)$$

with y determined by

$$\frac{\tanh y}{y} = \frac{v_W}{v}. \qquad (4.7)$$

Since $R(t)$ and its derivatives are positive for $t < \gamma v_W$, one can show that it follows from eqns. (4.3) and (4.4) that

$$\frac{\tanh y}{y} \geqq \alpha^{-1}. \qquad (4.8)$$

From (4.2) then follows *a fortiori*

$$\lim_{n \to \infty} \frac{1}{n} \log Q_n \leqq \log 2 - \frac{v\gamma}{2} + \frac{1}{2} \int_0^{v\gamma/\alpha} R(t)\, dt + \frac{1}{2} v\gamma R\left(\frac{v\gamma}{\alpha}\right) \cdot v\gamma R\left(\frac{v\gamma \tanh y}{y}\right) \quad (4.9)$$

For any fixed temperature $T > T_W$, i.e., if $T - T_W \geqq \varepsilon > 0$, with ε independent of γ, one therefore has

$$\lim_{n \to \infty} \frac{1}{n} \log Q_n = \log 2 - \frac{v\gamma}{2} + \frac{1}{2} \int_0^{v\gamma} R(t)\, dt + O(\gamma^2). \qquad (4.10)$$

Brout's original correction term to the Weiss approximation is thus rigorously justified in this temperature range, and the order in γ of the next term is correctly given by the unmodified γ-expansion.

In reference [4, section 7], we had developed a modified expansion, which avoids the obvious failure of the original expansion at T_W. The leading terms of this expansion agree with the result of Brout's "sphericalization" method [2] and with the result of Mühlschlegel and Zittartz [6], and others [3].

In reference [5] we showed that this result is obtained for the critical region by resumming the most divergent terms of the unmodified γ-expansion, provided that the function $R(t)$ defined by eqn. (3.7) remains finite or diverges no worse than logarithmically when t approaches $g(0)^{-1}$ from below. We then showed that the terms not included in the resummed result are *individually* of higher order in γ than the resummed result when $T > T_W$ in the critical region.

The critical region in the present paper is given by

$$\left| 1 - \frac{T}{T_W} \right| = O(\gamma^{2/3}), \qquad O\left(\gamma \log \frac{1}{\gamma} \right) \quad \text{and} \quad O(\gamma)$$

for the typical one-, two-, and three-dimensional model respectively, and the difference between the upper and lower bound is of order $\gamma^{4/3}$, $(\gamma \log(1/\gamma))^2$, and γ^2, respectively, in the critical region. These bounds are thus not yet strong enough to make the results of reference [5] rigorous.

The investigation of the thermodynamic limit, resulting in the inequality (4.5), indicates, however, the proper continuation of eqn. (4.10) of reference [5], which becomes meaningless for $v > v_c'$.

For the typical three-dimensional model, the proper treatment of the thermodynamic limit shows that the upper and lower bounds have a singularity at v_c' and v_c'' (given by eqns. (3.10–14) respectively), but the second derivatives with respect to v are still continuous.

Acknowledgements

It is a pleasure to thank Mrs. Maria E. V. Koelling for checking some of the calculations and Professor Colin J. Thompson for a valuable discussion.

BIBLIOGRAPHY

[1] J. L. Lebowitz and O. Penrose, *J. Math. Phys.*, **5**, 75 (1966).

[2] R. Brout, *Phys. Rev.*, **118**, 1009 (1960).

[3] Papers of these authors are listed and discussed in [4] and [5].

[4] A. J. F. Siegert and D. J. Vezzetti, *J. Math. Phys.*, **9**, 2173 (1968).

[5] C. J. Thompson, A. J. F. Siegert, and D. J. Vezzetti, *J. Math. Phys.*, **11**, 1018 (1970).

[6] B. Mühlschlegel and H. Zittartz, *Z. Physik*, **175**, 553 (1963).

[7] R. C. Tolman, "The Principles of Statistical Mechanics", Chap. VI, Section 51, eq. (51.13). Oxford University Press (1938).

[8] Kerson Huang, "Statistical Mechanics", Section 12.3, see especially Fig. 12.4, p. 264, Wiley and Sons, New York and London (1963).

Studying Spatially Cutoff $(\varphi^{2n})_2$ Hamiltonians

BARRY SIMON

*Departments of Mathematics and Physics,
Princeton University, Princeton (New Jersey)*

Abstract

We review here various methods of studying the properties of Hamiltonians of spatially cutoff two-dimensional self-coupled Boson fields (the so-called $P(\varphi)_2$ model). We discuss semiboundedness, selfadjointness and spectral properties. We do not discuss the methods for removing the cutoff.

0. INTRODUCTION

In this note we want to review the work that has been done on studying the operator $H_0 + V$, where H_0 is the Hamiltonian of a two-dimensional free Boson field of mass $m_0 > 0$ and $V = \int g(x) : P[\varphi(x)] : dx$; here $P(X)$ is a polynomial which is bounded below, $\varphi(x)$ is the free field, and $:\ :$ is Wick ordering (definitions in Section 1). Furthermore, $g \geqq 0$ and $g \in L^1 \cap L^2$. In the case $P(X) = X^4$, this theory was first developed by Glimm and Jaffe [10, 11, 12] who relied heavily on an important technical device of Nelson [33]. The first comprehensive results on general P are due independently to Rosen [34] and Segal [39]. The past two years have seen extensive study of the $P(\varphi)$ Hamiltonian in order to better understand and to simplify the proofs of Glimm and Jaffe. It is this work we will review.

We restrict ourselves to studying the spatial cutoff Hamiltonian and do not consider questions of the infinite volume limit. This is done for several reasons: (i) There exists an excellent review of the ideas and results about the infinite volume limit [13]. (ii) Most of the recent results involve the spatially cutoff theory. (iii) We feel that the spatially cutoff Hamiltonian is an interesting mathematical object and hope to interest mathematicians in its study. Some of the questions we will raise are not relevant to the infinite volume limit and are thus not relevant to the physics, but I think they are mathematically interesting, since they involve "differential operators in a continuous infinity of variables".

Thus we are not discussing:

1) Time automorphisms in the infinite volume limit (Guenin [22], Segal [40], Glimm and Jaffe [10]).

2) Volume lower bounds on the ground state energy (Glimm and Jaffe [12], Simon [44]).

3) Complex coupling constant behavior and asymptotic series (Simon and Hoegh-Krohn [48], Simon [45], Rosen and Simon [38]).

4) Properties of the Physical Vacuum (Glimm and Jaffe [12, 16, 17]).

5) Lorentz Automorphisms (Cannon and Jaffe [2], Rosen [37]).

6) Other models on which there are results (Glimm and Jaffe [14, 15], Hepp [23]).

In Section 1, we discuss Fock space in an abstract manner convenient for our purpose and introduce the "Q-space" representation of Fock space. In Sections 2 and 4 we discuss the main tools for studying spatially cutoff theories. In Section 3 we discuss lower-boundedness and in Section 5 selfadjointness of $H = H_0 + V$. In Section 6 we discuss what is known of the spectral properties of H.

1. FOCK SPACE AND Q-SPACE

We first write the Fock space construction familiar for the free field in a general mathematical setting [3, 41]. This setting clarifies exactly what is going on. Let $\mathcal{H}$ be a complex Hilbert space with a distinguished complex conjugation C. That is, we pick out an antilinear map C with $C^2 = 1$. The Fock space $\mathcal{F}$ over $\mathcal{H}$ is the symmetric tensor algebra over $\mathcal{H}$; that is $\mathcal{F} = \mathcal{F}_0 \oplus \mathcal{F}_1 \oplus \dots \oplus \mathcal{F}_n \oplus \dots$ where $\mathcal{F}_0 = \mathbb{C}$, $\mathcal{F}_1 = \mathcal{H}$, $\mathcal{F}_2 = \mathcal{H} \otimes_s \mathcal{H}, \dots$ where $\otimes_s$ denotes the symmetric tensor product. If $\mathcal{H} = L^2(\mathbb{R})$, then $\mathcal{F}_n = L_s^2(\mathbb{R}^n)$, the symmetric functions on $\mathbb{R}^n$. If A is an operator on $\mathcal{H}$, there is a natural operator $\Gamma(A)$ induced on $\mathcal{F}$ by $\Gamma(A) \restriction \mathcal{F}_n = A \otimes \dots \otimes A$. If U is unitary, so is $\Gamma(U)$, so Γ has a differential on the selfadjoint operator given by $\Gamma(e^{itA}) = e^{it\, d\Gamma(A)}$. By Leibnitz' rule, $d\Gamma(A) \restriction \mathcal{F}_n = = A \otimes 1 \otimes \dots \otimes 1 + 1 \otimes A \otimes \dots \otimes 1 + \dots + 1 \otimes \dots \otimes 1 \otimes A$. In particular, $d\Gamma(1) = N$, the number operator. If $\mathcal{H} = L^2(\mathbb{R}, dk)$ and $W: \mathcal{H} \to \mathcal{H}$ by $(Wf)(k) = = \omega(k)f(k)$, then $d\Gamma(W)$ is just the object one usually writes as $d\Gamma(W) = \int \omega(k) \cdot a^*(k)\, a(k)\, dk$, i.e.

$$(Wf)_n (k_1 \dots k_n) = \left[\sum_{i=1}^{n} \omega(k_i) \right] f(k_1 \dots k_n) \, .$$

One also has creation, annihilation and field operators in the abstract Fock space. Let $h \in \mathcal{H}$. Define $a^\dagger(h) = \sqrt{N} h \otimes \cdot$, i.e. if $\psi \in \mathcal{F}_n$ then $a^\dagger(h)\psi = (n+1)\Pi_s(h \otimes \psi)$ where Π_s is the projection onto the symmetric tensors. For example, if $\mathcal{H} = L^2(\mathbb{R}, dx)$, then $(a^\dagger(h)\phi)_n (x_1 \dots x_n) = (1/\sqrt{n}) \sum_{i=1}^{n} h(x_i) \phi(x_1, \dots \hat{x}_i \dots x_n)$. $a(h)$ is defined by $a(h) = [a^\dagger(Ch)]^*$ where C is the special complex conjugation on $\mathcal{H}$. Again, if $\mathcal{H} =$

$= L^2(\mathbb{R}, dx)$ and $Cf = \bar{f}$, then

$$(a(h)\psi)_n(x_1, \ldots x_n) = \sqrt{n+1} \int h(x_{n+1})\psi_{n+1}(x_1, \ldots x_{n+1})\, dx_{n+1}.$$

Of course, the reason for the $\sqrt{n}$ in the definition of $a^\dagger$ is that we obtain $[a(h), a^\dagger(g)] = \langle Ch, g \rangle \mathbf{1}$ (when applied to vectors in any $\mathscr{F}_n$). Finally, if $h \in \mathscr{H}$, and $Ch = h$, we define $\varphi(h) = 2^{-1/2}[a^\dagger(h) + a(h)]$. If we define $\varphi(h)$ as an operator on $\psi \in \mathscr{F}$ with $\psi_n = 0$ for n large, $\varphi(h)$ is selfadjoint by Nelson's analytic vector theorem and the $\{\varphi(h) \mid h \in \mathscr{H}\}$ commute in the sense that $e^{it\varphi(h)}$ and $e^{is\varphi(g)}$ commute for all t, s, h, g.

If $\mathscr{H} = \mathscr{H}_1 \oplus \mathscr{H}_2$, then $\mathscr{H} \otimes_s \mathscr{H} = \mathscr{H}_1 \otimes \mathscr{H}_1 \oplus \mathscr{H}_1 \otimes \mathscr{H}_2 \oplus \mathscr{H}_2 \otimes \mathscr{H}_2 = \mathscr{F}_2(\mathscr{H}_1) \oplus \mathscr{F}_1(\mathscr{H}_1) \otimes \mathscr{F}_1(\mathscr{H}_2) \oplus \mathscr{F}_2(\mathscr{H}_2)$ etc. In this way one gets a natural isomorphism of $\mathscr{F}(\mathscr{H}_1) \otimes \mathscr{F}(\mathscr{H}_2)$ and $\mathscr{F}(\mathscr{H})$. In a natural way, $\mathscr{F}(\mathscr{H}_1) \cong \mathscr{F}(\mathscr{H}_1) \otimes \mathscr{F}_0$ is imbedded in $\mathscr{F}(\mathscr{H})$.

In the practical case of two-dimensional field theories, we shift back and forth between two realizations of Fock space. Let $\mathscr{H}$ be $L^2(\mathbb{R}, dk)$; we think of $\mathscr{H}$ as square integrable functions on momentum space. Let $\mathscr{K}$ be the family of all functions f on $\mathbb{R}$ (thought of as x-space) with $\int |\hat{f}(k)|^2\, dk/\mu(k) < \infty$, where $\mu(k) = \sqrt{m_0^2 + k^2}$. The symbol $\hat{f}$ stands for the Fourier transform $\hat{f}(k) = (2\pi)^{-1/2} \int e^{-ikx} f(x)\, dx$. The inner product on $\mathscr{K}$ is given by $\langle f, g \rangle = \int (dk/\mu(k)) \hat{f}(k) \hat{g}(k)$. The map $F : \mathscr{K} \to \mathscr{H}$ by $(Ff)(k) = \mu(k)^{-1/2}\hat{f}(k)$ is a unitary map, $\Gamma(F) : \mathscr{F}(\mathscr{K}) \to \mathscr{F}(\mathscr{H})$. Given $f \in \mathscr{K}$, we define $(Cf)(x) = \bar{f}(x)$. Then $[(FCF^{-1})f](k) = f(-k)$. Thus the inner product is simple in $\mathscr{H}$ but the complex conjugation is simple in $\mathscr{K}$. When we write $\varphi(f)$ we will think of $f \in \mathscr{K}$ and will introduce the operator-valued distribution $\varphi(x)$ with $\int \varphi(x) f(x)\, dx = \varphi(f)$. When we write $a(g)$ we will think of $g \in \mathscr{H}$ and will introduce $a(k)$ by $\int a(k) g(k)\, dk = a(g)$. φ and a are related by

$$\varphi(x) = (4\pi)^{-1/2} \int e^{ikx}[a^\dagger(k) + a(-k)] \frac{dk}{\mu(k)^{1/2}}.$$

We identify $\mathscr{F}(\mathscr{H})$ and $\mathscr{F}(\mathscr{K})$ and call if $\mathscr{F}$.

Let μ be the map of $\mathscr{H} \to \mathscr{H}$ given by $(\mu f)(k) = \mu(k) f(k)$. Then the Hamiltonian for a free field is just $H_0 = d\Gamma(\mu)$.

For an interaction density we require the notion of Wick product. *Formally*

$$\varphi^n(x) = (4\pi)^{-n/2} \int \frac{dk_1}{\mu(k_1)^{\frac{1}{2}}} \cdots \frac{dk_n}{\mu(k_n)^{\frac{1}{2}}} e^{i(k_1 + \ldots + k_n)x} \prod_{i=1}^{n} (a^\dagger(k_i) + a(-k_i)).$$

$\varphi^n(x)$ is meaningless even if smeared in x, because an arbitrary product of a's and $a^\dagger$'s is not meaningful. The problem is that $a(k)$ is a densely defined operator but $a^\dagger(k)$ is not. Now we can define $a^\dagger(k_1) \ldots a^\dagger(k_j) a(-k_{j+1}) \ldots a(-k_n)$ as a quadratic form

$$\langle \phi, a^\dagger(k_1) \ldots a^\dagger(k_j) \ldots a(-k_n)\psi \rangle \equiv \langle a(k_j) \ldots a(k_1)\phi, a(-k_{j+1}) \ldots a(-k_n)\psi \rangle.$$

If any $a^\dagger$ came after an a we couldn't use this trick. This suggests that we define

$$:\varphi^n(x): = (4\pi)^{-n/2} \int \frac{dk_1 \ldots dk_n}{\mu(k_1)^{\frac{1}{2}} \ldots \mu(k_n)^{\frac{1}{2}}} e^{i(k_1 + \ldots + k_n)x} \times$$

$$\times \sum_{j=0}^{n} \binom{n}{j} a^\dagger(k_1) \ldots a^\dagger(k_j) a(-k_{j+1}) \ldots a(-k_n).$$

For the interaction we will take $V = \int g(x) \sum_{m=0}^{n} a_m : \phi^m(x): dx$. Here $g \in L^1 \cap L^2$; $g \geqq 0$ and n is even with $a_n > 0$. The important fact about two-dimensional space time is that V is an operator. Equivalently each $\int g(x): \varphi^n(x): dx$ is an operator.

Theorem 1.1 (Jaffe [25]; see also Glimm and Jaffe [10], Segal [42] and Simon and Hoegh-Krohn [48]):

In two-dimensional space time, $\int g(x): \varphi^n(x): dx =$

$$= \int \frac{\hat{g}(k_1 + \ldots + k_n)}{\mu(k_1)^{\frac{1}{2}} \ldots \mu(k_n)^{\frac{1}{2}}} \sum_{j=0}^{n} \binom{n}{j} a^\dagger(k_1) \ldots a^\dagger(k_j) \ldots a(-k_n) dk_1 \ldots dk_n$$

is an essentially selfadjoint operator on $F \equiv \{\psi \in \mathscr{F} | \psi_n = 0 \text{ for } n \text{ large}\}$ if $g \in L^2(\mathbb{R})$. The input of the proof is the fact that

$$\int \frac{|\hat{g}(k_1 + \ldots + k_n)|^2}{\mu(k_1) \ldots \mu(k_n)} dk_1 \ldots dk_n < \infty$$

if $g \in L^2$, and the N_τ estimates we discuss in Section 4.

We now discuss another realization of $\mathscr{F}$, first introduced by Segal [41], although its importance in constructive field theory is more recent. Returning to the abstract setting, let $\mathfrak{M}$ be the abelian von Neumann algebra generated by $\{e^{i\varphi(f)} | f \in \mathscr{H}\}$. By the standard Gelfand procedure, $\mathfrak{M}$ is isomorphically isometric to all continuous functions $C(Q)$, on a compact Hausdorff space Q. The Fock vacuum, Ω_0 induces a measure on Q; for let $\Pi: \mathfrak{M} \to C(Q)$ be the isometric isomorphism of $\mathfrak{M}$ and $C(Q)$. One obtains a positive linear functional on $C(Q)$ by $l(f) = \langle \Omega_0, \Pi^{-1}[f] \Omega_0 \rangle$ and thus a measure μ on Q with $\langle \Omega_0, A\Omega_0 \rangle = \int_Q \Pi(A) d\mu$ for each $A \in \mathfrak{M}$. Because $\mathfrak{M}$ is a von Neumann algebra, Q is a highly disconnected space; in fact, any $f \in L^\infty(Q, d\mu)$ is equal almost everywhere to a function in $C(Q)$! We thus forget about the topological structure of Q. We therefore have a map $\Pi: \mathfrak{M} \to L^\infty(Q, d\mu)$ with $\langle \Omega_0, A\Omega_0 \rangle = \int \Pi(A) d\mu$. Next, we use the fact that $\{A\Omega_0 | A \in \mathfrak{M}\}$ is dense in $\mathscr{F}$, the Fock space. For $A \in \mathfrak{M}$,

$$\|A\Omega_0\| = \langle \Omega_0, A^*A\Omega_0 \rangle = \int \Pi(A^*A) d\mu = \int |\Pi(A)(q)|^2 d\mu(q).$$

Thus $A\Omega_0 \to \Pi(A)$ defines an isometric map of a dense subset of $\mathscr{F}$ onto a dense subset of L^2. We conclude that this extends to a unitary map $U: \mathscr{F} \to L^2(Q, d\mu)$.

What have we accomplished by this construction? We have realized $\mathscr{F}$ as $L^2(Q, d\mu)$ in such a way that $e^{i\varphi(f)}$ becomes multiplication by a function on Q. In case $\mathscr{H}$ is one-dimensional, so that $N = d\Gamma(1)$ is just the harmonic oscillator, this passage from $\mathscr{F}$ to $L^2(Q)$ is almost the passage from a Hermite basis to the usual $L^2(Q, dq)$ representation. The difference is that $\mu(Q) = 1$. In case $\mathscr{H}$ is one-dimensional (up to sets of measure zero), Q is $\mathbb{R}$ and μ is the measure $\pi^{-1/2} e^{-x^2} dx$.

As one might expect, $\varphi(f)$ is an unbounded multiplication operator on Q. Since V is built out of φ's in an intuitive sense, it is not surprising that

Theorem 1.2 (essentially due to Nelson [33]; see also Segal [42] or Simon and
Hoegh-Krohn [48]):

There is a function V on Q so that the operator given by $\int dx\, g(x) : P(\phi):$ is multiplication by V. If $g \geq 0$ and $g \in L^1 \cap L^2$, then

$$\text{(a)} \quad V \in L^q(Q, d\mu) \quad \text{for any} \quad q < \infty .$$

$$\text{(b)} \quad e^{-tV} \in L^1 \quad \text{for any} \quad t > 0 .$$

That $V \in L^q$ follows from the fact that $\langle \Omega_0, V^{2n}\Omega_0 \rangle < \infty$, so that $V \in L^{2n}$ for all n. (b) is deeper and says that V is almost bounded from below. It is proven by approximating V with ultraviolet cutoff V_k, with V_k bounded below by a constant c_k, which goes to $-\infty$ as $k \to \infty$ and controlling $\|V_k - V\|_p$. For $\mu\{q \,|\, V(q) < c_k - 1\} \leq \|V_k - V\|_p^p$. For the details, see Simon and Hoegh-Krohn [48].

2. HYPERCONTRACTIVE ESTIMATES

One important technique in studying $H_0 + V$ is to employ smoothing properties of e^{-tH_0} on Q-space. These smoothing properties were discovered by Nelson [33], although not quite in the Q-space language. Nelson's ideas were developed by Glimm [8] and exploited extensively by Glimm and Jaffe [10, 11, 12]. The theorem we will shortly state was first explicitly stated by Segal [39, 43] but is implicit in Glimm [8]. First, some useful terminology (Simon and Hoegh-Krohn [48]):

Definition

Let (M, μ) be a measure space and measure with $\mu(M) = 1$. A semigroup e^{-tH_0} of selfadjoint operators on $L^2(M, \mu)$ is called hypercontractive if (i) $\|e^{-tH_0}\psi\|_p \leq \|\psi\|_p$ for all $\psi \in L^2 \cap L^p$ (and thus in L^p), and $t > 0$. (ii) $\|e^{-TH_0}\psi\|_4 \leq C\|\psi\|_2$ for some $T > 0$ and some C.

Theorem 2.1 (Nelson, Glimm, Segal):

Let $\mathscr{F}$ be an abstract Fock space and Q the associated Fock space. Let $\omega: \mathscr{H} \to \mathscr{H}$ be a selfadjoint operator on $\mathscr{H}$ with $\omega \geq c1 > 0$, and with $C\omega = \omega C$, where C is the distinguished complex conjugation. Then $e^{-t\,d\Gamma(\omega)}$ is a hypercontractive semi-group.

Our proof will be in three lemmas:

Lemma 2.1 (Glimm and Jaffe [11]).

If $\omega > 0$ and $C\omega = \omega C$, then $e^{-td\Gamma(\omega)}$ is positivity-preserving, i.e. if $f \geqq 0$ in Q-space, then $e^{-td\Gamma(\omega)}f \geqq 0$ in Q-space.

Lemma 2.2.

If (M, μ) is a probability measure space ($\mu(M) = 1$), if $T: L^2(M, \mu) \to L^2(M, \mu)$ is selfadjoint, positivity-preserving and $T1 = 1$, then $\|T\psi\|_p \leqq \|\psi\|_p$ for all p.

Lemma 2.3 (Nelson [33a]).

If $0 \leqq A \leqq c1$, with $c < 1$, then $\|\Gamma(A)\psi\|_p \leqq B\|\psi\|_2$ for some $p > 2$ and some B.

Remarks:

1. In [48] hypercontractive is defined so that $\|e^{-TH_0}\psi\|_4 \leqq \|\psi\|_2$. All consequences hold under the weaker hypothesis we gave above (and actually under even weaker hypotheses). In Theorem 2.1, the stronger contractive property can be proven but it is not particularly useful.

2. Lemma 2.3 and its proof, due to Nelson, are a considerable simplification of the usual proof. We would like to thank Professor Nelson for permission to quote it.

3. Lemma 2.3 only implies that $\Gamma(e^{-t\omega}) = e^{-td\Gamma(\omega)}$ is a bounded map from L^2 to L^p for some t and some $p > 2$. The implication that it is a bounded map from L^2 to L^4 for large t is a consequence of interpolation theorems of the type we shall discuss below.

Proof of Lemma 2.1 (Simon and Hoegh-Krohn [48]).

Given a finite dimensional subspace $V \subset \mathscr{H}$ and F with $\hat{F}$ continuous and bounded, let $\psi(F) = (\int_V F(h)\, e^{i\varphi(h)}dh)\,\Omega_0$. In terms of an orthonormal basis $h_1 \ldots h_n$ for V, $\psi(F) = (2\pi)^{n/2}\,\hat{F}(\varphi(h_1) \ldots \varphi(h_n))$. Thus $\psi(F) \geqq 0$ if and only if $\hat{F} \geqq 0$, i.e. if and only if F is a function of positive type. As V runs through all finite dimensional subspaces and F through all of C, the "coherent vectors", $\psi(F)$, span a dense subset of L^2. A simple computation proves that

$$e^{-td\Gamma(\omega)}\psi(F) = \left(\int_V F(h)\, e^{i\varphi(e^{-t\omega}h)}\, e^{-\frac{1}{2}\langle h,(1 - e^{-2t\omega})h\rangle}\, dh \right)\Omega_0.$$

If F is of positive type, so is $e^{-\frac{1}{2}\langle h, Ah\rangle}\, F$ for any positive matrix A, so that $e^{-td\Gamma(\omega)}\psi(F) \geqq 0$ if $\psi(F) \geqq 0$.

Proof of Lemma 2.2.

By use of interpolation theorems, we need only prove $\|Tf\|_1 \leqq \|f\|_1$ and $\|Tf\|_\infty \leqq$

$\leq \|f\|_\infty$. Since T is selfadjoint and $(L^1)^* = L^\infty$, we need only show that $\|Tf\|_1 \leq \|f\|_1$. Let $f \geq 0$, $f \in L^1$. Then $\|f\|_1 = \langle 1, f \rangle$. Since $Tf \geq 0$, $\|Tf\|_1 = \langle 1, Tf \rangle = \langle T1, f \rangle = \langle 1, f \rangle = \|f\|_1$. If f is real-valued, $f = f_+ - f_-$ with $f_+, f_- \geq 0$ and $\|f\|_1 = \|f_+\|_1 + \|f_-\|_1$. Thus $\|Tf\|_1 \leq \|Tf_+\|_1 + \|Tf_-\|_1 = \|f_+\|_1 + \|f_-\|_1 = 1$. In general, let $f = g + ih$ with g and h real. Since $\int_0^{2\pi} d\theta |a\cos\theta + b\sin\theta| = \sqrt{a^2 + b^2} \int_0^{2\pi} d\theta |\cos\theta| = C\sqrt{a^2 + b^2}$, we see that

$$\|Tf\|_1 = \int d\mu(q) \sqrt{(Tg)(q)^2 + (Th)(q)^2}$$

$$= C^{-1} \int d\theta d\mu(q) |T(g\cos\theta + h\sin\theta)| \leq C^{-1} \int d\theta d\mu(q) |g\cos\theta + h\sin\theta|$$

$$= \int d\mu \sqrt{g^2 + h^2} = \|f\|_1 .$$

Proof of Lemma 2.3 (Nelson).

Realize $\mathcal{H}$ as $L^2(\mathbb{R}, dx)$. Let $f \in \mathcal{F}_n$. Then $f = \int F(x_1, \ldots x_n) : \phi(x_1) \ldots \phi(x_n) : dx_1 \ldots dx_n \Omega_0$ for some symmetric $F \in L^2$ and $\|f\|^2 = (n!/2^n) \int |F(x_1 \ldots x_n)|^2 dx_1 \ldots dx_n$. Write $A_F = \int F(x_1 \ldots x_n) : \phi(x_1) \ldots \phi(x_n) :$. Then $\|f\|_4^4 = \langle \Omega_0, |A_F|^4 \Omega_0 \rangle = \|A_F^2 \Omega_0\|^2$. Write A_F as a sum $A_F = \sum_{i=1}^{2^n} b_i$, where each b_i is of the form

$$\int F(x_1 \ldots x_n) 2^{-n/2} a^\#(x_1) \ldots a^\#(x_n) \quad \text{and} \quad b_i \Omega_0 = 0 \quad \text{if} \quad i \neq 1 .$$

By a simple estimate $\|b_i[(N+1) \ldots (N+n)]^{-1}\| \leq 2^{-n/2}\|F\|_2$. Thus

$$\|f\|_4^4 = \|A_F(b_1\Omega_0)\|^2 \leq 2^{2n} \sup_{j=1,\ldots 2^n} \|b_j(b_1\Omega_0)\|^2 \leq$$

$$\leq (\|F\|_2^4 2^{-2n})(2^{2n})(2n)! \leq (\|f\|_2^4)(2^{2n})\binom{2n}{n} \leq 2^{4n}\|f\|_2^4 .$$

Thus $\|f\|_4 \leq 2^n\|f\|_2$ if $f \in \mathcal{F}_n$. By interpolation, we find Nelson's basic estimate $\|f\|_p \leq 4^{(1-2/p)n}\|f\|_2$ if $2 \leq p \leq 4$. Let $A \leq c1$ with $c < 1$. Pick $p > 2$ so that $4^{(1-2/p)}c < 1$ still holds. Then

$$\|\Gamma(A)f\|_p \leq \sum_{n=0}^\infty \|(\Gamma(A)f)_n\|_p$$

$$= \sum_{n=0}^\infty \|(A \otimes \ldots \otimes A)f_n\|_p$$

$$\leq \sum_{n=0}^\infty 4^{(1-2/p)n} \|(A \otimes \ldots \otimes A)f\|_2$$

$$\leq \|f\|_2 \sum_{n=0}^\infty [C^{4(1-2/p)}]^n = B\|f\|_2 .$$

It is useful to be able to extend the hypercontractive estimates from (p, p) and $(2, 4)$ to other p, q. One need only use the following interpolation theorems:

Riesz–Thorin Theorem: If $||Tf||_{p_1} \leq ||f||_{q_1}$ and $||Tf||_{p_0} \leq ||f||_{q_0}$, then $||Tf||_{p_t} \leq$ $\leq ||f||_{q_t}$, where $p_t^{-1} = tp_1^{-1} + (1 - t)_{p_0}^{-1}$; $q_t^{-1} = tq_1^{-1} + (1 - t) q_0^{-1}$.

Stein Interpolation Theorem: Let $T(z)$ be an analytic family of operators for $0 < \mathrm{Re}\, z < 1$, continuous in $0 \leq \mathrm{Re}\, z \leq 1$. If $||T(z)f||_{p_1} \leq ||f||_{q_1}$ for $\mathrm{Re}\, z = 1$ and $||T(z)f||_{p_0} \leq ||f||_{p_0}$ for $\mathrm{Re}\, z = 0$, then $||T(z)f||_{p_t} \leq ||f||_{q_t}$ for $\mathrm{Re}\, z = t$.

We then have:

Theorem 2.2 (Segal [39], Simon and Hoegh-Krohn [48]):

Let e^{-tH_0} be a hypercontractive semigroup. Then

(a) for any p, q with $1 < p, q < \infty$, there is a $T_{p,q}$ such that e^{-tH_0} is a bounded map of L^p into L^q if $t > T_{p,q}$.

(b) for any $1 < p < \infty$ there are $c, T, D > 0$ such that $||e^{-tH_0}f||_{p+ct} \leq e^{Dt}||f||_p$ if $0 < t < T$.

Proof:

(a) Applying the Riesz–Thorin theorem to e^{-TH_0} with $p_0 = 4, q_0 = 2$ and $p_1 =$ $= q_1 = \infty$, we see that $||e^{-TH_0}\psi||_{2p} \leq C||\psi||_p$ if $2 \leq p \leq \infty$. Thus $||e^{-nTH_0}\psi||_{2^{n+1}} \leq$ $\leq C^n||\psi||_2$. Since $p < q$ implies $||\psi||_p \leq ||\psi||_q$, this proves that e^{-nTH_0} is bounded from L^p to L^q if $p, q \geq 2$ and n is large. By duality, we have a simple result if $p, q \leq 2$. We handle the general result by composition.

(b) Apply Stein's theorem to $\exp(-zTH_0)C^{-z}$ with $p_0 = q_0 = 2$ and $p_1 = 4$, $q_1 = 2$. Then for $t < T$, $||e^{-tH_0}f||_{(2+\alpha t)} \leq e^{(\ln C)t}||f||_2$. By applying the Riesz–Thorin Theorem as in (a), we get the general p case.

3. TWO (PLUS ONE) PROOFS OF LOWER-BOUNDEDNESS

It is important that $H_0 + V$ be bounded from below. We will discuss various proofs of

Theorem 3.1 (Nelson, Glimm):

Let $H_0 + V$ be the form discussed in the introduction. Then $H_0 + V$ is bounded from below in the sense that for some C and all $\psi \in D(H_0) \cap D(V)$, $\langle \psi, (H_0 + V) \psi \rangle \geq$ $\geq - C||\psi||^2$.

For a theory with a box cutoff this result is due to Nelson [33], who employed the precursors of hypercontractive estimates and Feynman path integrals. Glimm [8] extended the result to the case we are considering. The basic trick of the first proof given below (which is a streamlined version of Nelson's proof) is due to Segal [43], with an extra trick (using monotonicity of log) due to Simon and Hoegh-Krohn [48].

(a) *Hypercontractive Proof*: Consider first $\exp(-TH_0)\exp(-2TV)\exp(-TH_0)$. By Theorem 2.1 $\exp(-tH_0)$ is a hypercontractive semigroup so that for the proper T, $\exp(-TH_0)$ is a bounded map of L^2 into L^4. By Theorem 1.2 $\exp(-2TV) \in L^4$, so that by Hölder's inequality $\exp(-2TV)$ is a bounded map of L^4 into L^2. Since $\exp(-TH_0)$ is a contraction from L^2 to L^2, $\exp(-TH_0)\exp(-2TV)\exp(-TH_0)$ is bounded from L^2 to L^2. Thus for some C, $\exp(-TH_0)\exp(-2TV)\exp(-TH_0) \leqq$ $\leqq C$ or $\exp(-2TV) \leqq C\exp(2TH_0)$. The log is a monotonic function on operators, so that $-2TV \leqq \log C + 2TH_0$, or $H_0 + V \geqq -(1/2T)\log C$.

(b) *The New Glimm–Jaffe Proof* [18]: The basis of this proof is the Duhamel formula:

$$\exp(-tH_2) = \exp(-tH_1) + \int_0^t \exp(-sH_1)(H_1 - H_2)\exp\left[-(t-s).H_1\right]dt, \quad (3.1).$$

which holds under suitable domain assumptions. It is the basis of the Dyson series. Suppose, formally, $A = \sum_{n=1}^\infty B_n$. Then iterating (3.1), $\exp(-tA) = \exp(-tB_1) - \int_0^t \exp(-s_1 B_1)(A - B_1)\exp\left[-(t-s_1)A\right]ds_1 = \exp(-tB_1) - \int_0^t \exp(-s_1 B_1)\cdot(A - B_1)\exp\left[-(t-s_1)B_2\right]ds_1 + \int_0^t ds_1 \int_0^t ds_2 \exp(-s_1 B_1)(A - B_1)\cdot\exp\left[s_2(B_1 + B_2)\right](A - B_1 - B_2)\exp\left[-(t-s_1-s_2)A\right] = \ldots$ It is the resulting infinite series that Glimm and Jaffe treat, where B_n is chosen so that $\sum_{n=1}^\infty B_n$ is $H_0 + V$ with an ultraviolet cutoff K_n which goes to infinity as $n \to \infty$. The individual terms are bounded as seen by using Feynman path integrals, a technique which we have not described.

(c) *Federbush* [5] has a proof of lower-boundedness for the box cutoff theory that may extend to the spatially cutoff case.

4. HIGHER-ORDER ESTIMATES

The basic higher-order estimates are:

$$H_0^2 + V^2 \leqq a(H_0 + V)^2 + b \qquad [(\varphi^4)_2 \ only] \tag{4.1}$$

$$N^j \leqq a(H_0 + V + b)^j \tag{4.2}$$

$$H_0^2 \leqq a(H_0 + V + b)^{2n} \quad \text{for} \quad (\varphi^{2n})_2 \tag{4.3}$$

$$H_0^{3-\varepsilon}N^j \leqq a(H_0 + V + b)^\alpha. \tag{4.4}$$

These estimates are to hold as inequalities of quadratic forms on $C^\infty(H_0) = \bigcap_n D(H_0)^n$. a, b, and α are constants which depend on the interaction, j and ε. ε is an arbitrary positive number. There are also estimates involving Lorentz generators and the momentum (see Rosen [35, 37]). Whether (4.1) extends to general $P(\varphi)_2$ theories is an open question—the anlogous estimate for $p^2 + x^2 + P(x)$ on

$L^2(\mathbb{R}, dx)$ does hold for arbitrary P with values in a sector avoiding the large negative reals (Simon [46]). Perturbation theory indicates that the vacuum for $H_0 + V$ is not in $D(H_0^3)$, so that the estimate in (4.4) with $\varepsilon = 0$ is probably false.

These estimates have the form of *a priori* estimates, that is, the manipulation in their proofs requires the estimates to justify them. A careful proof first proves (4.1–4) for an ultraviolet cutoff theory, proving the constants independent of the cutoff, and then passes to limits (see Rosen [35] and Section 4(b)). We will ignore these subtleties and all domain questions.

The higher order estimate (4.1) is due to Glimm and Jaffe [10]. The others are due to Rosen [35], generalizing a nonrelativistic result in Jaffe's thesis [26]. Rather than go through the detailed proofs, let us describe the techniques used.

(a) Abstract Nonsense

If A and B are positive selfadjoint operators, the statements $A \leq B$, $B^{-1/2}AB^{-1/2} \leq 1$, and $\|B^{-1/2}A^{1/2}\| \leq 1$ are equivalent.

(b) The First-Order Estimates

Introduce the operators $N_\tau = d\Gamma(\mu^\tau)$ where $N_0 = N$ and $N_1 = H_0$. Then the methods we used to show that $H_0 + V + b \geq 0$ also prove that $N_\tau + V + b \geq 0$ if $\tau > 0$, or $a(H_0 + V + b) \geq N_\tau$ if $a > 1$, $\tau \leq 1$, or $H_0 + V + b \geq W$ if W is the interaction associated with a polynomial of degree lower than that of the one defining V.

(c) N_τ-Estimates

The simple estimates from nonrelativistic quantum mechanics $\|a(N + 1)^{-1/2}\| \leq 1$ and $\|(N + 1)^{-1/2}a^\dagger\| \leq 1$ have been vastly generalized (see Glimm and Jaffe [13] for a very general statement). Two sample statements are:

Theorem 4.1 (N_τ-estimates of Glimm and Jaffe):
   (a) If $F(k_1, \ldots k_n) \in L^2(\mathbb{R}^n)$ and

$$A = \int F(k_1, \ldots k_n)\, a^\dagger(k_1) \ldots a^\dagger(k_j)\, a(k_{j+1}) \ldots a(k_n)\, dk_1 \ldots dk_n$$

then $(N + 1)^{-\alpha/2} A(N + 1)^{-\beta/2}$ are bounded operators as long as $\alpha + \beta \geq n$ and the norm is bounded by $C\|F\|_2$, where C depends only on α, β and n.
   (b) Let A be as above, where F obeys

$$\int |F(k_1, \ldots k_n)|^2 \, \frac{dk_1 \ldots dk_n}{\mu(k_1)^\tau \ldots \mu(k_n)^\tau} < \infty.$$

Then $(N_{\tau+1})^{-j/2} A(N_{\tau+1})^{-n+j/2}$ is bounded with norm less than $\|F(\mu_1, \mu_2 \ldots \mu_n)^{-\tau/2}\|_2$.

Proof:

The basic idea is that $a(k)$ is a densely defined operator and $[\int a(k)f(k)\,dk] \times$ $\times (N_\tau + 1)^{-1/2}$ has norm $\|f\mu^{-\tau/2}\|_2$. Thus $a(k)(N_\tau + 1)^{-1/2}$ has norm 1 as a map of $\mathscr{F}$ into $\mathscr{F} \otimes L^2(\mathbb{R}, \mu^{-\tau/2}\,dk)$. It follows that $a(k_1)\ldots a(k_j)(N_\tau + 1)^{-j/2}$ has norm 1 from $\mathscr{F}$ into $\mathscr{F} \otimes L^2(\mathbb{R}^j, (\mu_1\ldots\mu_j)^{-\tau/2}\,d^jk)$. (b) follows immediately. (a) follows from (b) by noting that $[(N+1), A] = (2j - n)\,A$.

Thus if $W = \int g(x):Q(\phi):dx$ and Q has degree m, then $W(N+1)^{-m/2}$ is bounded if $g \in L^2$, for $\int dk_1\ldots dk_n|g(k_1 + \ldots + k_n)|^2/\mu(k_1)\ldots\mu(k_n)$ is bounded. [For example, use the fact that $A_1 A_2 \ldots A_n \leq \sum_{i=1}^n (A_1 \ldots \hat{A}_i \ldots A_n)^{n/n-1}$, where $(A_1 \ldots \hat{A}_i \ldots A_n)$ means that A_i is missing in the product].

(d) The Pull-Through Formula

The pull-through formula is a technique discovered by Glimm and Jaffe in their treatment of Yukawa interactions [14]. Rosen adapted it in his proofs. $a(k)$ is a densely defined operator and $(H_0 + \mu(k))a(k) = a(k)H_0$. Letting $H = H_0 + V$, we see that for $z \in \mathbb{C}$, $(H + \mu + z)a(k) = a(k)(H + z) + [V, a(k)]$. Letting $R(z) = (H - z)^{-1}$ if $z \notin \sigma(H)$, we see that:

$$a(k)R(-z) = R(-z - \mu)a(k) + R(-z - \mu)[V, a(k)]R(-z) \qquad (4.5)$$

if $-z \notin \sigma(H)$ and $-z - \mu \notin \sigma(H)$. (Again this is an *a priori* estimate; we don't yet know that H is selfadjoint and that $\sigma(H)$ isn't all of $\mathbb{C}$.)

As a typical application, let us prove that

$$H_0 N_\tau \leq a(H + b)^2 \quad \text{if} \quad \tau < 0. \qquad (4.6)$$

By abstract nonsense we need only prove $\|N_\tau^{\frac{1}{2}} H_0^{\frac{1}{2}} R(-b)\psi\|^2 \leq a\|\psi\|^2$ for all ψ in some dense set. Now since $N_\tau = \int \mu^\tau a^+(k)a(k)\,dk$, $\|N_\tau^{\frac{1}{2}} H_0^{\frac{1}{2}} R(-b)\psi\|^2 =$ $= \int \mu^\tau \|a(k)H_0^{\frac{1}{2}} R(-b)\psi\|^2\,dk = \int \mu^\tau \|(H_0 + \mu(k))^{\frac{1}{2}} a(k)R(-b)\psi\|^2\,dk$.

By the pull-through formula, $(H_0 + \mu(k))^{\frac{1}{2}} a(k)R(-b)\psi = ① + ②$, where $① =$ $= (H_0 + \mu(k))^{\frac{1}{2}} R(-b - \mu)a(k)\psi$ and $② = (H_0 + \mu(k))^{\frac{1}{2}} R(-b - \mu)[V, a(k)] \cdot$ $\cdot R(-b)\psi$. Writing

$$\|①\| = \|(H_0 + \mu(k))^{\frac{1}{2}} R(-b - \mu)a(k)\psi\| = \|(H_0 + \mu(k))^{\frac{1}{2}} R^{\frac{1}{2}}(-b - \mu)\| \times$$

$$\times \|R^{\frac{1}{2}}(-b - \mu)(N_\tau + \mu(k) + 1)^{\frac{1}{2}}\| \times \|(N_\tau + 1 + \mu)^{-\frac{1}{2}} a(k)\psi\|,$$

we note that $H_0 + \mu \leq 2(H + b + \mu)$ and $N_\tau + 1 + \mu \leq 2(H + b + \mu)$ for suitable b independent of k, so that $\int \mu^\tau \|①\|^2\,dk \leq 2\int \mu^\tau \|a(k)(N_\tau + 1)^{-\frac{1}{2}}\psi\|^2\,dk =$ $= 2\|N_\tau^{\frac{1}{2}}(N_\tau + 1)^{-\frac{1}{2}}\psi\|^2 \leq 2\|\psi\|^2$. If $V = \int g(x):P(\varphi(x)):dx$, a simple computation shows that $[a(k), V] = \mu(k)^{-\frac{1}{2}}W$, where $W = \int g(x):P'(\varphi(x)):dx$ and $P'(X) = dP(X)/dX$. Since $\pm W \leq a(H + b)$ for suitable b, $\|R^{\frac{1}{2}}(-b - \mu)WR(-b)^{\frac{1}{2}}\|$ is bounded. As above, $\|(H_0 + \mu)^{\frac{1}{2}} R^{\frac{1}{2}}(-b - \mu)\|$ is bounded, so that

$$\int \mu^\tau \|②\|^2 \, dk \leqq C \int \mu^{\tau-1} \|R(-b-\mu)^{\frac{1}{2}}\psi\|^2 \, dk \leqq C'\|\psi\|^2 \int \mu^{\tau-1} \, dk \leqq C''\|\psi\|^2,$$

since $\tau < 0$.

(e) Hölder Inequalities

Let $n_1, n_2 \geqq 0$ with $n_1 + n_2 = n$, $n_1\tau_1 + n_2\tau_2 = n\tau$. By Hölder's inequality,

$$\left| \sum_{i=1}^{m} (a_i b_i) \right| \leqq \left(\sum_{i=1}^{m} |a_i|^{n/n_1} \right)^{n_1/n} \left(\sum_{i=1}^{m} |b_i|^{n/n_2} \right)^{n_2/n}.$$

Let $a_i = |c_i|^{n_1\tau_1/n}$ and $b_i = |c_i|^{n_2\tau_2/n}$. We see that

$$\left(\sum_{i=1}^{m} |c_i|^{\tau} \right)^n \leqq \left(\sum_{i=1}^{m} |c_i|^{\tau_1} \right)^{n_1} \left(\sum_{i=1}^{m} |c_i|^{\tau_2} \right)^{n_2}.$$

Since N_τ is multiplication by $\sum_{i=1}^{m} \mu(k_i)^\tau$ on $\mathscr{F}_m$, we have proven

Theorem 4.2 (Rosen [35]):
Let $n_1, n_2 \geqq 0$. Suppose $n_1 + n_2 = n$, $n_1\tau_1 + n_2\tau_2 = n\tau$. Then $N_\tau^n \leqq N_{\tau_1}^{n_1} N_{\tau_2}^{n_2}$.

For example, $N^2 \leqq N_{-1} H_0$, so that (4.6), proved above, implies $N^2 \leqq a(H+b)^2$. We thus have a proof of (4.2) if $j = 2$.

(f) The "Estimates for Free" Method

There is a simple trick, due to Simon (unpublished), for getting from N^j estimates to H_0 estimates. This method allows one to prove (4.3) and (4.4) from (4.2) and N_τ-estimates. Let us illustrate the method by proving (4.3) from (4.2). By an N_τ-estimate, if $V = \int g(x): P(x): dx$ and $\deg P = 2m$, then VN^{-m} is bounded. Thus since $N^m R(-b)^m$ is bounded (by (4.2)), $VR(-b)^m$ is bounded. Obviously, $HR(-b)^m = R(-b)^{m-1} - bR(-b)^m$ is bounded. Thus $H_0 R(-b)^m = (H-V)R(-b)^m$ is bounded, i.e. $H_0^2 \leqq a(H+b)^{2m}$. To prove (4.4), one writes

$$H_0^{3/2-\varepsilon} R(-b)^\alpha = H_0^{-1/2-\varepsilon}(H-V)^2 R(-b)^\alpha =$$

$$= H_0^{-1/2-\varepsilon}(H^2 + V^2 + 2VH + [H_0, V]) R(-b)^\alpha$$

and continues from there, using an N_τ-estimate to prove that $H_0^{-1/2-\varepsilon}[H_0, V] \times$ $\times H_0^{-1/2-\varepsilon}(N+1)^{-m+1}$ is bounded.

(g) The Double-Commutator Method

This is a trick due to Jaffe [26]. If $A \geqq 0$ and B is arbitrary, then $(A+B)^2 = A^2 + B^2 + AB + BA = A^2 + B^2 + 2A^{\frac{1}{2}}BA^{\frac{1}{2}} + [A^{\frac{1}{2}}, [A^{\frac{1}{2}}, B]]$. It is often easier to estimate $A^{\frac{1}{2}}BA^{\frac{1}{2}}$ and $[A^{\frac{1}{2}}, [A^{\frac{1}{2}}, B]]$ separately. For example, to prove (4.1) one writes

$$(H_0 + V)^2 = H_0^2 + V^2 + 2H_0^{\frac{1}{2}}VH_0^{\frac{1}{2}} + [H_0^{\frac{1}{2}}, [H_0^{\frac{1}{2}}, V]] =$$

$$= (\tfrac{1}{3}H_0^2 + V^2) + 2H_0^{\frac{1}{2}}(\tfrac{1}{6}H_0 + V)H_0^{\frac{1}{2}} + (\tfrac{1}{3}H_0^2 + [H_0^{\frac{1}{2}}, [H_0^{\frac{1}{2}}, V]]).$$

By a first-order estimate, $2H_0^{\frac{1}{2}}(\tfrac{1}{6}H_0 + V)H_0^{\frac{1}{2}} \geqq - bH_0 \geqq - \tfrac{1}{12}H_0^2 - d$. By an N_τ-estimate (when we are dealing with φ^4), $H_0^{-\frac{1}{2}-\varepsilon}[H_0^{\frac{1}{2}}, [H_0^{\frac{1}{2}}, V]]H_0^{-\frac{1}{2}-\varepsilon}$ is bounded, so that $\tfrac{1}{3}H_0^2 + [H_0^{\frac{1}{2}}, [H_0^{\frac{1}{2}}, V]] \geqq \tfrac{1}{3}H_0^2 + cH_0^{1+\varepsilon} \geqq - d'$. Thus $(H_0 + V)^2 \geqq \tfrac{1}{4}H_0^2 + V^2 - (d + d')$.

5. SIX PROOFS OF SELFADJOINTNESS

Selfadjointness of the Hamiltonian $H_0 + V$ in the $(\varphi^4)_2$ case was first proven by Glimm and Jaffe [10]. This was the critical step in the constructive field theory approach that first opened up a theory without cutoffs. The first proof for $(\varphi^{2n})_2$ is due to Rosen. In this section we want to discuss the various "independent" proofs of selfadjointness. Four proofs work for general $P(\varphi)_2$, and two for $(\varphi^4)_2$ only.

(a) The Hypercontractive Proof [39, 48]

This is a proof which employs the Trotter product formula and the hypercontractive estimates for e^{-tH_0}. In terms of a realization of Q-space as limits of $\mathbb{R}^n$ (which looks a little complicated), this is essentially Rosen's original $(\varphi^{2n})_2$ proof [34]. The abstract formulation is due to Segal [39], with some modifications and extensions due to Simon and Hoegh-Krohn [48].

In this proof, one first constructs a selfadjoint operator, H, and then proves that it is essentially selfadjoint on $D(H_0) \cap D(V)$ and $H \upharpoonright D(H_0) \cap D(V) = H_0 + V$.

Step (1): Pick V_n bounded, so that $V_n \to V$ in each L^p with $p < \infty$ and $e^{-tV_n} \to e^{-tV}$ in L^1 for each $t > 0$. V_n may be taken to be a function of V.

Step (2): Since V_n is bounded, H_n is a self-adjoint operator with $D(H_n) = D(H_0)$. Since the lower bound on $H_0 + V_n$ depends only on $\|e^{-8TV_n}\|_1$, the H_n are uniformly bounded from below. Equivalently, as operators on L^2, the e^{-tH_n} are uniformly bounded operators as n varies and t varies over compact subsets of $[0, \infty)$.

Step (3): We next note a useful general principle. Let $p < q < \infty$. If $\psi_n \to \psi$ in L^p, each $\psi_n \in L^q$ with $\sup \|\psi_n\|_q = C < \infty$, then $\psi \in L^q$ and $\|\psi\| < C$. For the unit ball in L^q is weak *-compact, so that $\psi_{n_i} \to \phi$ weakly in L^q for some subsequence and some ϕ with $\|\phi\|_q \leqq C$. But then since $L^p \supset L^q$ (so that $L^{p*} \subset L^{q*}$), $\psi_{n_i} \to \phi$ weakly in L^p. It follows that $\phi = \psi$.

Step (4): Let $p > q$; then each e^{-tH_n} maps L^p into L^q, and for some C independent of n, $||e^{-tH_n}\psi||_q \leqq C||\psi||_p$. C is also uniformly bounded for t in compact subsets of $[0, \infty)$. Let us prove this for $p = \infty, q = 4$. By the Trotter product formula, if $\psi \in L^\infty \subset L^2$, $e^{-tH_n}\psi = L^2\text{-lim}\ (e^{-tH_0/m}e^{-tV_n/m})^m\psi$. By the principle in step (3), we need only prove $||(e^{-tH_0/m}e^{-tV_n/m})^m\psi||_4 \leqq C||\psi||_\infty$ for each n and m with C independent of n and m. Consider the intermediate spaces, $L^\infty, L^{4m}, L^{4m/2}, L^{4m/3}, \ldots L^{4m/m}$. By Hölder's inequality, $||e^{-tV_n/m}\phi||_{4m/n+1} \leqq ||e^{-tV_n/m}||_{4m}||\phi||_{4m/n}$. Since $e^{-tH_0/m}$ is a contraction on each L^p, $||(e^{-tH_0/m}e^{-tV_n/m}\psi||_4 \leqq (||e^{-tV_n/m}||_{4m})^m||\psi||_\infty$. Since $(||e^{-tV_n/m}||_{4m})^m = (||e^{-4tV_n}||_1)^{1/4}$ and $\sup_n ||e^{-tV_n}||_1 < \infty$, $||e^{-tH_n}\psi||_4 \leqq C||\psi||_\infty$.

Step (5): For each $\psi \in L^\infty$, $e^{-tH_n}\psi$ is Cauchy in L^2. By the Duhamel formula (3.1), $e^{-tH_n}\psi - e^{-tH_m}\psi = \int_0^t e^{-sH_n}(V_m - V_n)e^{-(t-s)H_m}\psi ds$. By (4), $e^{-(t-s)H_m}\psi$ runs over a bounded set in L^4 as s runs between 0 and t and as m varies. $V_n - V_m$ maps L^4 into L^3, and since $||V_n - V_m||_{12} \to 0$,

$$\sup_{\substack{n \geqq m \\ s \in [0,t]}} ||(V_n - V_m)e^{-(t-s)H_m}\psi||_3 \to 0 \quad \text{as} \quad m \to \infty.$$

Finally, since e^{-sH_n} is uniformly bounded as a map of L^3 into L^2, $e^{-tH_m}\psi$ is Cauchy in L^2.

Step (6): For each $\psi \in L^2$, $L^2\text{-lim}_{n \to \infty} e^{-tH_n}\psi$ exists and defines a strongly continuous self-adjoint semigroup e^{-tH}. For $\psi \in L^\infty$, $e^{-tH_n}\psi \to e^{-tH}\psi$ in each $L^p(p < \infty)$. The first statement follows from (4), from the denseness of L^∞ in L^2 and from (2). The strong continuity of e^{-tH} follows from the uniformity of the limit locally in t. That the limit semigroup is of the form e^{-tH} for some self-adjoint H is a general theorem. $e^{-tH_n}\psi \xrightarrow{L^p} e^{-tH}\psi$ is proven as in step (5).

Step (7): Let $D = \{e^{-tH}\psi | \psi \in L^\infty, t > 0\}$. D is a core for H (domain of essential selfadjointness). This follows from the spectral theorem.

Step (8): Let $\psi \in L^\infty$. Then $H_n e^{-tH_n}\psi \xrightarrow{L^2} H e^{-tH}\psi$. For $e^{-tH_n}\psi = f_n(t)$ is analytic in the half plane $\{t | \operatorname{Re} t > 0\}$. By (2) the f_n are uniformly bounded. Since $f_n(t) \to e^{-tH}\psi$ for t real, the convergence is on $\{t | \operatorname{Re} t > 0\}$ and is uniform (Vitali's theorem). Thus, by the Cauchy integral formula, the derivatives converge.

Step (9): $D \subset D(H_0) \cap D(V)$ and $H \upharpoonright D = (H_0 + V) \upharpoonright D$. For let $\phi = e^{-tH}\psi$ with $\psi \in L^\infty$. Let $\phi_n = e^{-tH_n}\psi$. Then $\phi_n \to \phi$ in L^2 and L^4. Since $V_n \to V$ in L^4, $\phi \in D(V)$ and $V_n\phi_n \to V\phi$. By (8), $H_n\phi_n \to H\phi$. Thus, since $D(H_n) = D(H_0)$ and $H_n = H_0 + V_n$, $H_0\phi_n = (H_n - V_n)\phi_n \to (H - V)\phi$. Since H_0 is closed, $\phi \in D(H_0)$ and $H_0\phi = H\phi - V\phi$.

Step (10): $H_0 + V$ is essentially selfadjoint on $D(H_0) \cap D(V)$. For $H_0 + V \upharpoonright D(H_0) \cap D(V)$ is a symmetric extension of $H \upharpoonright D$, which is essentially self-adjoint by (7).

(b) The Higher-Order Estimates Proof [10, 35]

The original Glimm and Jaffe proof [10, 19] employed some special higher order estimates for φ^4 to prove that ultraviolet cutoff H's, known to be selfadjoint by methods of nonrelativistic quantum theory, had resolvents converging to the resolvent of a selfadjoint operator H. In a more general setting this is the idea of Rosen's second selfadjointness proof [35]. To avoid confusion we use H_I instead of V for the interaction. Glimm and Jaffe use $H_I(g)$ instead.

Step (1): We first introduce a "volume cutoff" free Hamiltonian $H_{0,V}$. Let $((x))$ be "the integer nearest x", i.e. $((x)) = n$ if n is an integer and $n - \frac{1}{2} < x \leqq n + \frac{1}{2}$. Let $k_V = (2\pi/V)((Vk/2\pi))$ and let μ_V be the multiplication in k space by the function $\mu_V(k) = \mu(k_V)$. Let $H_{0,V} = d\Gamma(\mu_V)$. In terms of $a^{\#}(k)$,

$$H_{0,V} = \sum \mu\left(\frac{2\pi n}{V}\right) \int_{(n-\frac{1}{2})V/2\pi}^{(n+\frac{1}{2})V/2\pi} a^+(k)\, a(k)\, dk = \int \mu(k_V)\, a^+(k)\, a(k)\, dk\,.$$

A simple N_τ-estimate shows that $(N + 1)^\alpha (H_0 - H_{0,V})(N + 1)^\beta \to 0$ as $V \to \infty$ if $\alpha + \beta = -1$.

Step (2): We introduce cutoff interactions $H_{I,V}$ and $H_{I,V,k}$. Define $\varphi_V(x)$ to be

$$\varphi_V(x) = (4\pi)^{-\frac{1}{2}} \int e^{ik_V \cdot x}\left[a^+(k) + a(-k)\right] \frac{dk}{\mu(k_V)^{\frac{1}{2}}} =$$

$$= \sum_{n=-\infty}^{\infty} (4\pi)^{-\frac{1}{2}}\, e^{2\pi i x n/V}\, \mu\left(\frac{2\pi n}{V}\right)^{-\frac{1}{2}} \int_{(n-\frac{1}{2})V/2\pi}^{(n+\frac{1}{2})V/2\pi} \left[a^+(k) + a(-k)\right] dk\,.$$

Let $H_{I,V} = \int g(x) : P(\varphi_V(x)) : dx$. If $P(x) = x^{2m}$, we can write $H_{I,V}$ as a sum $\sum_{n_1,\ldots,n_{2m}}^{\infty} (4\pi)^{-m} \ldots \ldots$ The sum over those n_i with $|n_i| \leqq k$ is denoted by $H_{I,V,k}$.

For the general P, we define $H_{I,V,k}$ by linearity. If $\deg P = 2m$, a simple N_τ-estimate proves that $\lim_{k \to \infty} (N + 1)^\alpha (H_{I,V,k} - H_{I,V})(N + 1)^\beta = 0$ and $\lim_{V \to \infty} (N + 1)^\alpha (H_{I,V} - H_I) \cdot (N + 1)^\beta = 0$ as long as $\alpha + \beta \leqq -m$.

Step (3): Recognize $H_{0,V}$ as being essentially the Hamiltonian of a system with discrete modes. Let $\mathcal{H}_V$ be that subspace of the one-particle space containing functions obeying $f(k) = f(k_V)$, i.e. functions constant on intervals

$$\left((n - \tfrac{1}{2})\frac{2\pi}{V}, (n + \tfrac{1}{2})\frac{2\pi}{V}\right].$$

Let $\mathcal{H}_V^{\perp}$ be the orthogonal complement of $\mathcal{H}_V$. Since $\mathcal{H}_V \oplus \mathcal{H}_V^{\perp} = \mathcal{H}$, $\mathcal{F} = \mathcal{F}_V \otimes \mathcal{F}_{V\perp}$, where $\mathcal{F}_V = \mathcal{F}(\mathcal{H}_V)$ and $\mathcal{F}_{V\perp} = \mathcal{F}(\mathcal{H}_V^{\perp})$.

Under this tensor product, $H_{0,V} = (H_{0,V} \restriction \mathcal{F}_V) \otimes 1 + 1 \otimes (H_{0,V} \restriction \mathcal{F}_{V\perp})$. $\mathcal{F}_V$ can be thought of as $l_2(-\infty, \infty)$ with the multiplication ω_V on l_2 by

$$(\omega_V a)_n = \mu\left(\frac{2\pi n}{V}\right) a_n.$$

Thus $H_{0,V} \restriction \mathcal{F}_V = d\Gamma(\omega_V \restriction \mathcal{H}_V)$ is essentially the Hamiltonian of a free system with discrete modes (actually we can realize $H_{0,V}$ as the Hamiltonian in a box of size V). Given k, writing $l^{(k)} = \{a \in l_2 \mid a_m = 0 \text{ if } |m| > k\}$, $l_2 = l^{(k)} \oplus (l^{(k)})^{\perp}$ and $\mathcal{F}_V = \mathcal{F}_{V,k} \otimes \mathcal{F}_{V,k\perp}$. We then have $\mathcal{F} = \mathcal{F}_{V,k} \otimes (\mathcal{F}_{V,k\perp} \otimes \mathcal{F}_{V\perp})$ and under this tensor product,

$$H_{0,V} = (H_{0,V} \restriction \mathcal{F}_{V,k}) \otimes 1 + 1 \otimes H_{0,V} \restriction (\mathcal{F}_{V,k\perp} \otimes \mathcal{F}_{V\perp})$$

and

$$H_{I,V,k} = (H_{I,V,k} \restriction \mathcal{F}_{V,k}) \otimes 1.$$

But $l^{(k)}$ is finite-dimensional ($l^{(k)} \cong \mathbb{C}^{2k+1}$), so that $\mathcal{F}_{V,k}$ is the Fock space of a finite number of degrees of freedom. We can realize $\mathcal{F}_{V,k}$ as $L^2(\mathbb{R}^{2k+1})$ so that $H_{0,V} \restriction \mathcal{F}_{V,k}$ becomes $\sum_{l=-k}^{k}(p_l^2 + \omega_l^2 q_l^2)$ and $H_{I,V,k} \restriction \mathcal{F}_{V,k}$ is a polynomial in q of positive degree (see e.g. [26, 50]).

Step (4): $H_{0,V} + H_{I,V,k} \restriction \mathcal{F}_{V,k}$ is unitarily equivalent to $-\Delta +$ a polynomial in $q_{-k}, \ldots q_k$ on $L^2(\mathbb{R}^{2k+1})$ with the polynomial bounded from below. By theorems from nonrelativistic quantum theory, it is essentially selfadjoint on

$$\mathscr{S}(\mathbb{R}^{2k+1}) \subset D(H_{0,V} \restriction \mathcal{F}_V) \cap D(H_{I,V,k} \restriction \mathcal{F}_V).$$

It follows that $H_{0,V} + H_{I,V,k} \equiv H_{V,k}$ is essentially selfadjoint on

$$D(H_{0,V}) \cap D(H_{I,V,k})$$

since $H_{0,V,k} \restriction \mathcal{F}_{V,k\perp} \otimes \mathcal{F}_{V\perp}$ is just a $d\Gamma$ operation.

Step (5): The higher-order estimates are really estimates of the form $(N+1)^j \leq a(H_{V,k} + b)^j$ with a and b independent of V and k (see the remarks on the *a priori* nature of the h.o.e.). One also has $(H_{V,k} + b) \leq c(H_0 + d)^m$ by N_τ-estimates.

Step (6): $(H_{V,k} + b)^{-m}$ is norm-Cauchy as $k \to \infty$, for

$$(H_{V,k} + b)^{-m} - (H_{V,k'} + b)^{-m} =$$

$$= \sum_{j=1}^{m} (H_{V,k} + b)^{-j}(H_{I,k',V} - H_{I,k,V})(H_{V,k'} + b)^{-m+j-1},$$

so that

$$\|(H_{V,k} + b)^{-m} - (H_{V,k'} + b)^{-m}\| \leq a^{m+1} \sum_{j=1}^{m} \|(N + 1)^{-j}(dH_{I,k,V})(N + 1)^{-m+j-1}\|$$

by the higher-order estimates.

Step (7): $(H_{V,k} + b)^{-m}$ converges in norm as $k \to \infty$ to some positive operator R_V^{-m}, so $(H_{V,k} + b)^{-1}$ converges to R_V^{-1}. Since

$$(H_{V,k} + b)^{-1} \geq C^{-1}(H_0 + d)^{-m},$$

R_V^{-1} has an unbounded selfadjoint inverse which we denote by H_V. In the same way, $(H_V + b)^{-1}$ converges to $(H + b)^{-1}$ for some selfadjoint operator H.

Step (8): By a limiting argument, H obeys $H_0 \leq a(H + b)^{2m}$ and $N^{2m} \leq a(H + b)^{2m}$. It follows that any ψ of the form $\psi = (H + b)^{-2m}\psi \in D(H_0) \cap D(H_I)$. By a limiting argument, $H\psi = H_0\psi + V\psi$.

Step (9): $\text{Ran}(H + b)^{-2m}$ is a core for H, so as in the hypercontractive proof, $H_0 + V$ is essentially selfadjoint on $D(H_0) \cap D(V)$.

(c) Gidas–Federbush Proof

Gidas [31] has found a proof of selfadjointness by combining Federbush's techniques [5] with the ideas of (a) and (b) above.

(d) Konrady's Proof [$(\varphi^4)_2$ Only]

In case $P(\varphi) = \varphi^4$, Konrady [30] has found a very simple proof of selfadjointness. Konrady uses two simple estimates (which are precisely the estimates originally used by Glimm and Jaffe [10]):

i) (The lower-boundedness) $H_0 + V + b \geq \frac{1}{2}N$ for some b.
ii) (An N_τ-estimate) $\pm [N, [N, V]] \leq cN^2 + d$ for some c and d.

He also uses the following "perturbation theorem" of Kato [27]:

Theorem 5.1:
Let A be selfadjoint and let B be symmetric with $D(B) \supset D(A)$. Suppose

$$\|B\psi\|^2 \leq a\|A\psi\|^2 + b\|\psi\|^2 \quad \text{for all} \quad \psi \in D(A).$$

Then
a) If $a < 1$, $A + B$ is selfadjoint on $D(A)$ and essentially selfadjoint on any core of A.
b) If $a = 1$, $A + B$ is essentially selfadjoint on any core for A.

Konrady first looks at $H_0 + eN^2 + V$, where e is chosen so that $V^2 \leq \frac{1}{2}e^2N^4$ (by an N_τ-estimate). Then $(H_0 + eN^2) + V$ is essentially selfadjoint on $C^\infty(H_0) =$

$= \cap_n D(H_n^0)$ by part (a) of Theorem 5.2. If we can prove

$$e^2 N^4 \leqq (H_0 + eN^2 + V)^2 + f^2 \tag{5.1}$$

by (b) $H_0 + V = (H_0 + eN^2 + V) - eN^2$ is essentially selfadjoint on $C^\infty(H_0)$. Thus we need only prove (5.1) or $0 \leqq (H_0 + V)^2 + eN^2(H_0 + V) + e(H_0 + V)N^2 + f^2$. By a double commutator,

$$eN^2(H_0 + V) + e(H_0 + V)N^2 = 2N(H_0 + V + b)N + [N, [N, H_0 + V]] -$$

$$- 2bN^2 \geqq N^3 - 2bN^2 - cN - d \quad \text{(by estimates (i) and (ii))} \geqq - f^2$$

for some f. This proves (5.1) and thus selfadjointness.

(e) The Masson-McClary Proof ($(\varphi^4)_2$ Only)

Masson and McClary [32] have a proof that $H_0 + V$ is essentially self-adjoint on $D(H_0) \cap D(V)$ in case $P(x) = x^4$.

In essence, they do the following: Let ψ be a vector in $C^\infty(H_0)$ with $\psi_n = 0$ for n large enough, i.e. $\psi \in \bigoplus\limits_{n=1}^{N} \mathscr{F}_n$ for some N. Then $\psi \in D(H_0) \cap D(V)$ and $F \cap C^\infty(H_0)$; the set of such ψ, is dense in $\mathscr{F}$. Let $D_\psi = \{\sum_{n=0}^\infty a_n H^n \psi \,|\, N = 1, 2, 3, \ldots\}$ and let $\mathscr{H}_\psi = \bar{D}_\psi$. Using the lower bound on H and the theory of Jacobi matrices, they prove that $H \upharpoonright D_\psi$ is selfadjoint as an operator on $\mathscr{H}_\psi$. By a simple argument employing the fact that H is essentially selfadjoint if and only if

$$\mathrm{Ker}(H^* \pm i) = \{0\} \,,$$

$H_0 + V$ is seen to be essentially selfadjoint on $F \cap C^\infty(H_0)$.

(f) The New Glimm and Jaffe Proof

Using their method of "high energy perturbation theory" which we discussed in Section 2(b), Glimm and Jaffe [18] have found a selfadjoint operator H in the general $P(\varphi)_2$ case. In [16] they prove $H = H_0 + V$.

6. SPECTRAL PROPERTIES OF H

The most interesting open questions about the spatially cutoff theory involve the spectral properties of H, essentially proving that it has no continuous singular spectrum. In studying these questions, the $(\varphi^2)_2$ theory is a useful guide. While it is trivial from a physical point of view (only a mass shift in the infinite-volume limit), it is "exactly soluble" in the sense that it is unitarily equivalent to $d\Gamma(\mu_g) - E_g$ for a suitable operator μ_g, where

$$E_g = \inf \sigma(H_0 + V). \tag{6.1}$$

We will use the definition (6.1) for general $P(x)$. In fact, as shown by various authors [6, 7, 4, 37], $\mu_g = \sqrt{-\Delta + m_0^2 + g(x)}$ if $V = \int g(x):\varphi^2(x):dx$. $H_0 + V$ can even be analyzed if g is negative so long as $\inf_x g(x) > -m_0^2$. Rosen [37] has analyzed the spectral properties of $H_0 + V$ in this $(\varphi^2)_2$ case in detail, and his results are often suggestive of what might be true in the general $P(\varphi)_2$ case.

(a) The Discrete and Essential Spectrum

Recall that a point $\lambda \in \sigma(H)$ is called *discrete* if λ is an isolated point of $\sigma(H)$ and an eigenvalue of finite multiplicity. If $\lambda \in \sigma(H) \setminus \sigma_{\mathrm{disc}}(H)$ we say that λ is in the essential spectrum of H. For the general $P(\varphi)_2$ case:

Theorem 6.1:

Let $V = \int g(x):P(\varphi):dx$ with the usual assumptions on g and P. Let $H = H_0 + V$ and let E_g be given by (6.1). Let m_0 be the mass in H_0. Then:

(a) (Glimm and Jaffe) H has purely discrete spectrum in $[E_g, E_g + m_0)$, in particular, E_g is an eigenvalue.

(b) $\sigma_{\mathrm{ess}}(H) = [E_g + m_0, \infty)$.

Remarks:

(a) was first proven in [11] for $(\varphi^4)_2$, but the method extended to $(\varphi^{2n})_2$ once Rosen had proved selfadjointness in that case. H_0 is not critical. If ω is any real operator on the one-particle space and $\omega \geq c1, c > 0$, then $d\Gamma(\omega) + V$ has a purely discrete spectrum in $[\inf \sigma(d\Gamma(\omega) + V), \inf + c)$ (see for example [48]). Gross [20, 21] has proven the existence of an eigenvector for E_g using only the fact that e^{-tH} obeys a hypercontractive estimate from L^2 to L^4.

(b) follows from (a) and Theorem 6.4 below. It does not hold for arbitrary $d\Gamma(\omega) + V$; the fact that $\omega = \mu(k)$ has a purely continuous spectrum is critical.

Proof:

Consider the operators $H_{V,k} = H_{0,V} + H_{I,V,k}$ used in the higher-order estimates proof of selfadjointness (Section 5b). We saw (in step (3)) that

$$H_{V,k} = (H_{V,k} \upharpoonright \mathscr{F}_{V,k} \otimes 1) + (1 \otimes H_{0,V} \upharpoonright \mathscr{F}_{V\perp} \otimes \mathscr{F}_{V,k\perp})$$

under the decomposition $\mathscr{F} = \mathscr{F}_{V,k} \otimes (\mathscr{F}_{V\perp} \otimes \mathscr{F}_{V,k\perp})$. In step (4) of the higher-order estimates proof of selfadjointness, we saw that $H_{V,k} \upharpoonright \mathscr{F}_{V,k}$ had a purely discrete spectrum. On $\mathscr{F}_{V\perp} \otimes \mathscr{F}_{V,k\perp}$, $H_{0,V}$ is 0 on $\mathbb{C} \subset \mathscr{F}_{V\perp} \otimes \mathscr{F}_{V,k\perp}$ and bigger than m_0 on $\mathbb{C}^\perp$. Letting $E_{g,V,k} = \inf \sigma(H_{V,k})$, it follows that $H_{V,k}$ has a purely discrete spectrum on $\mathscr{F}_{V,k} \otimes \mathbb{C} \subset \mathscr{F}$ and spectrum only in $[m_0 + E_{g,V,k}, \infty)$ on $(\mathscr{F}_{V,k} \otimes \mathbb{C})^{\perp} = {}$ $= \mathscr{F}_{V,k} \otimes \mathbb{C}^{\perp}$. Thus, $H_{V,k}$ has a purely discrete spectrum on $[E_{g,V,k}, E_{g,V,k} + m_0)$.

Now abstract analysis takes over. Letting A_α stand for $H_{V,k} - E_{V,k}$ we know that A_α has a purely discrete spectrum in $[0, m_0)$ and that A_α converges to $H - E_g \equiv A$ in the sense of norm convergence of resolvents (step (7) in the higher order estimates proof). It is easy to see that a self-adjoint operator B has a purely discrete spectrum in (a, b) if and only if $f(B)$ is compact for any continuous function f with support in (a, b). For such functions $f(A_\alpha) \to f(A)$, so that A has a purely discrete spectrum in $[0, m_0)$.

(b) Properties of the Vacuum

We know that $H_0 + V = H$ has an eigenfunction with eigenvalue the bottom of $\sigma(H)$. Such eigenfunctions are called vacuum vectors for H. An important result is:

Theorem 6.2 (Glimm and Jaffe):
   The vacuum for H is unique, i.e. the eigenvalue E_g is nondegenerate.

Remark:
   This was first proven in [11], using a generalization of the Perron–Frobenius theorem. Using an idea of Segal [43], Simon and Hoegh-Krohn [48] constructed a simpler proof along the Glimm–Jaffe line. It is this proof we sketch below.

Proof:
   Let V_n be the cutoff potentials in the hypercontractive proof of selfadjointness (Section 5a). By using hypercontractive techniques and Duhamel's formula, one proves that

$$(H_0 + V_n - E)^{-1} \to (H_0 + V - E)^{-1}$$

in norm and that

$$(H_0 + V - V_n - E)^{-1} \to (H_0 - E)^{-1}$$

in norm. As a result, using the Trotter product formula:

$$e^{-tH} = \underset{n \to \infty}{s\text{-lim}} \left[\underset{m \to \infty}{s\text{-lim}} (e^{-tV_n/m} e^{-tH_0/m})^m \right], \tag{6.2a}$$

$$e^{-tH_0} = \underset{n \to \infty}{s\text{-lim}} \left[\underset{m \to \infty}{s\text{-lim}} (e^{+tV_n/m} e^{-tH/m})^m \right]. \tag{6.2b}$$

Since e^{-tH_0} and e^{-tV_n} are positivity preserving, (6.2a) implies that e^{-tH} is positivity preserving. By (6.2b), e^{-tH_0} is in the von Neumann algebra, N, generated by $\{e^{-tH}\}_{t \geq 0}$ and $\{\psi \mid \psi \in L^\infty(Q)\}$. Let $A \in N'$ the commutant of N. Then

$$e^{-tH_0} A\Omega_0 = A(e^{-tH_0}\Omega_0) = A\Omega_0 ,$$

so that $A\Omega_0 = c\Omega_0$ for some constant. But then for any $\phi \in L^\infty$, $A\phi = A\phi\Omega_0 (\Omega_0$ is the function $1) = \phi(A\Omega_0) = c\phi\Omega_0 = c\phi$. Since L^∞ is dense, $A = c1$. Thus N acts irreducibly on L^2.

Let ψ be an eigenvector for H with $H\psi = E_g\psi$. Then $e^{-tH_0}\psi = \|e^{-tH_0}\|\psi$. Since e^{-tH_0} is positivity preserving, this is true if ψ is replaced by $\mathrm{Re}\,\psi$ or $\mathrm{Im}\,\psi$. Therefore we can suppose ψ to be real-valued without loss. Since e^{-tH_0} is positivity preserving,

$$\langle |\psi|, e^{-tH_0}|\psi| \rangle \geqq \langle |\psi|, e^{-tH_0}\psi \rangle \geqq \|e^{-tH_0}\| \langle |\psi|, \psi \rangle \geqq \|e^{-tH_0}\| \|\psi\|^2.$$

It follows that $e^{-tH_0}|\psi| = \|e^{-tH_0}\| |\psi|$, so that $|\psi| \pm \psi$ are also eigenvectors for e^{-tH_0}. Thus without loss we can suppose that ψ is almost everywhere nonnegative, and $\psi \not\equiv 0$.

Let $M = \{f \,|\, \langle \psi, |f| \rangle = 0\} = \{f \,|\, f\psi = 0 \text{ a.e.}\}$, since $\psi \geq 0$. M is a closed subspace of $L^2(Q)$ which is clearly left invariant by $L^\infty(Q)$. Moreover, if

$$f \in M_+ = \{f \in M \,|\, f \geqq 0\},$$

then $e^{-tH}f \geqq 0$ so $\langle \psi, |e^{-tH}f| \rangle = \langle \psi, e^{-tH}f \rangle = \langle e^{-tH}\psi, f \rangle = C\langle \psi, f \rangle = 0$. Thus e^{-tH_0} leaves M_+ and hence M invariant. Since $M \neq \mathscr{H}$ ($\psi \neq 0$) and $L^\infty \cup \{e^{-tH}\}_{t \geqq 0}$ acts irreducibly, $M = \{0\}$, i.e. ψ is almost everywhere nonzero and positive.

If E_g were not a simple eigenvalue, we could find two orthogonal, everywhere strictly positive functions, which is impossible, so that E_g is a simple eigenvalue.

From the proof of the last theorem, the higher-order estimates and the hypercontractive estimates, one concludes:

Theorem 6.3:

The vacuum Ω_g is an almost everywhere strictly positive function on Q-space. It is in each L^p space ($p < \infty$) and is a C^∞ vector for N, the number operator. For every $\varepsilon > 0, \Omega_g \in D(H_0^{3/2-\varepsilon})$.

Remark:

Perturbation theory suggests that $\Omega_g \notin D(H_0^{3/2})$.

(c) Other Bound States; Bound States in the Continuum

Almost nothing is known about whether bound states occur in general in the gap $(E_g, E_g + m_0)$ or above the continuum limit, and, if they do occur in the gap, whether they are finite or infinite in number. What is known is:

(i) In the $(\varphi^2)_2$ theory, Rosen [36] has found that if $g \geq 0$, no bound states appear. Bound states do appear if g is negative somewhere. There are finitely many bound states below $E_g + m_0$ and infinitely many above the continuum limit.

(ii) Using a perturbation theory argument and the $(-1)^N$ symmetry, Simon [47] has shown that there are $(\varepsilon\varphi^4 - \varphi^2)_2$ theories with bound states above the continuum in $[E_g + m_0, E_g + 2m_0)$. Using his perturbation theory argument, one can also find such theories with bound states in $[E_g, E_g + m_0)$.

Nothing else is known, but on the basis of (i) and (ii), one can conjecture or ask:

Conjecture: If g has compact support, there are at most finitely many bound states in $[E_g, E_g + m_0]$.

Conjecture: There are no bound states in $[E_g + 2m_0, \infty)$ if $P(x)$ has degree larger than 2. (This is almost surely true, but looks very hard to prove).

Question: If $P(x)$ has only positive coefficients, are there any bound states in $(E_g, E_g + m_0)$? The $(\varphi^2)_2$ case may not be reliable here, since $:\varphi^2:$ is formally a positive object minus an infinite constant, while $:\varphi^4:$ has some infinite negative operators.

Question: Is there at least a gap above E_g independent of g in case $P(x) = x^4$? This would have very important consequences in the $g \to 1$ limit [13].

Question: Is there no gap above E_g independent of g in case $P(x) = x^4 - 2m_0x^2$? That such a gap should not exist in this case is suggested by some broken symmetry ideas of Goldstone as interpreted by Wightman [49].

(d) Absolutely Continuous Spectrum

Theorem 6.4 (Hoegh-Krohn, Kato and Mugibayashi, Yakymiv):

$\sigma_{\text{a.c.}}(H) = [E_g + m_0, \infty)$ if g obeys the additional condition of being C^2, with $g', g'' \in L^2$.

Remarks:

1. Hoegh-Krohn [24] and Yakymiv [51] proved the general $P(\varphi)_2$ case using Rosen's higher-order estimates. Independently, Kato and Mugibayashi [28] proved the $(\varphi^4)_2$ result. They were unaware of Rosen's estimates and proved the $N^2 \leqq \leqq (H + b)^3$ estimate in the $(\varphi^4)_2$ case. Given Rosen's estimates, their results can be extended. The basic technique of the proof is due to Kato and Mugibayashi [29] and is a modification of ideas from potential scattering.

2. We will sketch the proof which shows that $[E_g + m_0, \infty) \subset \sigma_{\text{a.c.}}(H)$. That $\sigma_{\text{a.c.}}(H) \cap (-\infty, E_g + m_0) = \varnothing$ is a consequence of Theorem 6.1 (a).

Proof:

We will first prove that for $\psi \in D(H^{\frac{1}{2}}), f \in \mathscr{F}_1$,

$$\lim_{t \to \pm \infty} e^{iHt} e^{-iH_0 t} a^{\#}(f) e^{iH_0 t} e^{-iHt} \psi \equiv a_{\pm}^{\#}(f)$$

exists, where $\#$ stands for $\dagger$ or nothing.

Step (1): Think of f as a function on k-space and suppose first that f is smooth with compact support in $\mathbb{R} \setminus \{0\}$ and $\psi \in C^{\infty}(H)$. A simple argument, using the fact that $C^{\infty}(H) \subset D(NH_0)$, allows us to differentiate

$$F(t) = e^{iHt} e^{-iH_0 t} a^{\#}(f) e^{iH_0 t} e^{-iHt} \psi$$

yielding

$$F'(t) = i e^{iHt} [V, a^{\#}(e^{\pm i\mu t}f)] e^{-iHt} \psi \,,$$

where we have used

$$e^{-iH_0 t} a^{\#}(f) e^{iH_0 t} = a^{\#}(e^{\pm i\mu t}f),$$

and the $\pm$ sign depends on whether $\#$ is adjoint or not. We show that $\|F'(t)\| \in L^1$. The higher-order estimates imply that

$$\|F'(t)\| \leq c \|[V, a^{\#}(e^{i\mu t}f)](N + 1)^{-m}\| \|(H + b)^m \psi\|,$$

where $m \equiv \frac{1}{2}\deg P$. By an N_τ-estimate, the first factor is bounded by a constant multiple of the square root of

$$\int \frac{dk_1 \dots dk_{n-1}}{\mu(k_1) \dots \mu(k_{n-1})} \left[\int \hat{g}(k_1 + \dots + k_n) e^{i\mu(k_n)t} f(k_n) \frac{dk_n}{\mu(k_n)} \right]^2 .$$

Since f vanishes near $k = 0$, we can write

$$e^{i\mu(k_n)t} f(k_n) = -\frac{f(k_n)}{t^2} \left(\frac{\mu(k_n)}{k_n} \frac{d}{dk_n} \right)^2 e^{i\mu(k_n)t}$$

and integrate by parts in the k_n integral. Since $f, g, f', g', f'', g'' \in L^2$, we can bound $\|F'(t)\|$ by Ct^{-2}. Since $\|F'(t)\| \in L^1$, $F(t)$ is norm-Cauchy. This proves that $\lim_{t \to \pm \infty} F(t)$ exist if f and ψ obey the special conditions.

Step (2): By the estimates

$$\|a^{\#}(f)(N + 1)^{-\frac{1}{2}}\| \leq \|f\|_2, \qquad \|(N + 1)^{\frac{1}{2}} (H + b)^{-\frac{1}{2}}\| \leq c$$

for suitable c, we see that

$$\|e^{iHt} e^{-iH_0 t} a^{\#}(f) e^{iH_0 t} e^{-iHt} \psi\| \leq c \|f\|_2 \|(H + b)^{\frac{1}{2}} \psi\| \leq c,$$

so that the limits $\lim_{t \to \pm \infty}$ exist if $f \in L^2(k)$ and $\psi \in D((H + b)^{\frac{1}{2}})$.

Step (3): Since

$$[H, e^{iHt} e^{-iH_0t} a^{\#}(f) e^{iH_0t} e^{-iHt}] = e^{iHt} e^{-iH_0t} a^{\#}(\pm \mu f) e^{iH_0t} e^{-iHt} +$$

$$+ e^{iHt}[V, a^{\#}(e^{\pm i\mu t}f)] e^{-iHt}, \quad \text{if} \quad \mu f \in L^2 \quad \text{and} \quad \psi \in C^{\infty}(H_0),$$

we can obtain uniform bounds on $\|H e^{iHt} a^{\#}(e^{\pm \mu f}) e^{-iHt}\psi\|$, and so, since H is closed, $a_{\pm}^{\#}(f)\psi \in D(H)$. In this way, when $f \in \mathscr{S}$ we conclude that $a_{\pm}^{\#}(f)$ leave $C^{\infty}(H)$ invariant. On $C^{\infty}(H)$ we conclude that

$$[a_{\pm}^{\dagger}(f), a_{\pm}(g)] \psi = -\langle \bar{f}, g \rangle \psi. \tag{6.3}$$

Step (4): Let Ω_g be the vacuum for H. Then $a_{\pm}(f)\Omega_g = 0$ for

$$\|a_{\pm}(f)\Omega_g\| = \lim_{t \to \infty} \|a(e^{\pm i\mu(t)}f)\Omega_g\| = 0.$$

That $a(e^{\pm i\mu(t)}f)\phi \to 0$ if $\phi \in D(N^{\frac{1}{2}})$ follows from the fact that it goes to 0 if

$$\phi \in \overset{N}{\underset{n=1}{\oplus}} \mathscr{F}_n \text{ for any } N \text{ by the Riemann–Lebesgue lemma and}$$

$$[a(e^{\pm i\mu(t)}f)\phi_n](k_1, \ldots, k_n) = \sqrt{n+1} \int dk_{n+1}\phi_{n+1}(k_1, \ldots, k_{n+1}) f(k_{n+1}) e^{\pm i\mu(k_n)t}.$$

Step (5): Since $a_{\pm}(f)\Omega_g = 0$ and $[a_{\pm}^{\dagger}(f), a_{\pm}(g)] = -\langle \bar{f}, g \rangle$ on $C^{\infty}(H)$, we can form $\mathscr{F}_{\Omega_g}^{\pm}$, the closed linear span of the $\{a_{\pm}^{\dagger}(f_1) \ldots a_{\pm}^{\dagger}(f_n)\Omega_g\}$ and map $\mathscr{F}_{\Omega_g}^{\pm}$ onto $\mathscr{F}$ by

$$U_{+}(a_{+}^{\dagger}(f_1) \ldots a_{+}^{\dagger}(f_n)\Omega_g) = (a^{\dagger}(f_1) \ldots a^{\dagger}(f_n)\Omega_g).$$

U_{+} is unitary.

Step (6): $e^{iHs} a_{+}^{\#}(f) e^{-iHs} = a_{+}^{\#}(e^{\pm i\mu(k)}f)$ by an elementary computation, namely

$$e^{iHs}(e^{iHt} e^{-iH_0t} a^{\#}(f) e^{iH_0t} e^{-iHt}) e^{-iHs} = e^{iH(t+s)} e^{-iH_0(t+s)} a^{\#}(e^{\pm i\mu f}) e^{iH_0(t+s)} e^{-iH(t+s)}.$$

Step (7): Thus

$$U_{+}[e^{iHs} a_{+}^{\dagger}(f_1) \ldots a_{+}^{\dagger}(f_n)\Omega_g] = e^{iE_g t} a^{\dagger}(e^{i\mu t}f_1) \ldots a^{\dagger}(e^{i\mu t}f_n)\Omega_0 =$$

$$= e^{i(H_0 + E_g)t} U_{+}[a_{+}^{\dagger}(f_1) \ldots a_{+}^{\dagger}(f_n)\Omega_g].$$

Thus $U_{+}e^{iHs} = e^{i(H_0 + E_g)t} U_{+}$ or $H \upharpoonright \mathscr{F}_{\Omega_g}^{+}$ is unitarily equivalent to $H_0 + E_g$. Since

$$\sigma_{\text{a.c.}}(H_0) = [m_0, \infty), \text{ it follows that } \sigma_{\text{a.c.}}(H) \supset [E_g + m_0, \infty).$$

By Remark 2 this proves the theorem.

The above construction leads naturally to several questions. We first note that $a_{\pm}(f)\Omega = 0$ for any eigenfunction Ω of H (see step (4)). Thus

Question: Let $\mathscr{F}_{\pm} = \{\psi \,|\, a_{\pm}(f)\psi = 0 \text{ for all } f \in L^2(k)\}$. Is $\mathscr{F}_{\pm} = \{\psi \,|\, \psi \text{ is an eigenfunction of } H\}$?

Letting $\mathcal{F}^{\pm}$ be the closed linear span of $\{a_{\pm}^{\dagger}(f_1)\ldots a_{\pm}^{\dagger}(f_n)\,\psi \,|\, \psi \in \mathcal{F}_{\pm} \cap C^{\infty}(H)\}$, as in step (7) above, one can construct maps $U^{\pm}:\mathcal{F}^{\pm} \to \mathcal{F} \otimes \mathcal{F}_{\pm}$, so that $U^{\pm}(H \restriction \mathcal{F}^{\pm})(U^{\pm})^{-1} = H_0 \otimes 1 + 1 \otimes H \restriction \mathcal{F}_{\pm}$. One can ask:

Question (Asymptotic Completeness): Is $\mathcal{F}^{+} = \mathcal{F}^{-} = \mathcal{F}$?

(e) Singular Continuous Spectrum

In the study of nonrelativistic Hamiltonian operators, the proof of the absence of singular continuous spectrum is one of the hardest spectral questions. It has only recently been solved for atomic Hamiltonians [1]. There is almost no information in the case of $P(\varphi)_2$-spatially cutoff Hamiltonians, although the natural conjecture is $\sigma_{\text{sing}} = \varnothing$. Two remarks to be made are:

(i) In case g has compact support, Rosen [36] has proved $\sigma_{\text{sing}} = \varnothing$ in the soluble $(\varphi^2)_2$ model.

(ii) If the answers to both questions at the end of the discussion (d) above were "yes", then $\sigma_{\text{sing}} = \varnothing$. However, from our experience with nonrelativistic quantum theory, we suspect that proving $\mathcal{F}^{+} = \mathcal{F}^{-} = \mathcal{F}$ will be harder than proving $\sigma_{\text{sing}} = \varnothing$, and may well require first showing that $\sigma_{\text{sing}} = \varnothing$.

REFERENCES

[1] E. Balslev and J. Combes, *Commun. Math. Phys.,* **22**, 280 (1971).

[2] J. Cannon and A. Jaffe, *Commun. Math. Phys.,* **17**, 261 (1970).

[3] J. Cook, *Trans. Am. Math. Soc.,* **74**, 222–245 (1953).

[4] W. Eachus, Thesis, Syracuse University (1970).

[5] P. Federbush, *J. Math. Phys.,* **12**, 2050–2051 (1971).

[6] K. O. Friedrichs, *Comm. Pure Appl. Math.,* **6**, 1–72 (1953).

[7] J. Ginibre and G. Velo, *Commun. Math. Phys.,* **18**, 65–81 (1970).

[8] J. Glimm, *Commun. Math. Phys.,* **8**, 12–25 (1968).

[9] J. Glimm, *Commun. Math. Phys.,* **5**, 343 (1967).

[10] J. Glimm and A. Jaffe, *Phys. Rev.,* **176**, 1945–1951 (1968).

[11] J. Glimm and A. Jaffe, *Ann. Math.,* **91**, 362–401 (1970).

[12] J. Glimm and A. Jaffe, *Acta math.,* **125**, 203–267 (1970).

[13] J. Glimm and A. Jaffe, "Field Theory Models", In: *Les Houches Lectures,* 1970, C. DeWitt and R. Stora, Eds. Gordon and Breach, New York (1971).

[14] J. Glimm and A. Jaffe, *Ann. Phys,* **60**, 321–383 (1970)

[15] J. Glimm and A. Jaffe, *J. Func. Anal.*, **7**, 323–358 (1971).

[16] J. Glimm and A. Jaffe, *London Math. Soc., London Lectures,* 1971.

[17] J. Glimm and A. Jaffe, *New York Univ.—Harvard Univ. Preprint* (1972).

[18] J. Glimm and A. Jaffe, *Commun. Math. Phys.,* **4**, 253–258 (1971).

[19] J. Glimm and A. Jaffe, *Commun. Pure Appl. Math.,* **22**, 401–414 (1969).

[20] L. Gross, *Bull. Am. Math. Soc.,* **77**, 343–347 (1971).

[21] L. Gross, *J. Func. Anal.,* **10**, 52–109 (1972).

[22] M. Guenin, *Commun. Math. Phys.,* **3**, 120–132 (1966).

[23] K. Hepp, "Théorie de la renormalisation", Springer Lecture Notes in Physics (1969).

[24] R. Hoegh-Krohn, *Commun. Math. Phys.,* **18**, 109 (1970).

[25] A. Jaffe, *J. Math. Phys.,* **7**, 1251–1255 (1966).

[26] A. Jaffe, Thesis, Princeton Univ. (1966).

[27] T. Kato, "Perturbation Theory for Linear Operators", Springer Verlag, New York (1966).

[28] Y. Kato and N. Mugibayashi, *Prog. Theor. Phys.,* **45**, 628 (1971).

[29] Y. Kato and N. Mugibayashi, *Prog. Theor. Phys.,* **30**, 103 (1963).

[30] J. Konrady, *Commun. Math. Phys.,* **22**, 295 (1971).

[31] B. Gidas, *Univ. Michigan Preprint* (1971).

[32] D. Masson and W. McClary, *Commun. Math. Phys.,* **21**, 71 (1971).

[33] E. Nelson, In: "Mathematical Theory of Elementary Particles", R. Goodman and I. Segal, Eds. M.I.T. Press (1966).

[33a] E. Nelson, *Proc. Berkeley—Am. Math. Soc. Summer Institute,* 1971.

[34] L. Rosen, *Commun. Math. Phys.,* **16**, 157–183 (1970).

[35] L. Rosen, *Commun. Pure Appl. Math.,* **24**, 417 (1971).

[36] L. Rosen, *J. Math. Anal. Appl.,* **38**, 276–311 (1972).

[37] L. Rosen, *J. Math. Phys.,* **13**, 918–927 (1972).

[38] L. Rosen and B. Simon, *Trans. Am. Math. Soc.,* **165**, 365–379 (1972).

[39] I. Segal, *Ann. Math.,* **92**, 462–481 (1970).

[40] I. Segal, *Proc. Nat. Acad. Sciences,* **57**, 1178–1183 (1967).

[41] I. Segal, *Trans. Am. Math. Soc.,* **81**, 106–134 (1956).

[42] I. Segal, *J. Func. Anal.,* **4**, 404–456 (1969).

[43] I. Segal, *Bull. Am. Math. Soc.,* **75**, 1390–1395 (1969).

[44] B. Simon, *J. Func. Anal.,* **10**, 251–258 (1972).

[45] B. Simon, *Phys. Rev. Lett.,* **25**, 1583 (1970).

[46] B. Simon, *Ann. Phys.,* **58**, 76–136 (1970).

[47] B. Simon, *Proc. AMS* (to appear).

[48] B. Simon and R. Hoegh-Krohn, *J. Func. Anal.,* **9**, 121–180 (1972).

[49] A. S. Wightman, *Physics Today,* **22**, 53 (1969).

[50] A. S. Wightman, "Cargèse Lectures", Gordon and Breach (1967).

[51] Y. Yakymiv, *Kiev Univ. Preprint* (1971).

References to Functional Analysis

(a) For Nelson's Theorem:

E. Nelson, *Ann. Math.,* **67** 572–615 (1959); A. E. Nussbaum, *Ark. Mat.,* **6**, 179 (1965).

(b) For the Q-space construction:

I. Gelfand, D. Raikov, and G. Shilov, "Commutative Normed Rings", Chelsea Pub. Co., Bronx, N.Y. (1964).

(c) For interpolation theorems:

E. Stein, *Trans. Am. Math. Soc.,* **83**, 482–492 (1956).

(d) For the Trotter Product Formula:

H. Trotter, *Proc. Am. Math. Soc.* **10**, 545–551 (1959).

P. Charnoff, *J. Func. Anal.,* **2**, 238–242 (1968).

(e) For Jacobi Matrices:

M. Stone, "Linear Transformations in Hilbert Space", *Am. Math. Soc.* (1932).

Note added in proof. The past year has seen considerable progress in the study of the $P(\varphi)_2$-model. The most exciting results involve the infinite volume limit and use ideas of Nelson [33a]. This work is outside the scope of this review, but see (1) J. Dimock, J. Glimm, and T. Spencer, *NYU Preprint* (1972); (2) E. Nelson, *Princeton Preprint* (1972); (3) F. Guerra, *Phys. Rev. Lett.,* **28**, 1213 (1972); (4) F. Guerra, L. Rosen, and B. Simon, *Commun. Math. Phys.,* **27**, 10–22 (1972) and *Princeton Preprint* (1972).

Some refinements of the results in this review have also appeared:

(I) The $L^1 \cap L^2$ condition on g can be replaced with $L^{1+\varepsilon} + L^2$ ($g \geqq 0$, of course). Actually, without changing methods, $L^1 \cap L^{1+\varepsilon}$ can be treated (see F. Guerra et al., *Princeton Preprint*).

(II) Lemma 2.3 has been improved by Nelson (*Princeton Preprint*) to show that $\Gamma(c): L^p \to L^q$ if and only if $c \leqq (p-1)^{1/2}/(q-1)^{1/2}$ and the bound is always 1.

(III) A useful general lower bound for spatially cutoff field theories appears in Guerra et al. (*Commun. Math. Phys.,* **27**, 10–22 (1972) and to appear): For any $\alpha > 1$, there is a T so that

$$\exp(-TE(H_0 + V)) \leqq \langle \Omega_0, \exp(-T(H_0 + \alpha V))\Omega_0 \rangle$$

(IV) Gidas [31] has extended Federbush's lower bound method to spatial cutoff as opposed to box cutoff (see p. 205).

(V) The self-adjointness theorem for hypercontractive semigroups has been extended to prove essential self-adjointness on $C^\infty(H_0) \cap D(V)$ (see B. Simon, *Math. Ann.,* to appear).

(VI) Several papers on the theory of hypercontractive semigroups have appeared:

Segal (*Inv. Math.* (1972)), Semenov (*Kiev Preprint*), Faris (*Battelle Preprint*). These illustrate and extend the theory.

(VII) The answer to the first question on p. 218 is probably "yes" for some polynomials with positive coefficients. For example in a $:q^4:$ anharmonic oscillator, in the lowest order nonvanishing perturbation theory, the mass gap is decreasing!

Scaling Laws and Universality — or Statistical Mechanics Is Not Dead !

H. EUGENE STANLEY*

*Physics Department, Massachusetts Institute of Technology,
Cambridge, Massachusetts* 02139 (*USA*)

Abstract

We present an introduction for the non-specialist to some of the current problems of phase transitions and critical phenomena. Then we develop from a simple and unified point of view the scaling law hypotheses for thermodynamic functions, static correlation functions, and dynamic correlation functions. In particular, we make clear the relation between homogeneity assumptions and "scaling functions", and we show how new experimental tests of the scaling hypotheses might be carried out. We discuss a simple geometric interpretation of the scaling hypothesis and show how this has led to the recent calculation of the thermodynamic scaling functions. Further we show how the original two-exponent relations (such as $dv = 2 - \alpha$) acquire correction terms that turn them into thermodynamic function scaling laws (such as $\alpha + 2\beta + \gamma = 2$). The universality hypothesis is introduced, and recent calculations supporting its validity are discussed. The question of how critical-point exponents vary with D and d is discussed, where D and d denote the spin space and lattice dimensionality, respectively. Finally, a brief introduction to a purely phenomenological "dynamic cluster approximation" is presented.

Contents

* Supported by the National Science Foundation, the O.N.R., and the A.F.O.S.R.

INTRODUCTION

There has been a lot of talk in the US—and among some of the participants of this conference—to the effect that physics is dead, or at least that statistical mechanics and field theory are dead!

I could not disagree more with this claim, and I hope to show you why. I shall focus upon the area of phase transitions and critical phenomena, a field with a history extending back at least a century (to Andrews' discovery of the critical point of carbon dioxide) but which still generates fresh ideas every year. I choose this field to illustrate my thesis because this is one of two fields in which my own research interests lie... I could as well have chosen the area of cooperative phenomena in biological systems, where perhaps the thesis would be easier still to prove.

Since a greater proportion of you are field theorists than "statistical mechanists", I shall not assume that you are already familiar with the concepts I shall use, and I shall accordingly develop the subject essentially "from the beginning". Since many of you are familiar with scaling concepts from field theory, I shall emphasize this aspect of research in the critical phenomena field. Although phase transitions have developed by a careful interplay of theory and experiment, I shall—for the sake of pedagogy and to save time—concentrate mainly on the theoretical aspects here.

Let me organize my introduction to phase transitions about three simple questions:
(i)   "WHAT HAPPENS?"
(ii) "WHY STUDY?" (i.e., "SO WHAT?")
(iii) "WHAT DO WE ACTUALLY DO?"

1. "WHAT HAPPENS?"

What happens near the critical point is easily explained by means of an example. Consider a simple magnet. The most striking macroscopic property of a simple magnet is that it is magnetized—it can, e.g., pick up thumbtacks. So let us measure the number of thumbtacks (the "magnetization"—cf. Figure 1) as we heat the magnet at a uniform rate. As we heat the magnet, the thumbtacks fall off one by one smoothly and continuously. As we continue to heat, the thumbtacks fall off at a greatly increasing rate, in fact a rate that goes to infinity (cf. Figure 1) as we approach a certain temperature at which there are no thumbtacks left stuck. This temperature is called the critical temperature, denoted by T_c, and phenomena associated with the critical temperature, such as that just described, are called critical phenomena.

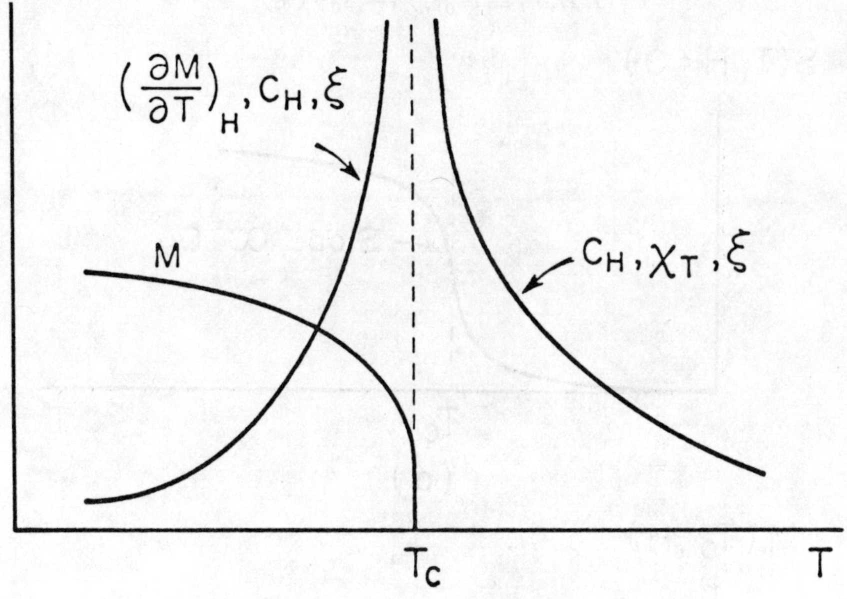

Figure 1

Behavior (schematic) of some common physical quantities near the critical point.

Examples of other critical phenomena (cf. Figure 2) are the singularities in two important "response functions",

   (i)   the constant field specific heat,

C_H = response in heat content to a change in temperature

$$= T\left(\frac{\partial S}{\partial T}\right)_H, \tag{1.1}$$

where here $S = S(T, H)$ denotes the entropy of the system, and

   (ii)   the isothermal susceptibility

$$\chi_T = \text{response of magnetization to a change in magnetic field}$$

$$= \left(\frac{\partial M}{\partial H}\right)_T, \tag{1.2}$$

where $M = M(T, H)$ is the magnetization and H is the magnetic field.

Thermodynamic functions are macroscopic in nature. What is going on at a microscopic level to account for the anomalies in the thermodynamic functions? Simply said, the motion of increasing numbers of particles is becoming correlated; e.g., the correlation function,

$$C_2(T, H, r) \equiv \langle s_0 s_r \rangle - \langle s_0 \rangle \langle s_r \rangle, \tag{1.3}$$

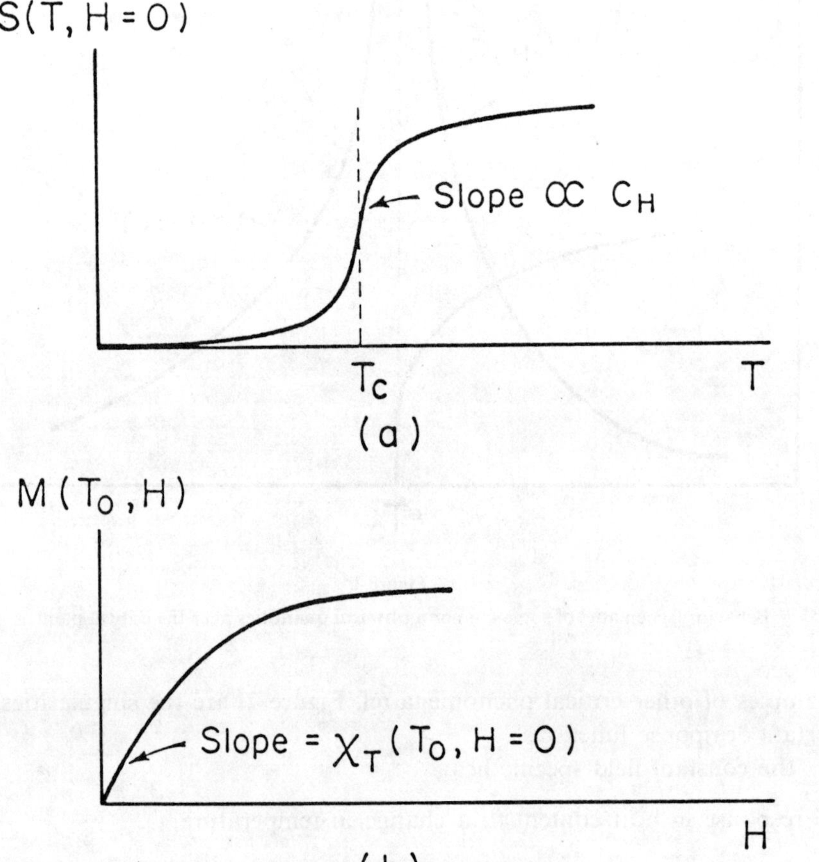

Figure 2

Definitions of two response functions, the constant-field specific heat $C_H(T, H = 0)$, and the isothermal susceptibility evaluated in zero field $\chi_T(T, H = 0)$. Both functions are singular at the critical point of a simple magnet.

which describes the degree of correlation among the constitutent magnetic moments s_r of our magnet, is becoming long-range. Here the angular brackets denote thermal averages.

In fact, the "moments" of the correlation function,

$$\mu_a(T, H) \equiv \int |r|^a C_2(T, H, r)\, dr \,, \tag{1.4}$$

are found to diverge for all positive a and even for a limited range of negative a (roughly $a \gtrsim -2$)! In particular, the zero-field isothermal susceptibility is directly proportional to the zeroth moment

$$\chi(T, H = 0) \propto \mu_0(T, H = 0) = \int C_2(T, 0, r)\, dr \,, \tag{1.5}$$

so the divergence of the susceptibility can be seen to be directly related to the increase in range of $C_2(T, H, r)$.

As an aside, let me say that a few years ago (as a beginning graduate student) I proposed, on the basis of certain numerical calculations, that in a class of two-dimensional ($d = 2$) systems, one could have a divergent susceptibility. This proposal was greeted with confusion at first, because people thought that a spontaneous magnetization would have to appear at the same temperature as the susceptibility diverged, and it was known rigorously that one could not have a nonzero spontaneous magnetization in this class of $d = 2$ systems. However, we emphasized (Stanley and Kaplan, 1966, 1967; Stanley, 1968 a and b) that the two singularities needn't be related, as the spontaneous magnetization is given by

$$M^2 \propto \lim_{r \to \infty} \langle s_0 s_r \rangle \tag{1.6}$$

and should be termed "infinite range order", while the susceptibility, given by eqn. (1.5), reflects the existence of "long-range order" that needn't be infinite in range! Suppose, for example, that

$$C_2(T, 0, r) \sim r^{-\lambda T} \,. \tag{1.7}$$

Then by substituting (1.7) into (1.6) we see that $M = 0$ for all $T > 0$, but on substituting (1.7) into (1.5) we see that $\chi = \infty$ for small T (for $T < 2/\lambda$).

This proposal was greeted with due skepticism at the time, justifiable perhaps because it involved the concept of a phase transition to an altogether new and different low-temperature phase—one with sufficient long-range order for χ to diverge but with no infinite-range order (so that $M = 0$) (cf. Figure 3). In fact, about the only theorist who would really tolerate the proposal was Freeman Dyson, and he did a crude low-temperature calculation that supported the general picture (F. J. Dyson, unpublished). Experimentalists were less intolerant, and the proposal of phase transitions in $d = 2$ systems stimulated numerous investigators to consider doing experiments on two-dimensional magnets, or the closest approximation thereto that could be made. An entire industry of "two-dimensional magnetism" has since

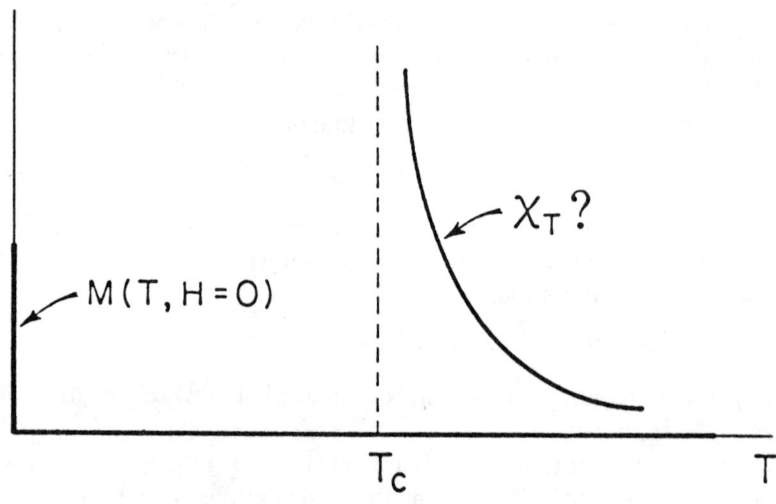

Figure 3

The situation for $d = 2$ materials for spin dimensionality $D > 1$.

grown up, and three sessions of the international congress on magnetism last summer were devoted to this subject (see the review by Birgeneau, 1973). It should be said that the original question of whether χ indeed diverges for the ideal, isotropic, $d = 2$ system remains unresolved to this date!

2. "WHY STUDY?"

There are, of course, many answers to the question "why study critical phenomena?", but certainly the simplest answer (and in fact almost an answer to why anything is studied) is that there exist rather striking discrepancies between the predictions of closed-form theories and our findings in the real world. For example, in Figure 4 we compare the temperature dependences of the functions discussed in Section 1 with the predictions of the mean field theory of cooperative phenomena. As Kac, Brout, and others have emphasized (Kac, 1968; Brout, 1965), the mean-field theory (mft) corresponds to a model Hamiltonian,

$$\mathcal{H}^{\text{mft}} \equiv -\frac{J}{N} \sum_{i=1}^{N} \sum_{j=1}^{N} s_i s_j, \tag{2.1}$$

in which each spin s_i interacts with all other spins in the system with an equal exchange (J/N). Such an interaction Hamiltonian is unlikely to be a realistic model to describe the effects noted above, such as the sudden dramatic increase in the correlation length.

In fact, one feature that phase transitions in a wide range of system Hamiltonians

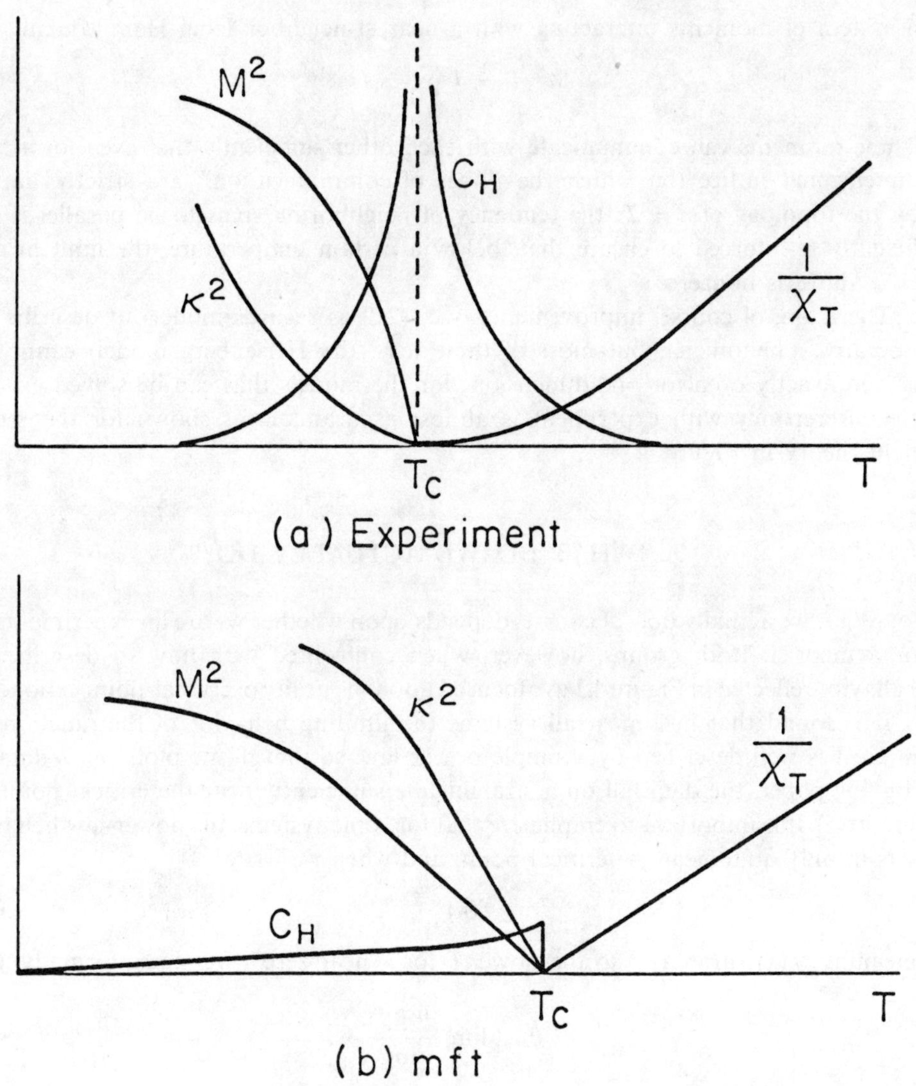

Figure 4

Behavior (schematic) of common physical quantities near the critical point (a), compared with the predictions of the mean field theory (b).

have in common is the fact that they all seem to involve the "propagation of order" from one particle to the next via relatively short-range forces. That a system with nearest-neighbor forces *only* can even exhibit a phase transition in the first place has fascinated and intrigued the greatest minds of this century. For example, consider a system of moments interacting with a nearest-neighbor Ising Hamiltonian

$$\mathscr{H} = - J \sum_{\langle ij \rangle}^{*} s_i s_j. \tag{2.2}$$

These moments can communicate with each other sufficiently that even for a two-dimensional lattice (on which the "lines of communication" are strictly limited by the topology of $d = 2$), the tendency of neighboring spins to be parallel is sufficiently re-inforced to ensure that, below a certain temperature, the limit in eqn. (1.7) above is nonzero.

There are, of course, improvements over (2.2) as regards models to describe co-operative phenomena, but most of these (e.g., the Heisenberg model) cannot be solved exactly even for one dimension; for the models that can be solved exactly, the discrepancy with experiment is almost as dramatic as shown for the mean-field theory in Figure 4.

3. "WHAT DO WE ACTUALLY DO?"

"What we actually do", of course, depends upon whether we are an experimentalist or a theorist. Both groups, however, when confronted with how to describe the behavior reflected in Figure 4, have focused upon the utility of critical-point exponents.

It is found that in almost all systems the limiting behavior of the functions of interest is well described by a simple power law, so that if one plots one's data on log–log paper, the data fall on a straight line sufficiently near the critical point (cf. Figure 5). It is important to emphasize that for some systems, the power-law behavior sets in only quite near the critical point, and when we write

$$f(x) \sim x^{\theta}, \tag{3.1a}$$

meaning "$f(x)$ varies as x to the power θ for x near zero", we mean, formally, that

$$\theta \equiv \lim_{x \to 0} \frac{\log f(x)}{\log x}. \tag{3.1b}$$

In general, there is a different exponent for each function and each path of approach to the critical point. The thermodynamic functions considered above all concern approaches to the critical point ($T = T_c, H = 0$), in which $H = 0$ and $\tau \equiv T - T_c \to 0$. Accordingly, we define the three exponents α, β, and γ in Table 1 to describe, respectively, the behavior of the specific heat, the spontaneous magnetization, and the isothermal susceptibilty along this path.

The second column gives a typical range of experimental values for the exponents, while the third column gives the values predicted by the mean-field theory. The reader would do well to verify from inspection of Figure 4 that the mean-field ex-

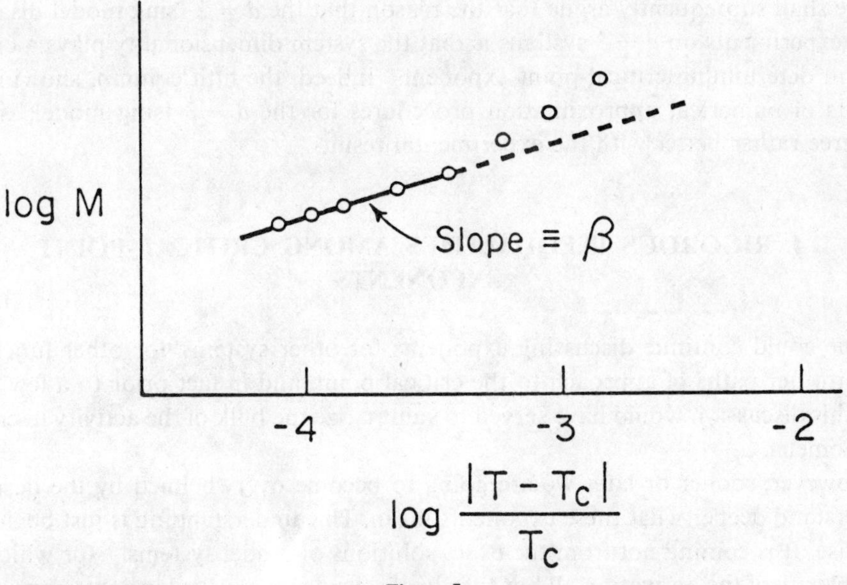

Figure 5

Sketch of definition of the critical-point exponent $\beta[M \sim |\tau|^{\beta}]$ to illustrate the general definition (3.1).

ponents are correctly listed. That the agreement is far from perfect is consistent with the discrepancies between Figures 4a and 4b (experiment and mft).

Theorists have endeavored to calculate the exponents for as many models as possible. For example, the fourth column shows the exponents calculated for the $d = 2$ Ising model, for which many of the zero-field properties are known exactly. We see that these exponents agree no better with experiment than do the mean field exponents—in fact, they err in the opposite direction, with the experimental numbers lying in between the mft and $d = 2$ Ising predictions.

TABLE 1

Definitions and typical values of selected critical-point exponents for
a simple magnet.

	Definition	Experiment	mft	$d = 2$ Ising	$d = 3$ Ising
α', α	$C_H \sim (-\tau)^{-\alpha'}$ $\sim \tau^{-\alpha}$	-0.1 to 0.2	0	0	$\simeq 1/8$
β	$M \sim (-\tau)^{\beta}$	0.2 to 0.4	$1/2$	$1/8$	$\simeq 5/16$
γ', γ	$\chi_T \sim (-\tau)^{-\gamma'}$ $\sim \tau^{-\gamma}$	1.1 to 1.5	1	$7/4$	$\simeq 5/4$
$\alpha' + 2\beta + \gamma'$		$\simeq 2$	2	2	$\simeq 2$

We shall subsequently argue that the reason that the $d = 2$ Ising model disagrees with experiments on $d = 3$ systems is that the system dimensionality plays a crucial role in determining critical-point exponents. Indeed, the fifth column, showing the results of numerical approximation procedures for the $d = 3$ Ising model, is seen to agree rather better with the experimental results.

4. RIGOROUS INEQUALITIES AMONG CRITICAL-POINT EXPONENTS

One could continue discussing exponents for other systems, for other functions, or for other paths of approach to the critical point, and in fact prior to a few years ago this discussion would have served to summarize the bulk of the activity in critical phenomena.

However, sooner or later we are going to become overwhelmed by the desire to understand deeper what these exponents mean. This understanding is just beginning to arise. It is coming not from the exact solutions of model systems—for which the complexity of the derivation all but totally obscures any physical insights concerning the magnitude of the exponent obtained—but rather from an altogether different approach. This approach began historically with the introduction of rigorous relations among the critical-point exponents—relations which took the form of inequalities and generally involved three exponents. The inequality involving α', β, and γ' is simply

$$\alpha' + 2\beta + \gamma' \geqq 2 \tag{4.1}$$

and was first proposed to be an equality by Essam and Fisher (1963) on the basis of some calculations on an artificial system. Relation (4.1) was proved, as an inequality, later the same year by Rushbrooke (Rushbrooke, 1963), and is generally called the Rushbrooke inequality. Rushbrooke derived (4.1) for a magnetic system by a simple and elegant application of the relation

$$C_H - C_M = T\left(\frac{\partial M}{\partial T}\right)_H^2 \bigg/ \chi_T , \tag{4.2}$$

which is a transcription to magnetic language of a fundamental equation for a simple fluid that is familiar to all who have studied thermodynamics; it is appropriate that Rushbrooke himself should have seen how to exploit this equation in a fashion that was to lead to a whole new era in the study of critical phenomena, for the fluid analog of (4.2) appears as the first equation on the first page of Rushbrooke's classic text on statistical mechanics that he had written some 14 years earlier (Rushbrooke, 1949). As one who has just completed the proofreading of a book (Stanley, 1971), I can vouch that an author, confronted with the possibility of errors that will "hound him till his death," reads and re-reads the first pages time and time again.

I shall pause from time to time to set examples ("exercises") for the reader. I have chosen these to be sufficiently simple for the entire set to be done on the back of a napkin or two over a leisurely lunch.

Exercise 1

Derive (4.1) from (4.2).

The inequality (4.1) is only representative of a host of "rigorous" inequalities among the critical-point exponents; a very recent review appears in Sec. 9C of Griffiths (1972). I use the word "rigorous" with reservations in front of this audience, because some of the inequalities are in fact far from rigorous, but the arguments used would be accepted by most reasonable men—if not by the stubborn. I might also add that the proofs of some of the other inequalities lack the simplicity and elegance that characterizes the proof of the Rushbrooke inequality.

You will notice from Table 1 that the exponents for all the systems shown satisfy the Rushbrooke inequality— but just barely. In fact, when the inequality was first proposed, there were numerous examples of experiments whose results appeared to contradict (4.1). This caused some concern among many professors, because the work involved in making an experimental determination of an exponent was often so much that one could not simply re-run the experiment. Moreover, the people who had actually performed the experiments had completed their Ph.D.'s and were off on postdoctoral fellowships in various exotic places. The thought must have entered several minds to follow the pattern of US automobile manufactures and to recall certain Ph. D.'s, but fortunately there was another "out"—since the inequalities generally involved three exponents, and since any *one* of the laboratories generally was not responsible for measuring all three, there was always the chance that "the other laboratory" had erred in its experimental determination. Eventually all was sorted out—whether by merely multiplying error bars or by repeating the entire experiment, and it now appears that essentially all data satisfy (4.1), even if "just barely"!

5. STATIC AND DYNAMIC SCALING HYPOTHESES

That exactly-soluble systems appeared to satisfy (4.1) [and most of the other "rigorous" inequalities] as *equalities*, and the fact that most experimental systems and numerical calculations also were not inconsistent with the possibility that (4.1) holds as an *equality*, did not go unnoticed. However, all attempts to rigorously prove (4.1)—or any of the other relations—as equalities have been singularly unsuccessful.

It is in the finest tradition of theoretical physics that when one cannot solve the original problem one seeks to replace it with a simpler problem that one can solve (Kac, 1964). In this instance, the "breakthrough" occurred in 1965—two years after Rushbrooke—by many investigators working independently (Widom, 1965; Domb and Hunter, 1965; Kadanoff, 1966; Patashinskii and Pokrovskii, 1966; Griffiths, 1967). This work generally goes by the name of the "homogeneity" or "scaling law" approach to critical phenomena.

There are almost as many approaches to the scaling problem as there are investigators. The approach I shall tell you about here is one developed in collaboration with Mr. Alex Hankey of MIT, with, more recently, some not inconsiderable assistance from Professor Tom Chang of North Carolina State University. We believe our approach has the virtue of being extremely simple and systematic. In particular, we obtain all the predictions of scaling for both the static and the dynamic regime from the same unified approach. Also, we have been able to obtain additional predictions using our approach, which are of particular interest and will be mentioned at the appropriate time.

Our approach makes systematic use of a class of functions which we call "generalized homogeneous functions" (GHF's). A function $f(x, y, z, \ldots)$ is a GHF if we can find functions $g_x(\lambda)$, $g_y(\lambda)$, $\ldots$, $g_f(\lambda)$ such that for all positive λ,

$$f[g_x(\lambda)x, g_y(\lambda)y, \ldots] = g_f(\lambda)f[x, y, \ldots], \tag{5.1}$$

where the functions $g_i(\lambda)$ are arbitary except that they possess inverses. It is elementary to show (Chang, Hankey, and Stanley, 1972; Cooper, 1968) that (5.1) is equivalent to the statement that there exist numbers $a_x, a_y, \ldots, a_f$ such that

$$f(\lambda^{a_x} x, \lambda^{a_y} y, \ldots) = \lambda^{a_f} f(x, y, \ldots). \tag{5.2}$$

Homogeneous functions form a subset of GHF's for which $a_x = a_y = \ldots$. Thus, for example, $f(x, y) = xy^2 + y^4$ and $x^2 + y^2$ are both GHF's and the latter is a homogeneous function as well; the scaling powers are $a_x = 1/2$, $a_y = 1/4$ in the first example, and $a_x = a_y = 1/2$ in the second example, with $a_f = 1$ for both cases. Clearly, we can multiply all scaling powers by an arbitrary number p, because any function that satisfies (5.2) with scaling powers $a_x, a_y, \ldots, a_f$ also satisfies (5.2) with scaling powers $pa_x, pa_y, \ldots, pa_f$ and, *unless* $a_f = 0$, we can choose p such that the scaling power of the function is unity.

Although the set of functions which are GHF's is vastly larger than the set of functions that are homogeneous, GHF's nevertheless form a rather small class of functions. However, the class of functions that can be closely approximated by some GHF near the origin $x = y = \ldots = 0$ is considerably larger (with notable exceptions such as functions possessing essential singularities, and functions with numerous cross terms such as $xy^2 + x^2y^3 + x^3y$). This is the key to the apparent success of the scaling hypothesis, which is taken up in the next section.

The scaling hypothesis is just that: a hypothesis. It involves GHF's, and because of the properties of GHF's, it can be made about a variety of functions. We shall make a scaling hypothesis about three different classes of functions, thermodynamic functions (TF), static correlation functions (SCF), and dynamic correlation functions (DCF). We shall show below that these statements are not entirely independent of one another, and that in fact,

$$\text{DCF Hypothesis} \Rightarrow \text{SCF Hypothesis} \Rightarrow \text{TF Hypothesis}. \tag{5.3}$$

Thus it is at least logically possible that, say, the TF and SCF Hypotheses are true for a system but the DCF Hypothesis is not; but the converse is *not* possible. (This was not always appreciated—see, e.g., the discussion in the Proceedings of the 1968 International Conference on Statistical Mechanics at Kyoto.) We shall prove relation (5.3) after we state the hypotheses:

TF Hypothesis: Close to the critical point $\tau = H = 0$, the singular part of the Gibbs potential per spin $G(\tau, H)$ is "asymptotically" a GHF.

SCF Hypothesis: Close to the critical point and for large $|r|$, the static correlation function $C_2(\tau, H, r)$ is a GHF.

DCF Hypothesis: Close to the critical point and for large $|r|$ and t, the dynamic correlation function $\mathscr{C}_2(\tau, H, r, t)$ is a GHF.

Here the dynamic correlation function is defined in analogy to the definition of the static correlation function (cf. eqn. (1.3)).

According to the definition of a GHF (eqn. (5.1)), we can write out equations which the three functions mentioned above must satisfy:

TF Hypothesis:

$$G(\lambda^{a_\tau} \tau, \lambda^{a_H} H) = \lambda G(\tau, H); \tag{5.4a}$$

SCF Hypothesis:

$$C_2(\lambda^{b_\tau} \tau, \lambda^{b_H} H, \lambda^{b_r} r) = \lambda\, C_2(\tau, H, r); \tag{5.4b}$$

DCF Hypothesis:

$$\mathscr{C}_2(\lambda^{b_\tau} \tau, \lambda^{b_H} H, \lambda^{b_r} r, \lambda^{b_t} t) = \lambda\, \mathscr{C}_2(\tau, H, r, t). \tag{5.4c}$$

Note that we have chosen the same scaling parameters for the dynamic as for the static correlation function, since

$$C_2(\tau, H, r) = \mathscr{C}_2(\tau, H, r, t = 0), \tag{5.5}$$

and we know that the scaling parameters of a given function are unique up to a multiplicative factor p.

6. PROPERTIES OF GHF's

To work out the implications of the three hypotheses, we need only apply in a straightforward fashion certain of the properties of GHF's. These we summarize in the form of theorems, the proofs of which follow from the following simple observation that if

$f(x, y, \ldots)$ is a GHF with scaling powers $a_x, a_y, \ldots,$ and a_f, and if
$g(x, y, \ldots)$ is a GHF with scaling powers $a_x, a_y, \ldots, a_g,$ then

$$\text{the product } fg \text{ is a GHF with scaling power } a_f + a_g; \tag{6.1a}$$

$$\text{the quotient } f/g \text{ is a GHF with scaling power } a_f - a_g; \tag{6.1b}$$

and

$$\text{the sum } f \pm g \text{ is a GHF if and only if } a_f = a_g. \tag{6.1c}$$

Theorem 1

All partial derivatives of a GHF are also GHF's, and the scaling power is determined in terms of the original scaling powers. In particular, the partial derivative $f_x(x, y, \ldots)$ is a GHF with scaling power $1 - a_x$,

$$f_x(\lambda^{a_x} x, \lambda^{a_y} y, \ldots) = \lambda^{1 - a_x} f_x(x, y, \ldots). \tag{6.2}$$

Theorem 2

All Legendre transforms of a GHF are also GHF's, and the scaling powers of the new "conjugate" variables are determined in terms of the original scaling powers. For example, the function

$$\bar{f}(\bar{x}, y, \ldots) = f(x, y, \ldots) - x\bar{x} \tag{6.3}$$

denotes the Legendre transform of $f(x, y, \ldots)$ in which the conjugate variable $\bar{x} \equiv f_x$ replaces x as independent variable. Then properties (6.1a) and (6.1c) imply that

$$\bar{f}(\lambda^{1 - a_x} x, \lambda^{a_y} y, \ldots) = \lambda^{a_f} f(x, y, \ldots). \tag{6.4}$$

Because of Theorems 1 and 2, the TF Hypothesis implies that every thermodynamic function is a GHF, since there is no function that cannot be obtained from the Gibbs potential by some combination of Legendre transforms and partial derivatives.

Theorem 3

All Fourier transforms of a GHF are also GHF's, with the scaling powers of the transformed variables being determined in terms of the original scaling parameters. Specifically, if

$$\hat{f}(\hat{x}, y, \ldots) \equiv \int dx f(x, y, \ldots) e^{i\hat{x} \cdot x}, \tag{6.5}$$

then

$$\hat{f}(\lambda^{-a_x} \hat{x}, \lambda^{a_y} y, \ldots) = \lambda^{1 + da_f} \hat{f}(\hat{x}, y, \ldots), \tag{6.6}$$

where d is the dimensionality of the variable x.

Using Theorem 3, we can complete the proof of the logical relation (5.3). The first implication sign follows directly from the fact that $C_2(\tau, H, r) = \mathscr{C}_2(\tau, H, r, t = 0)$, since if a function $f(x, y, \ldots)$ is a GHF, then so is the function $f(x = 0, y, \ldots)$.

The second implication sign follows from the following argument: Theorem 3 implies that the "structure factor"

$$S(\tau, H, \boldsymbol{q}) \equiv \int d\boldsymbol{r}\, C_2(\tau, H, \boldsymbol{r})\, e^{i\boldsymbol{q}\cdot\boldsymbol{r}} \tag{6.7a}$$

is a GHF,

$$S(\lambda^{b_\tau}\tau, \lambda^{b_H}H, \lambda^{-b_r}q) = \lambda^{1+db_r}S(\tau, H, q). \tag{6.7b}$$

But from (1.5) we see that the isothermal susceptibility is just the $q = 0$ Fourier component of the correlation function,

$$\chi_T(\tau, H) \sim S(\tau, H, q = 0), \tag{6.8}$$

apart from factors, such as kT, that do not involve the independent variables which are being scaled, so that χ_T is a GHF with the same scaling power as S. Now $\chi_T(\tau, H) = -(\partial^2 G/\partial H^2)_T$, and if the integration factors are consistent with scaling, then it follows that $G(\tau, H)$ is a GHF,

$$G(\lambda^{b_\tau}\tau, \lambda^{b_H}H) = \lambda^{1+db_\tau+2b_H}\, G(\tau, H). \tag{6.9}$$

Thus $G(\tau, H)$ is a GHF and (5.3) is justified. By comparing (6.9) and (5.4a), note that the scaling parameters a_τ, a_H in (5.4a) are expressible in terms of the b's of (5.4b) and (5.4c), and comparison of (5.4a) and (6.9) results in

$$a_\tau = b_\tau/(1 + db_r + 2b_H); \quad a_H = b_H/(1 + db_r + 2b_H). \tag{6.10}$$

We have seen that Theorems 1–2 provide the "bridge" between the GHF hypothesis for $G(\tau, H)$ and a GHF property for all thermodynamic functions, and that Theorem 3 provides a bridge between the GHF hypothesis for the correlation function and the GHF properties of the thermodynamic functions. We shall see that Theorem 4 provides a "bridge" between the GHF approach to scaling and those approaches that begin with assumptions about 'scaling functions'.

Theorem 4

A function $f(x, y, \ldots)$ is a GHF with scaling powers $a_x, a_y, \ldots, a_f$ if there exists some function $g_\pm(u, v, \ldots)$ such that

$$f(x, y, z, \ldots) = |x|^{a_f/a_x} g_{\mathrm{sgn}\,x}(y/|x|^{a_y/a_x}, z/|x|^{a_z/a_x}, \ldots) \tag{6.11a}$$

or, equally, if there exists some function $h_\pm(u, v, \ldots)$ such that

$$f(x, y, z, \ldots) = |y|^{a_f/a_y} h_{\mathrm{sgn}\,y}(x/|y|^{a_x/a_y}, z/|y|^{a_z/a_y}, \ldots) \tag{6.11b}$$

and so forth for all the independent variables. Conversely, if $f(x, y, \ldots)$ is a GHF, then (6.11a) and (6.11b) follow, where the functions $g_\pm(u, v, \ldots)$ and $h_\pm(u, v, \ldots)$ are given in terms of $f(x, y, \ldots)$ by the simple relations

$$g_\pm(u, v, \ldots) = f(\pm 1, u, v, \ldots) \tag{6.12a}$$

$$h_\pm(u, v, \ldots) = f(u, \pm 1, v, \ldots). \tag{6.12b}$$

An immediate "corollary" to Theorem 4 is that all GHF's have power-law singularities at the origin, when it is approached along one of the principal axes. The exponent for the path of approach $|x| \to 0$ (with $y = z = 0$) is given by a_f/a_x, the ratio of the scaling power of the function to the scaling power of the variable that is approaching zero, since from (6.11a) and (6.12a) we have

$$f(x, 0, 0, \dots) = |x|^{a_f/a_x} f(1, 0, 0, \dots). \tag{6.13}$$

We are now armed with enough theorems to obtain all the predictions of the scaling hypotheses (5.4 a, b, c) that we might wish.

The remaining discussion will therefore center on simple applications of the theorems to obtain exponent relations and "scaling functions".

Exercise 2

Prove Theorems 1–4, including the corollary to Theorem 4, eqn. (6.13).

7. RELATIONS AMONG THE CRITICAL-POINT EXPONENTS

7.1. Thermodynamic Functions

To begin with, we shall show that the Rushbrooke inequality (4.1) with an equal sign is a direct consequence of the TF Hypothesis.

Since

$$C_H \sim (\partial S/\partial T)_H \sim -(\partial^2 G/\partial T^2)_H, \tag{7.1a}$$

it follows from Theorem 2 and (5.4a) that C_H is a GHF with scaling power $a_C = 1 - 2a_\tau$, so that because α' is defined along the path $\tau \to 0$,

$$-\alpha' = (1 - 2a_\tau)/a_\tau. \tag{7.1b}$$

Since

$$M = -(\partial G/\partial H)_T, \tag{7.2a}$$

we have that $a_M = 1 - a_H$ and hence

$$\beta = (1 - a_H)/a_\tau, \tag{7.2b}$$

and since

$$\chi_T \equiv (\partial M/\partial H)_T = -(\partial^2 G/\partial H^2)_T, \tag{7.3a}$$

we have $a_\chi = 1 - 2a_H$ and

$$-\gamma' = (1 - 2a_H)/a_\tau. \tag{7.3b}$$

Combining (7.1), (7.2), and (7.3), we have the "Rushbrooke equality",

$$\alpha' + 2\beta + \gamma' = 2. \tag{7.4}$$

Note from (6.13) that in the present formulation we have the same values for exponents, regardless whether $\tau \to 0^-$ (T approaches T_c from below) or $\tau \to 0^+$ (T approaches T_c from above); i.e., $\alpha = \alpha'$, $\gamma = \gamma'$, etc. This restriction can be lifted easily within the present approach (Chang, Hankey, and Stanley, 1972) and I invite you to show this on your luncheon napkins.

Note also that the Rushbrooke equality is intuitively obvious within the scaling framework, because it simply says that the exponent $(\beta - 1)$ for the temperature derivative of the magnetization (i.e., the "rate of thumbtacks falling off of our uniformly heated magnet"—cf. Figure 1),

$$(\partial M/\partial T)_H = (\partial^2 G/\partial T \partial H) \sim |\tau|^{\beta-1}, \tag{7.5}$$

is precisely *halfway* between the exponents for the functions $(\partial^2 G/\partial T^2)_H$ and $(\partial^2 G/\partial H^2)_T$, so that $\beta - 1 = 1/2(-\alpha' - \gamma')$.

Clearly, we can find the scaling power and hence the critical-point exponents of any thermodynamic function in terms of the two unknown scaling powers a_τ and a_H, and Table 2 lists these for a variety of common functions. Then one can

TABLE 2

Predictions of the GHF hypothesis (5.4a) for the power law dependences for all four thermodynamic potentials and for a variety of other thermodynamic functions obtained by partial differentiation of the potentials. These expressions illustrate the utility of Theorem 2 and the Corollary to Theorem 4 in affording an expression of all thermodynamic-function critical-point exponents in terms of two unspecified "scaling powers" a_τ and a_H.

$$G(\tau, H) \sim |\tau|^{1/a_\tau} \qquad\qquad U(\sigma, M) \sim |\sigma|^{1/a_\sigma}$$
$$\sim |H|^{1/a_H} \qquad\qquad\qquad \sim |M|^{1/a_M}$$
$$A(\tau, M) \sim |\tau|^{1/a_\tau} \qquad\qquad E(\sigma, H) \sim |\sigma|^{1/a_\sigma}$$
$$\sim |M|^{1/a_M} \qquad\qquad\qquad \sim |H|^{1/a_H}$$

$$\sigma(\tau, H) \sim G^{(1,0)}(\tau, H) \sim |\tau|^{a_\sigma/a_\tau} \Rightarrow 1 - \alpha = a_\sigma/a_\tau = (1 - a_\tau)/a_\tau$$
$$\sim |H|^{a_\sigma/a_H} \Rightarrow \psi = a_\sigma/a_H = (1 - a_\tau)/a_H$$

$$M(\tau, H) \sim G^{(0,1)}(\tau, H) \sim |\tau|^{a_M/a_\tau} \Rightarrow \beta = a_M/a_\tau = (1 - a_H)/a_\tau$$
$$\sim |H|^{a_M/a_H} \Rightarrow \delta^{-1} = a_M/a_H = (1 - a_H)/a_H$$

$$\chi_T(\tau, H) \sim G^{(0,2)}(\tau, H) \sim |\tau|^{a_\chi/a_\tau} \Rightarrow -\gamma' = -\gamma = a_\chi/a_\tau = (1 - 2a_H)/a_\tau$$
$$\sim |H|^{a_\chi/a_H} \Rightarrow 1 - \delta^{-1} = a_\chi/a_H = (1 - 2a_H)/a_H$$

$$C_H(\tau, H) \sim G^{(2,0)}(\tau, H) \sim |\tau|^{a_C/a_\tau} \Rightarrow -\alpha' = -\alpha = a_C/a_\tau = (1 - 2a_\tau)/a_\tau$$
$$\sim |H|^{a_C/a_H} \Rightarrow -\phi = a_C/a_H = (1 - 2a_\tau)/a_H$$

$$C_M(\tau, M) \sim A^{(2,0)}(\tau, M) \sim |\tau|^{a_C/a_\tau} \Rightarrow -\alpha' = -\alpha = a_C/a_\tau = (1 - 2a_\tau)/a_\tau$$
$$\sim |M|^{a_C/a_M} \Rightarrow -\phi\delta = a_C/a_M = (1 - 2a_\tau)/a_M$$

proceed to eliminate the unknown scaling powers a_τ, a_H from triplets of exponents, obtaining the "scaling laws" of the form of the Rushbrooke equality (7.4).

It is convenient to express the scaling powers in terms of two exponents. For example, if we choose the first two exponents, α, β, we find from (7.1b) and (7.2b) that

$$a_\tau = \frac{1}{2 - \alpha} \tag{7.6a}$$

and

$$a_H = 1 - \frac{\beta}{2 - \alpha}. \tag{7.6b}$$

Additional expressions analogous to (7.6) are shown in Table 3.

TABLE 3

Expressions for the scaling parameters a_τ, a_H
in terms of pairs of critical-point exponents;
these relations may be verified using Table 2.

Exponents	a_τ	a_H
(α, β)	$(2 - \alpha)^{-1}$	$(2 - \alpha - \beta)/(2 - \alpha)$
(α, γ)	$(2 - \alpha)^{-1}$	$(2 - \alpha + \gamma)/2(2 - \alpha)$
(α, δ)	$(2 - \alpha)^{-1}$	$\delta/(\delta + 1)$
(β, γ)	$(2\beta + \gamma)^{-1}$	$(\beta + \gamma)/(2\beta + \gamma)$
(β, δ)	$[\beta(\delta + 1)]^{-1}$	$\delta/(\delta + 1)$
(γ, δ)	$(\delta - 1)/\gamma(\delta + 1)$	$\delta/(\delta + 1)$

Note that it is mathematically impossible to have a "two-exponent equality" involving only two critical point exponents. We shall return to this point shortly (cf. Section 8).

7.2. Static Correlation Functions

For the static correlation function, we have *three* scaling parameters, b_τ, b_H, and b_r, and although the a's can be determined from a knowledge of the b's (cf. eqn. (6.10)), there is no way of obtaining the b's from a knowledge of the a's. Accordingly, we shall see that the critical-point exponents for the correlation function are not determined exclusively in terms of exponents for the thermodynamic functions.

The exponent η is defined by the relation

$$S(\tau = 0, H = 0, q) \sim q^{-2 + \eta}, \tag{7.7}$$

where S is the structure factor defined in (6.7a), and by Theorem 3 it has scaling power $1 - db_q$, where $b_q = -b_r$. Accordingly, we have from (7.7) that

$$-2 + \eta = \frac{1 - db_q}{b_q} = \frac{1 + db_r}{-b_r},\tag{7.8}$$

or

$$\eta = 2 - d - 1/b_r.\tag{7.9}$$

A particularly useful quantity is the generalized correlation length $\xi_a(\tau, H)$ defined by

$$\xi_a(\tau, H) \equiv [\mu_a(\tau, H)/\mu_0(\tau, H)]^{1/a},\tag{7.10}$$

where $\mu_a(\tau, H)$ is defined in (1.4) and μ_0 is just the susceptibility (cf. (1.5)). Now, μ_a is just the $q = 0$ component of the Fourier transform of $r^a C_2$, which has scaling power $1 + ab_r$. Hence by Theorem 4, μ_a has a scaling power $a_f + db_r = 1 + ab_r + db_r$, and μ_a/μ_0 has scaling power ab_r, so that the correlation length itself has scaling power b_r,

$$\xi_a(\lambda^{b_r} r, \lambda^{b_H} H) = \lambda^{b_r} \xi_a(\tau, H).\tag{7.11}$$

As we approach the critical point along the two orthogonal paths, we obtain

$$\xi_a \sim |\tau|^{b_r/b_\tau} \equiv |\tau|^{-v_a}\tag{7.12}$$

and

$$\xi_a \sim |H|^{b_r/b_H} \equiv |H|^{-\mu_a}.\tag{7.13}$$

Since v_a, μ_a are independent of a, we drop the subscript in what follows. We can solve (7.9), (7.12), and (7.13) for the new scaling parameters b_τ, b_H, and b_r, with the results

$$b_\tau = [v(d - 2 + \eta)]^{-1},\tag{7.14}$$

$$b_H = [\mu(d - 2 + \eta)]^{-1},\tag{7.15}$$

and

$$b_r = [-(d - 2 + \eta)]^{-1}.\tag{7.16}$$

Numerical values of the b's for a few selected models are given in Table 4.

Using (6.10) we can solve for the a's in terms of the correlation function exponents,

$$a_\tau = \frac{1/v}{2/\mu + \eta - 2}\tag{7.17a}$$

and

$$a_H = \frac{1/\mu}{2/\mu + \eta - 2},\tag{7.17b}$$

and by using Table 2 we can express any thermodynamic function exponent in terms of the correlation function exponents (the converse of this statement is false!). For example, the susceptibility exponent γ is given in terms of the a's by eqn (7.3b), whence on using (7.17) we have

$$\gamma = (2 - \eta)v.\tag{7.18}$$

TABLE 4

Numerical values of the scaling parameters a_τ, a_H for thermodynamic functions, and b_τ, b_H, and b_r for static correlation functions, for some model systems thought to obey the TF and SCF hypotheses. The notation $\simeq$ indicates that the number is based upon numerical approximation methods.

Model	a_τ	a_H	b_τ	b_H	b_r	b_M
Mean-field theory (mft)	1/2	3/4	—	—	—	—
$d = 2$ Ising model ($D = 1$)	1/2	15/16	4	15/2	-4	1/2
$d = 3$ Ising model ($D = 1$)						
(taking $v = 0.638$ and $\eta = 0.041$)	$\simeq 8/15$	$\simeq 5/6$	$\simeq 1.506$	$\simeq 2.3526$	$\simeq -0.9606$	$\simeq 0.4708$
$d = 3$ plane rotator model ($D = 2$)						
(taking $v = 0.680$, $\eta = 0.040$)	$\simeq 1/2$	$\simeq 5/6$	$\simeq 1.414$	$\simeq 2.3567$	$\simeq -0.9615$	$\simeq 0.4721$
$d = 3$ classical Heisenberg model ($D = 3$)						
($v = 0.717$, $\eta = 0.040$, $\gamma = 1.405$,						
$\Delta = 1.77$)	$\simeq 10/21$	$\simeq 5/6$	$\simeq 1.341$	$\simeq 2.3736$	$\simeq -0.9615$	$\simeq 0.4891$
$d = 3$ spherical model ($D = \infty$)	1/3	5/6	1	5/2	-1	1/2

Similarly, one finds (cf. Table 2),

$$\frac{\delta - 1}{\delta} = (2 - \eta)\mu. \tag{7.19}$$

Rigorous inequalities analogous to eqns. (7.18) and (7.19) have recently been proved as inequalities (subject to certain assumptions),

$$\gamma' \geqq (2 - \eta)v' \tag{7.20}$$

and

$$\frac{\delta - 1}{\delta} \geqq (2 - \eta)\mu \tag{7.21}$$

by Liu, Joseph, and Stanley (1972). The second of these, (7.21), actually *followed* (historically) the corresponding exponent equality, (7.19), in contrast to, say, the Rushbrooke inequality (4.1), which *preceded* the scaling equality (7.4).

For those of you who are energetic (and wouldn't mind an exponent inequality possibly being named after you), I can recommend sitting down and deriving all possible exponent relations that are equalities under the scaling hypothesis, and then investigating which of these can be *proved* as inequalities.

Inequality (7.20) is interesting for another reason: Fisher (1969) had earlier proved a relation quite similar to (7.20),

$$(2 - \eta)v \geqq \gamma. \tag{7.22}$$

Hence (7.20) and (7.22) combine to yield

$$\frac{\gamma'}{v'} \geqq \frac{\gamma}{v}, \tag{7.23}$$

which is a rigorous inequality between *low*-temperature and *high*-temperature exponents! Moreover, if the anticipated symmetry between low- and high-temperature exponents should hold, or actually only if

$$\frac{\gamma'}{v'} = \frac{\gamma}{v}, \tag{7.24}$$

then it follows from (7.20) and (7.22) that

$$\gamma = (2 - \eta)v; \qquad \gamma' = (2 - \eta)v'. \tag{7.25}$$

That is to say, we have obtained one of the "scaling laws" (the exponent relations arising from the scaling hypothesis are generally called scaling laws) directly from a plausible assumption concerning symmetry of exponents!

7.3. Dynamic Correlation Functions

In order to make predictions concerning critical-point exponents for dynamic phenomena, it is convenient to transform from the DCF Hypothesis of eqn. (5.4c) to the dynamic structure factor, defined in analogy to the static structure factor of (6.7a),

$$\mathscr{S}(\tau, H, \boldsymbol{q}, \omega) \equiv \int_{-\infty}^{\infty} dt \int d\boldsymbol{r}\, \mathscr{C}_2(\tau, H, \boldsymbol{r}, t)\, e^{i(\boldsymbol{q}\boldsymbol{r} - \omega t)}. \tag{7.26}$$

According to Theorem 3, the Fourier transform of a GHF is also a GHF; hence (5.4c) implies that

$$\mathscr{S}(\lambda^{a_\tau}\tau, \lambda^{a_H} H, \lambda^{a_q}\boldsymbol{q}, \lambda^{a_\omega}\omega) = \lambda^{1 - db_q - b_\omega}\,\mathscr{S}(\tau, H, \boldsymbol{q}, \omega), \tag{7.27}$$

where b_q, the scaling power of q, is not a new parameter but is given simply by

$$b_q = - b_r \tag{7.28}$$

and b_ω, the scaling power of ω, is simply

$$b_\omega = - b_t. \tag{7.29}$$

If the dynamic structure factor has a large central peak centered about $\omega = 0$ which dominates the function as the critical point is approached, then a parameter of particular significance is the half-width ω_c, which approaches zero. This is defined by the equation

$$\int_{-\omega_c}^{\omega_c} \mathscr{S}(\tau, H, q, \omega)\,d\omega = \frac{1}{2} \int_{-\infty}^{\infty} \mathscr{S}(\tau, H, q, \omega)\,d\omega \tag{7.30}$$

or, equivalently,

$$\int_{-\tilde{\omega}_c}^{\tilde{\omega}_c} \mathscr{S}(1, \tilde{H}, \tilde{q}, \tilde{\omega})\,d\tilde{\omega} = \frac{1}{2} \int_{-\infty}^{\infty} \mathscr{S}(1, \tilde{H}, \tilde{q}, \tilde{\omega})\,d\tilde{\omega}, \tag{7.31}$$

where we have multiplied both sides of (7.30) by $\tau^{(b_f - b_\omega)b_\tau}$ with $b_f \equiv 1 - db_q - b_\omega$, the scaling power of the dynamic structure factor (cf. eqn. (7.27)). Clearly, $\tilde{\omega}_c$ depends upon the variables $\tilde{H}$, $\tilde{q}$ and hence it is a function of them:

$$\tilde{\omega}_c \equiv \omega_c / \tau^{b\omega/b_\tau} = \tilde{\omega}_c(\tilde{H}, \tilde{q}). \tag{7.32}$$

This shows that ω_c is GHF of the same scaling power as ω,

$$\omega_c(\lambda^{b_\tau}\tau, \lambda^{b_H}H, \lambda^{b_q}q) = \lambda^{b_\omega}\omega_c(\tau, H, q). \tag{7.33}$$

Exercise 3

Work out the critical-point exponents for ω_c as $\tau \to 0$ and $H \to 0$.

Exercise 4

Show that ω_c is a homogeneous function of the variables q and $\kappa \equiv 1/\xi$.

Exercise 5

Further, show that the "shape function" defined by

$$F(\omega/\omega_c) = \omega_c \mathscr{S}(\tau, H, q, \omega) \Big/ \int_{-\infty}^{\infty} \mathscr{S}(\tau, H, q, \omega)\,d\omega \tag{7.34}$$

is a function of q/κ only.

Exercises 4 and 5 show that our formulation implies the conventional Halperin–Hohenberg formulation of dynamic scaling (Halperin and Hohenberg, 1969).

8. RELATIONS *NOT* PREDICTED BY THE SCALING HYPOTHESIS: 2-EXPONENT RELATIONS

This section concerns the exponent relations

$$dv = 2 - \alpha \tag{8.1a}$$

and

$$\frac{\delta - 1}{\delta + 1} = \frac{2 - \eta}{d}. \tag{8.1b}$$

In the form the scaling hypothesis was first proposed, relations (8.1) were predicted to be consequences (see, especially, Kadanoff, 1966).

And so began a long sequence of exchanges between those who do hard, lengthy numerical calculations and those who propose "scaling hypotheses" to interpret the results of this work: for the "numerical experiments" persisted in giving results that yielded the left-hand sides of (8.1) being just about 2 percent larger than the right-hand sides. For example, in the case of the $d = 3$ Ising model, $\alpha = 1/8$ so that (8.1a) predicts $v = 5/8 = 0.625$, while the numbers persisted in giving a number somewhat larger (about 0.638 ± 0.002—see Moore et al., 1969).

What about "rigorous inequalities"? Here Josephson (1967) and Gunton and Buckingham (1968), Stell (1968), and, more rigorously, Buckingham and Gunton (1969), proved versions of (8.1) with the $=$ sign replaced by a $\geqq$ sign. Hence the numbers were consistent with the rigorous inequalities, but not with the "scaling" equalities of (8.1).

Accordingly, considerable "encouragement" was given the numerical experimentalists to come up with numbers that supported scaling. The argument for believing eqns. (8.1) was strengthened by the fact that the two principal exactly-soluble model Hamiltonians agreed with eqns. (8.1): the $d = 2$ Ising model ($v = 1$, $\alpha = 0$, $\delta = 15$, and $\eta = 1/4$) and the $d = 3$ spherical model ($v = 1$, $\alpha = -1$, $\delta = 5$, and $\eta = 0$). However, the argument for believing the numbers was that the numerical approximation methods used had never given wrong answers in the past; this statement is *itself* only approximately true.

Accordingly, you cannot imagine our excitement when, very recently, Alex Hankey and I observed that homogeneity alone led to *correction terms* to eqns. (8.1) which, when the appropriate numbers for the exponents were plugged in, produced precisely the missing two percent decrement! That agreement became quite understandable a few hours later, when Alex observed that the correction terms were just such as to turn the "aberrant inequalities" into highly-respected thermodynamic equalities. For example, in evaluating the analog of (8.1a), we find

$$2 - \alpha = 1/a_\tau = 1 + db_r + 2b_H = -db_r/b_\tau + (2b_M - 1)/b_\tau, \qquad (8.2)$$

where we have used (6.10) to relate a_τ to the b's, and we have defined

$$b_M = 1 + db_r + b_H. \qquad (8.3)$$

Clearly, unless $b_M = 1/2$, we have

$$dv = 2 - \alpha + X, \qquad (8.4)$$

where

$$X \equiv (d - 2 + \eta)v - 2\beta \qquad (8.5)$$

is the appropriate "correction term". But if we use the scaling equality $\gamma = (2 - \eta)v$, we find that (8.4) and (8.5) combine to give

$$2 - \alpha = 2\beta + \gamma, \qquad (8.6)$$

which is just the "Rushbrooke equality" of (7.4).

Similarly, we showed that (8.1b) is replaced by

$$(\delta - 1)/(\delta + 1) = \gamma/(2\beta\delta - \gamma) \qquad (8.7)$$

which, on rearrangement, yields the "Widom equality",

$$\beta(\delta - 1) = \gamma. \qquad (8.8)$$

However, again if we assume that $b_M = 1/2$, we find that we obtain (8.1b) instead of (8.7). Thus both (8.1a) and (8.1b) are valid if and only if $b_M = 1/2$ *in addition to homogeneity*.

The next step was to inquire whether *perhaps* $b_M = 1/2$? Certainly $b_M = 1/2$ for the two exactly soluble systems, the $d = 2$ Ising and the $d = 3$ spherical models. What does $b_M = 1/2$ correspond to? According to (6.1a) the function M^2 has scaling power $2b_M$, and we know that the second term in the definition (1.3) of $C_2(T, H, r)$ does indeed scale as M^2. Hence, since C_2 has scaling power unity, one might anticipate that for some reason M^2 should also have scaling power unity. However, there is no *a priori* reason why C_2 should have the same scaling properties as M^2 and it is our contention that it need not.

If indeed (8.1a) and (8.1b) are *not* valid, then is it not reasonable to ask if the rigorous inequalities corresponding to eqns. (8.1),

$$dv \geq 2 - \alpha \tag{8.9}$$

and

$$(\delta - 1)/(\delta + 1) \geq (2 - \eta)/d, \tag{8.10}$$

might not be "looser" than they need be? This question was answered by Liu, Joseph and Stanley (1972), who showed that one could in fact produce inequalities which were in a sense "tighter" than (8.10).

It indeed turned out that Stell (1970) and Snider (1971) had realized some of our points independently, though not, in our opinion, in as straightforward a fashion.

9. SCALING FUNCTIONS

9.1. Thermodynamic Equation of State near the Critical Point

The equation of state is a functional relation among three variables, such as $M = M(T, H)$. In this section we apply Theorem 4 to show that the equation of state near the critical point is predicted to be of a certain form. Thus we shall see that the assumption that a function is a GHF leads to predictions concerning the function for *all* values of its argument near the critical point—not just its critical-point exponents.

Let us first apply Theorem 4 to the function $H(\tau, M)$. Then eqn. (6.11a) implies that

$$H(\tau, M) = |\tau|^{a_H/a_\tau} \mathscr{F}^{(1)}_{\mathrm{sgn}\,\tau}(M/|\tau|^{a_M/a_\tau}), \tag{9.1a}$$

where the first thermodynamic scaling function $\mathscr{F}^{(1)}_{\pm}(u)$ is given by (6.12a),

$$\mathscr{F}^{(1)}_{\pm}(u) = H(\tau = \pm 1, u) \tag{9.1b}$$

and has two branches, according as whether $\tau > 0 (T > T_c)$ or $\tau < 0 (T < T_c)$. This was the first scaling function to be introduced.

Equation (9.1a) can be written in the form

$$H_\tau = \mathscr{F}^{(1)}_{\mathrm{sgn}\,\tau}(M_\tau) \tag{9.2}$$

where we have defined the magnetic *field* scaled with respect to temperature as the "scaled variable" on the left-hand side of (9.2),

$$H_\tau \equiv H/|\tau|^{a_H/a_\tau} = H/|\tau|^{\beta\delta}, \tag{9.3a}$$

and we have defined the *magnetization* scaled with respect to temperature as the scaled variable on the right-hand side of (9.2),

$$M_\tau \equiv M/|\tau|^{a_M/a_\tau} = M/|\tau|^{\beta}. \tag{9.3b}$$

Thus (9.2) says that H_τ, the scaled magnetic field, is a two-valued function of M_τ, the scaled magnetization.

This prediction has been confirmed by experiments on various materials over the past few years, and because the *converse* of Theorem 4 holds, the fact that H_τ is a two-valued function of M_τ *only* implies that $H(\tau, M)$ is a GHF and indeed provides an *experimental confirmation* of the scaling hypothesis for thermodynamic functions (5.4a).

One can equally well apply (6.11b) instead of (6.11a), with the result that

$$H_s(\tau, M) = |M|^{a_H/a_M}\, \mathscr{F}^{(2)}_{\mathrm{sgn}\,M}\, (\tau/|M|^{a_\tau/a_M}), \tag{9.4a}$$

where the second thermodynamic scaling function is given by

$$\mathscr{F}^{(2)}_{\pm}(u) = H(\tau, M = \pm\, 1) \tag{9.4b}$$

and again has two branches, according whether $M > 0$ or $M < 0$. However, the function $H(\tau, M = \pm\, 1)$ is an odd function of magnetization M, so that the two branches of the function $\mathscr{F}^{(2)}_{\pm}$ of (9.14b) are symmetrical, and in fact data are normally taken only for one sign of H (and hence only one sign of M). Thus all experimental data are found in practice to fall on a single curve.

Eqn. (9.14a) may be written in the form

$$H_M = \mathscr{F}^{(2)}_{\mathrm{sgn}\,M}(\tau_M), \tag{9.5}$$

where we have defined the scaled variables

$$H_M \equiv H/|M|^{a_H/a_M} = H/|M|^{\delta} \tag{9.6a}$$

as being the magnetic field, scaled with respect to magnetization, and

$$\tau_M \equiv \tau/|M|^{a_\tau/a_M} = \tau/|M|^{1/\beta} \tag{9.6b}$$

as being the temperature, scaled with respect to magnetization.

Experimental data have also been plotted in accordance with (9.5), which has the advantage over (9.2) that data fall on a single curve; both plots are schematically

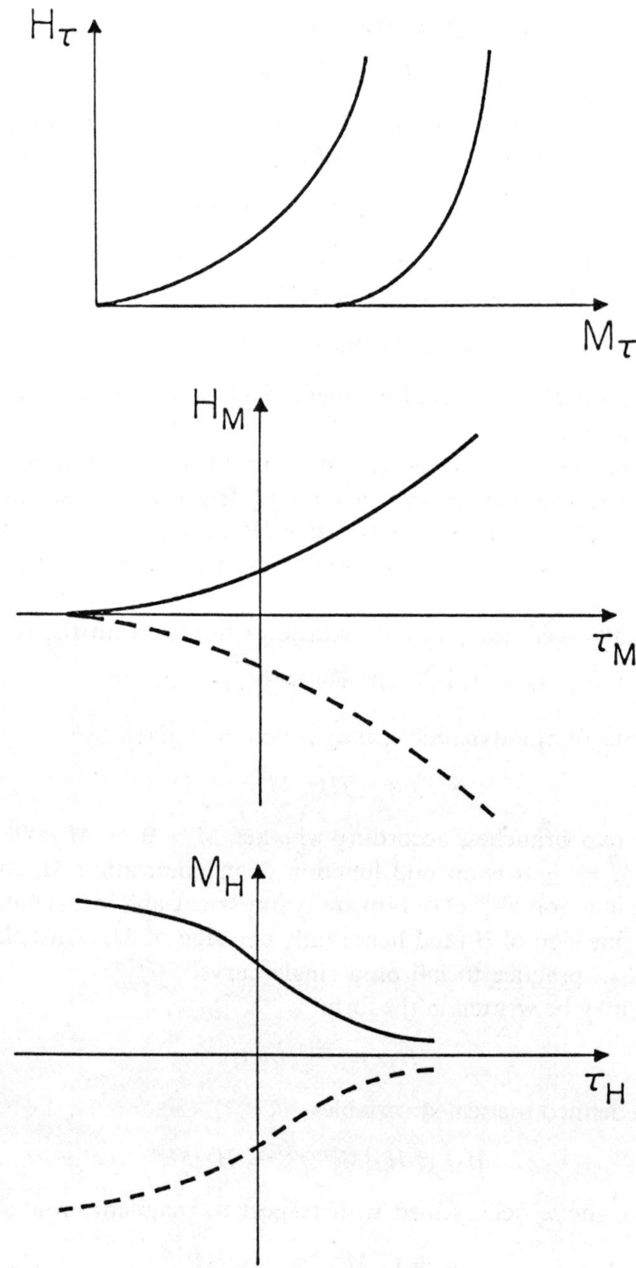

Figure 6

Sketches of the three possible "scaling functions" for interpretation of *MHT* equation-of-state data near the critical point. A dashed line indicates that data is not taken for this "branch" of the scaling function due to *MH* symmetry.

sketched in Figure 6. A drawback of using $\mathscr{F}^{(1)}_{\pm}$ to plot data is that the critical temperature $\tau = 0$ corresponds to $H_\tau = M_\tau = \infty$ (cf. eqns. (9.3)), and hence critical isotherm data cannot be plotted! A similar problem arises with using $\mathscr{F}^{(2)}_{\pm}$; the data for M very small correspond to very large values of H_M and τ_M (cf. eqns (9.6)) and hence in order to get all the data on a single plot, the use of log–log paper must be resorted to, with its consequent lack of precision.

A third way of plotting M–H–T equation-of-state data does not suffer from either of these defects. We apply eqn. (6.11b) to the function $M(\tau, H)$, with the result

$$M_s(\tau, H) \equiv |H|^{a_M/a_H} \, \mathscr{F}^{(3)}_{\operatorname{sgn} H}(\tau/|H|^{a_\tau/a_H}), \tag{9.7a}$$

where the third thermodynamic scaling function is given by

$$\mathscr{F}^{(3)}_{\pm}(u) = M_s(\tau, H = \pm 1). \tag{9.7b}$$

As with $\mathscr{F}^{(2)}_{\pm}(u)$, the fact the M is an odd function of H means that data are normally taken for only one sign of H and all data taken fall on a single curve (cf. Figure 6). We write (9.7a) in the form

$$M_H = \mathscr{F}^{(3)}_{\operatorname{sgn} H}(\tau_H), \tag{9.8}$$

where

$$M_H \equiv M/|H|^{a_M/a_H} = M/|H|^{1/\delta} \tag{9.9a}$$

and

$$\tau_H \equiv \tau/|H|^{a_\tau/a_H} = \tau/|H|^{1/\beta\delta}. \tag{9.9b}$$

All three scaling functions are summarized in Table 5. Figure 7 shows the data of Ho and Litster (1969) on the ferromagnetic insulator $CrBr_3$ plotted according to the prescription of (9.8), and we see that all the data fall quite smoothly on a single curve.

TABLE 5

(a) Summary of scaling functions predicted by the relations (9.2), (9.5), and (9.8) for the M–H–T equation of state; (b) concerns the S–H–T equation of state. Here $\sigma \equiv S(\tau, H) - S(0, 0)$.

(a) Function	Dependent variable	Independent variable
$H_\tau = \mathscr{F}^{(1)}_{\operatorname{sgn} \tau}(M_\tau)$	$H_\tau \equiv H/|\tau|^{\beta\delta}$	$M_\tau \equiv M/|\tau|^{\beta}$
$H_M = \mathscr{F}^{(2)}_{\operatorname{sgn} M}(\tau_M)$	$H_M \equiv H/|M|^{\delta}$	$\tau_M \equiv \tau/|M|^{1/\beta}$
$M_H = \mathscr{F}^{(3)}_{\operatorname{sgn} H}(\tau_H)$	$M_H \equiv M/|H|^{1/\delta}$	$\tau_H \equiv \tau/|H|^{1/\beta\delta}$

(b) Function	Dependent variable	Independent variable
$\tau_\sigma \equiv \mathscr{G}^{(1)}_{\operatorname{sgn} \sigma}(H_\sigma)$	$\tau_\sigma \equiv \tau/|\sigma|^{a_\tau/a_\sigma}$	$H_\sigma \equiv H/|\sigma|^{a_H/a_\sigma}$
$\tau_H \equiv \mathscr{G}^{(2)}_{\operatorname{sgn} H}(\sigma_H)$	$\tau_H \equiv \tau/|H|^{a_\tau/a_H}$	$\sigma_H \equiv \sigma/|H|^{a_\sigma/a_H}$
$\sigma_\tau \equiv \mathscr{G}^{(3)}_{\operatorname{sgn} \tau}(H_\tau)$	$\sigma_\tau \equiv \sigma/|\tau|^{a_\sigma/a_\tau}$	$H_\tau \equiv H/|\tau|^{a_H/a_\tau}$

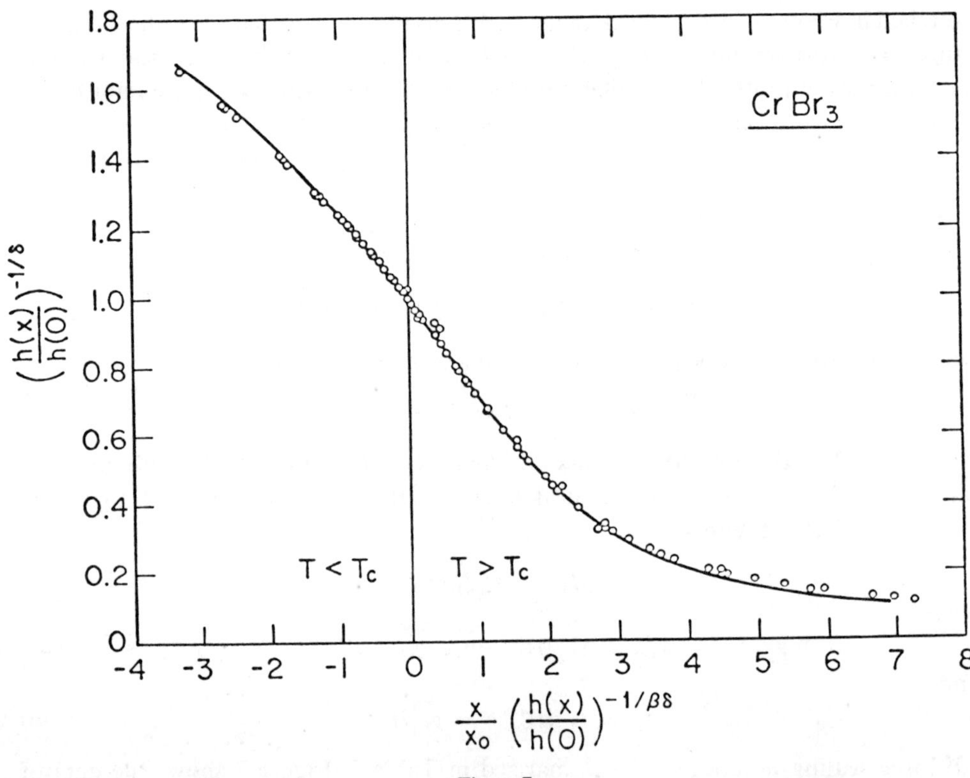

Figure 7

The 99 CrBr₃ data points of Ho and Litster (1969) plotted according to the prescription of Eqn. (9.8). Note that all data fall reasonably evenly spaced when plotted on a linear scale (as opposed to a log–log scale). The solid curve was obtained from the numerical calculations of Milošević and Stanley (1972) for the $S = 1/2$ Heisenberg model, using the method of high-temperature series expansions (see Section 9.2).

Clearly, the same procedure can be applied to any triplet of three variables, and should prove useful when interpreting measurements of the entropy as a function of T and H (cf. Table 5) that are now being contemplated (T. Arrott, private communication).

9.2. Numerical Confirmation of the Scaling Hypothesis and Calculation of Scaling Functions

Recently it has become possible to numerically test the TF Hypothesis (5.4a) for the Heisenberg model, using the method of high-temperature series expansions. The main idea is to utilize another property of GHF's: if we know the values of a GHF along any path encircling the origin, then we know it everywhere.

Exercise 6

Prove the preceding statement.

Milošević and Stanley (1972) have calculated the function $H(\tau, M)$ on a family of paths of the form shown in Figure 8; these paths essentially "encircle" the origin, since a line from the origin $T = T_c$, $M = 0$ to any arbitrary point in the $T-M$ plane must pass through the path shown (because of $H-M$ symmetry, we need not show the symmetrical path in the fourth quadrant). They have utilized each path to calculate

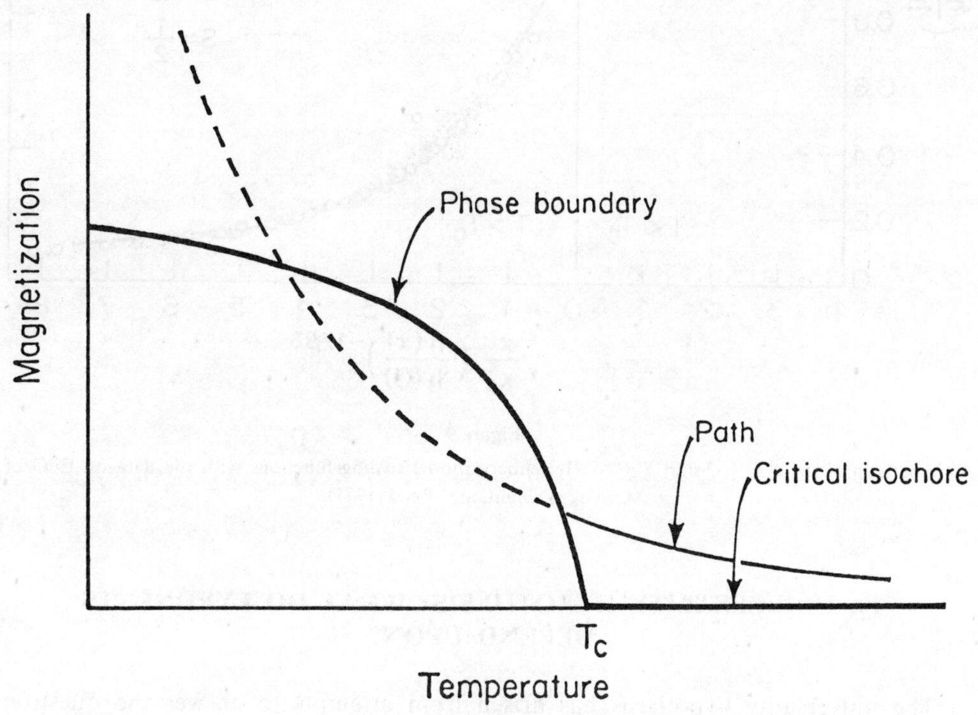

Figure 8

The curve on which the function $H(\tau, M)$ is calculated, used in calculating the scaling function from high-temperature series expansions.

the scaling function $\mathscr{F}^{(2)}(x)$ of (9.5), from which they have also calculated the function $\mathscr{F}^{(3)}(x)$ of (9.8). The scaling functions are found to be independent of the path chosen, which leads to a very strong consistency check on the GHF hypothesis. Comparison with experimental data is shown in Figures 9 and 10. It is to be emphasized that there are *no* adjustable parameters whatsoever in the theoretical curves shown.

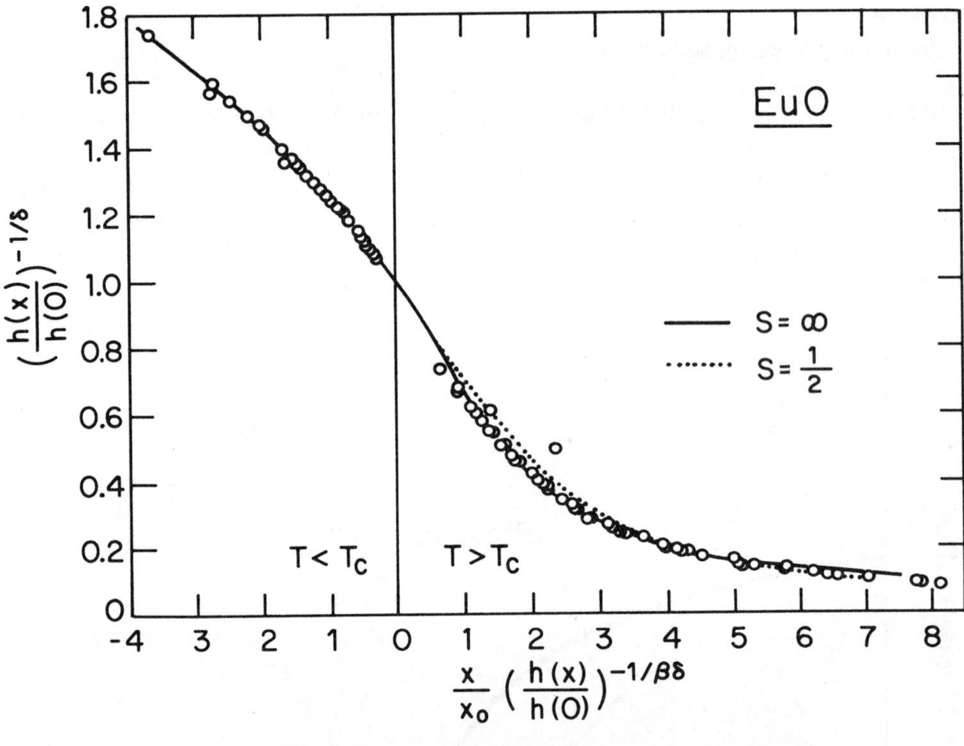

Figure 9

Comparison of the $S = 1/2$ and $S = \infty$ Heisenberg model scaling functions with the data on EuO of Menyuk, Dwight, and Reed (1971).

10. UNIVERSALITY HYPOTHESIS: WHAT DO EXPONENTS DEPEND UPON?

The universality hypothesis has arisen from attempts to answer the question *"On what features of an interaction Hamiltonian do critical-point exponents depend?"*

10.1. A General Interaction Hamiltonian

For the sake of concreteness, let us introduce a general Hamiltonian which encompasses, as special cases, a very large proportion of the models that exhibit critical phenomena. This Hamiltonian utilizes the concept of classical spin vectors, situated on sites r and r' in a d-dimensional lattice, and interacting with energy parameters $J^{\alpha}_{r-r'}$ that depend upon the vector $(r - r')$. The important thing is

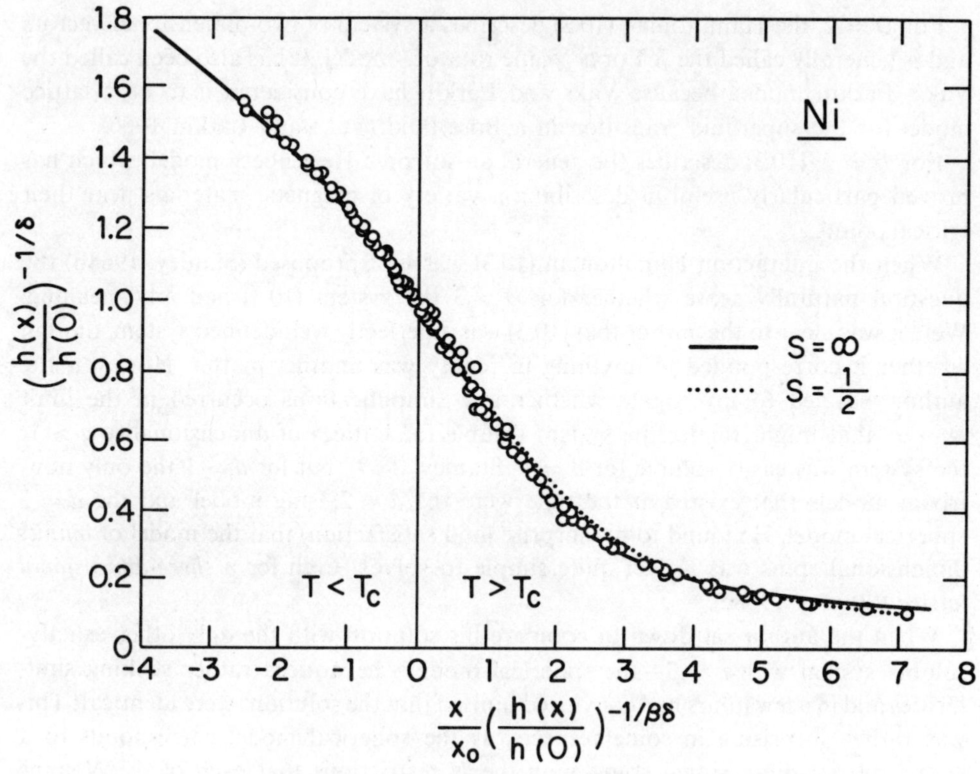

Figure 10

Comparison of the Heisenberg model calculations with the measured scaling function for
of Kouvel and Comly, 1968).

that the spin vectors are taken to be unit vectors spanning a D-dimensional space,
i.e.,

$$S \equiv (S^1, S^2, \ldots, S^D), \tag{10.1}$$

with

$$(S^1)^2 + (S^2)^2 + \ldots + (S^D)^2 = 1. \tag{10.2}$$

Thus the Hamiltonian under consideration is

$$H = -\sum_{r,r'} \sum_{\alpha=1}^{D} J^{(\alpha)}_{r-r'} \, S^\alpha_r S^\alpha_{r'}. \tag{10.3}$$

Note that when $D = 1$, this Hamiltonian reduces to the Ising model of eqn. (2.2),
since the spins become simply one-dimensional "sticks", capable of assuming the
two discrete orientations $+1$ (up) and -1 (down).

For $D = 2$, the Hamiltonian (10.3) describes a system of two-dimensional vectors and is generally called the XY or a "plane rotator" model. It has also been called the Vaks–Larkin model, because Vaks and Larkin have considered it to be a lattice model for the superfluid transition in a Bose fluid (Vaks and Larkin 1966).

For $D = 3$, (10.3) describes the general anisotropic Heisenberg model, which has proved particularly useful in describing a variety of magnetic materials near their critical points.

When the interaction Hamiltonian (10.3) was first proposed (Stanley, 1968b) the question naturally arose whether for $D > 3$ the system (10.3) had any meaning. Well, it was clear to the author that (10.3) was a perfectly well-defined system, though whether it corresponded to anything in reality was another matter. However, the author was led to investigate whether any simplifications occurred in the limit $D \to \infty$ that might render the system soluble for lattices of dimensionality $d > 1$; the system was easily soluble for $d = 1$ (Stanley 1969), but for $d > 1$ the only non-trivial models that existed at the time were the $d = 2$ Ising model and the $d = 3$ spherical model. He found to his surprise (and satisfaction) that the model of *infinite* dimensional spins was in fact quite simple to solve—even for a *three-dimensional lattice* (Stanley 1968c).

When the author sat down to compare his solution with the only other exactly-soluble system with $d = 3$ — the spherical model — he noticed rather striking similarities and in a few hours had convinced himself that the solutions were identical! This was rather surprising in some respects, as the spherical model corresponds to a system of *one*-dimensional spins, with the N restrictions that each of the N spins be of unit length being replaced by the single constraint that

$$S_1^2 + S_2^2 + \ldots + S_N^2 = N. \tag{10.4}$$

It is still not very intuitively clear why an Ising Hamiltonian like (2.2) together with the "spherical constraint" (10.4) should in fact have the same partition function as the $D \to \infty$ limit of the system (10.3), where the spins are D-dimensional vectors *of unit length*! Perhaps some of you in the audience whose intuitions about higher dimensions have been sharpened by your field-theoretical studies can explain this one to me.

10.2. Parameters in (10.3) upon which Exponents Might Depend

A program of study on which we have recently embarked is aimed at establishing which parameters in the general-model Hamiltonian (10.4) are important for determining the values of critical-point exponents, and which are "irrelevant". To this end, we considered

   (1) the lattice dimensionality, d;
   (2) the spin dimensionality, D;

(3) the "lattice anisotropy" or "nonuniformity of the interaction", i.e., the dependence of $J^\alpha_{r-r'}$ upon the direction of $r - r'$ (There are many important materials for which the interaction strength in one crystal direction is different than in another direction, so that this is not "of merely academic interest");

(4) the "spin-space anisotropy", or dependence of $J^\alpha_{r-r'}$ upon α (Probably no material is *perfectly* isotropic and there are certainly some materials for which the anisotropy is believed to play a very important role);

(5) the range of interaction, or dependence of $J^\alpha_{r-r'}$ upon $r - r'$ (Again, there are probably no materials for which the commonly made assumption that only nearest neighbors of one another interact, and that all other pairs of spins are completely coupled, is valid);

(6) the spin-quantum number, S. The Hamiltonian (10.3) was for classical spins, which correspond to the $S = 1/2$ Ising model and the $S \to \infty$ limits of the other systems. Materials in nature have a wide range of spin-quantum number S, and accordingly we must consider what happens to these models for general quantum number S.

The conclusion of our work is that the only features of (10.3) that are important are lattice dimensionality d and the symmetry of the ground state (which can be measured by an "effective dimensionality" (Jasnow and Wortis, 1968) of the spin). This conclusion is sometimes called by a rather pretentious sounding name, "the universality hypothesis", and it has the implication—if believed—that one need only consider *two* parameters, d and D. Hence (10.3) may be replaced by

$$\mathscr{H}_V \equiv -J \sum_{r,\delta} \sum_{\alpha=1}^{D} S^\alpha_r S^\alpha_{r+\delta}, \tag{10.5}$$

where here δ denotes a vector between nearest-neighbor pairs of lattice sites.

10.3. Nature of the "Evidence" Favoring the Universality Hypothesis

It is perhaps useful to discuss the evidence provided to rule out the possible dependence of exponents upon items (3) and (5), lattice anisotropy and further-neighbor interactions. This work is being done by Gerald Paul for his Ph.D. thesis (Paul and Stanley, 1971a, b, 1972a, b). He considers the interaction Hamiltonians

$$- \mathscr{H}_{\text{lat.anis.}} = \sum_{\langle ij\rangle}^{xy} J_{xy} S^z_i S^z_j + \sum_{\langle ij\rangle}^{z} J_z S^z_i S^z_j \equiv J_x \left[\sum_{\langle ij\rangle}^{xy} S^z_i S^z_j + R \sum_{\langle ij\rangle}^{z} S^z_i S^z_j \right] \tag{10.6}$$

and

$$-\mathscr{H}_{nnn} = J_1 \sum_{ij}^{nn} \mathbf{S}^{(D)}_i \cdot \mathbf{S}^{(D)}_j + J_2 \sum_{ij}^{nnn} \mathbf{S}^{(D)}_i \cdot \mathbf{S}^{(D)}_j \equiv J_1 \left[\sum_{ij}^{nn} \mathbf{S}^{(D)}_i \cdot \mathbf{S}^{(D)}_j + R' \sum_{ij}^{nnn} \mathbf{S}^{(D)}_i \cdot \mathbf{S}^{(D)}_j \right], \tag{10.7}$$

where $R \equiv J_z / J_{xy}$ and $R' \equiv J_2 / J_1$.

In eqn. (10.6) the first summation is over pairs of nearest-neighbor sites whose relative displacement vector r_{ij} has no z component, while the second summation is over all other pairs of nearest-neighbor sites. Note that in the limit $R \to 0$, the sc and fcc lattices reduce to noninteracting planes ($d = 2$); in the limit $R \to \infty$, the sc lattice reduces to an array of noninteracting chains ($d = 1$) and the fcc lattice reduces to a bcc lattice ($d = 3$). The predictions of the universality hypothesis for (10.6) are indicated in Figure 11.

In eqn. (10.7) we have isotropically interacting D-dimensional classical spins. The first and second summations are over nearest-neighbor and next-nearest-neighbor pairs of lattice sites. We treat the sc, bcc, and fcc lattices which reduce in the limit $R' \to \infty$ to *two noninteracting* fcc, sc, and sc lattices, respectively. Thus we expect nearest-neighbor only exponents in the limits $R \to 0$ *and* $R \to \infty$.

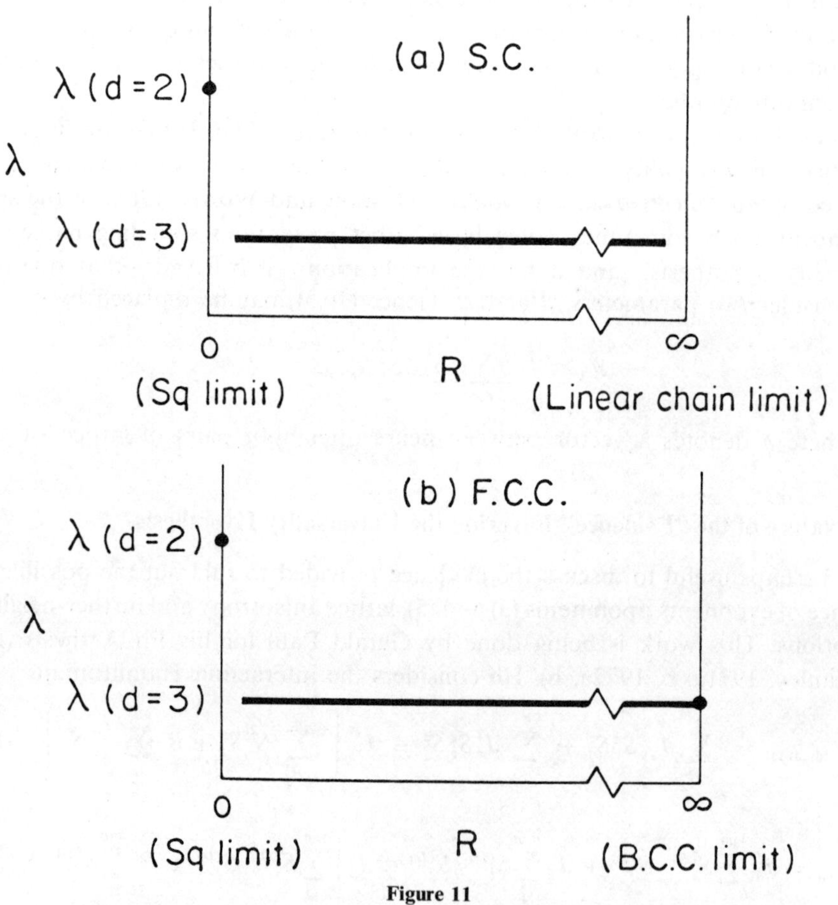

Figure 11

Predictions of the universality hypothesis for $\mathcal{H}_{\text{lat.anis.}}$ for sc and fcc lattices. The "arbitrary exponent" λ is predicted to change discontinuously at $R = 0$ and $R = \infty$.

An example that will serve to illustrate how changing R does not seem to change the value of the critical-point exponent from its $R = 1$ value is seen by looking at Figure 12, where we show a sequence of estimates plotted against $1/n$ (for ease of extrapolation to $n = \infty$). By contrast, the critical temperature most definitely does vary smoothly with R, and this dependence is shown in Figure 13.

For the case of $\mathscr{H}_{nnn}$, eqn. (10.7), the evidence is not very compelling unless one systematically studies all spin dimensionalities D, and then makes use of the fact that the limiting case $D = \infty$ is exactly soluble and so an arbitrary number of expansion coefficients can be found. One finds that the initial estimates indeed look as if exponents are going to depend on $R' = J_2/J_1$, but that when one systematically looks at all D together (cf. Figure 14) one sees that there is amazingly similar behavior of the estimates

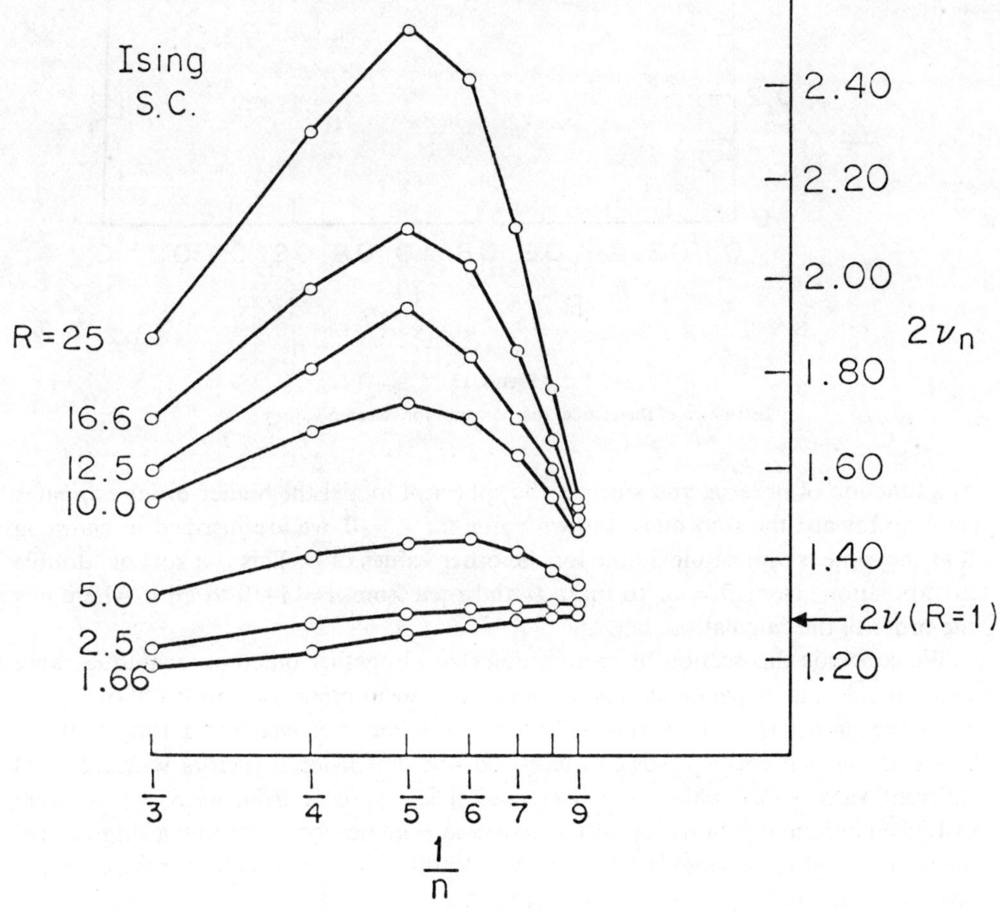

Figure 12

Estimates for γ for different values of $R \equiv J_z/J_{xy}$.

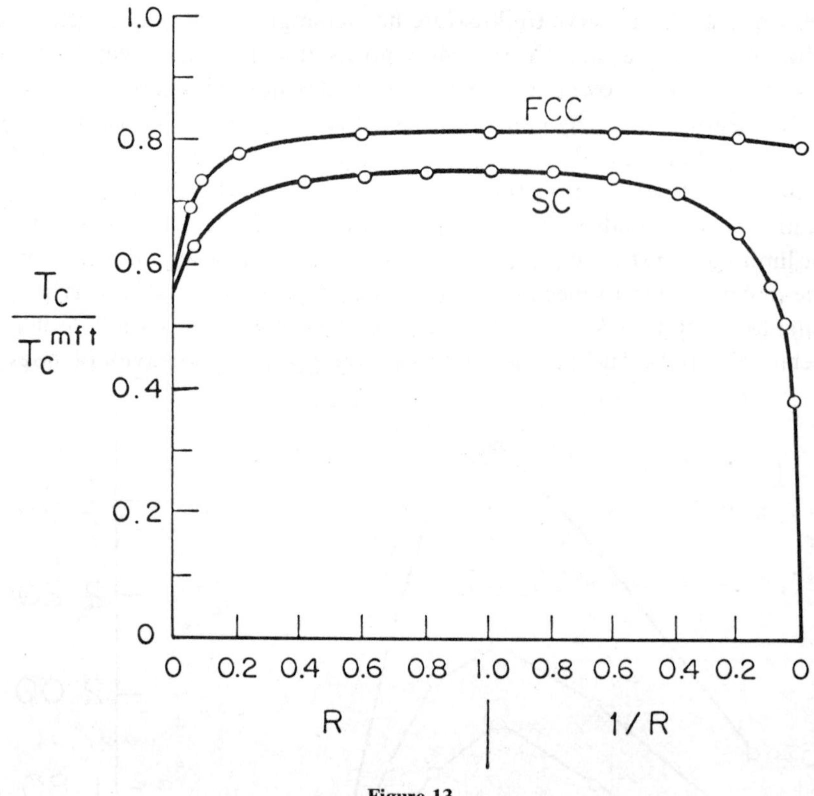

Figure 13

Estimates of the critical temperature for selected values of R.

as a function of order n, and since in the spherical model the higher-order estimates tend up toward the rigorously known value for $r' = 0$, we are justified in claiming that the same is almost inevitable for the other values of D. This is a sort of "double extrapolation" from $D = \infty$ to finite D and then from $n = 1$–10 to all n, where n is the order of the calculation.

We conclude this section by mentioning that altogether different techniques were used to rule out dependence upon spin-quantum number S (item (6) above). For the Ising model ($D = 1$), Jasnow and Wortis (1968) believed for a time that for $S > 1/2$—which corresponds to quantized one-dimensional vectors with $(2S + 1)$ different values—the value of γ decreased a few percent from its $S = 1/2$ value of 1.25. Subsequently, however, they discovered evidence for a "confluent singularity" on top of the physical singularity which was leading to the misleading extrapolations and now they believe that $\gamma = 5/4$ for all S.

The case $D = 3$ is just the reverse: here there appear to be not "confluent singu-larities" but rather other singularities in the complex-temperature plane that are

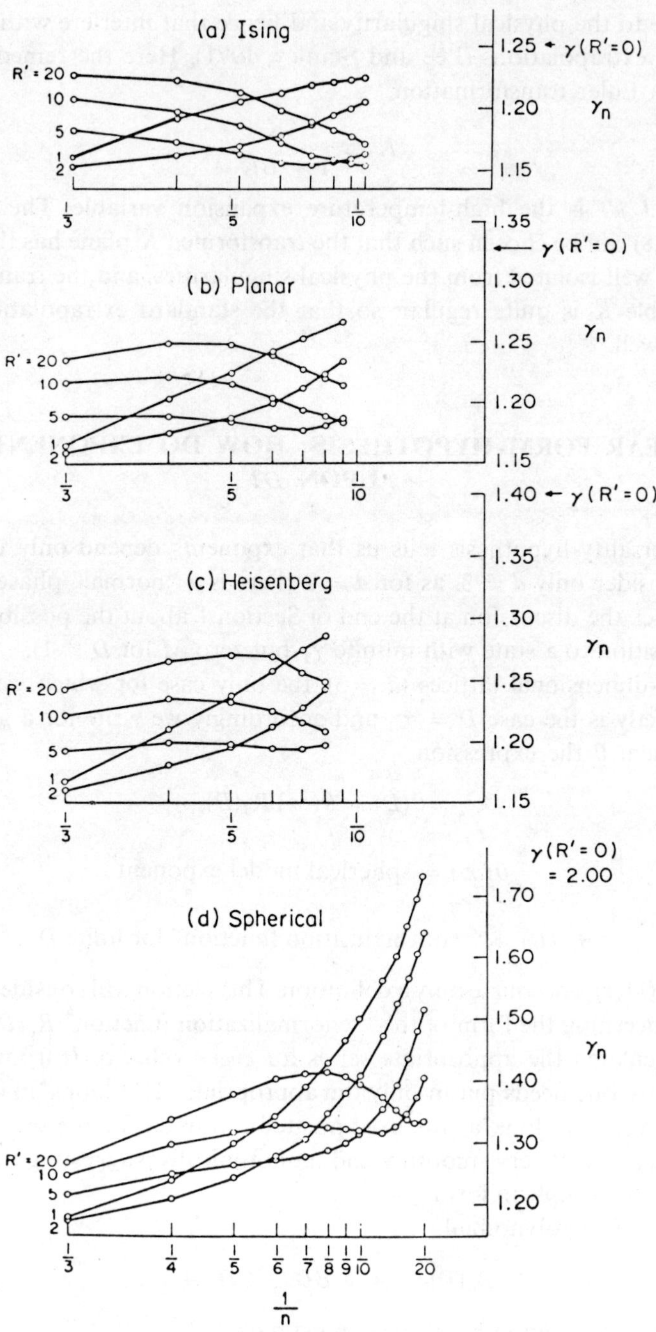

Figure 14

Estimates of γ for different values of $R' \equiv J_2/J_1$.

located close to the physical singularity and hence that interfere with the regularity of the series extrapolations (Lee and Stanley, 1971). Here the remedy has been to introduce an Euler transformation

$$\overline{K} \equiv \frac{AK}{1 + BK} \tag{10.8}$$

where $K \equiv J/kT$ is the high-temperature expansion variable. The parameters A and B in (10.8) can be chosen such that the transformed $\overline{K}$ plane has the nonphysical singularities well isolated from the physical singularities, and the transformed series in the variable $\overline{K}$ is quite regular, so that the standard extrapolation procedures work quite well.

11. BILINEAR FORM HYPOTHESIS: HOW DO EXPONENTS DEPEND UPON D?

The universality hypothesis tells us that exponents depend only upon D and d. Here we consider only $d = 3$, as for $d = 2$ there is a "normal" phase transition for $D = 1$ *only* (cf. the discussion at the end of Section 1 about the possibility of a novel type of transition to a state with infinite χ_T but zero M for $D > 1$).

For three-dimensional lattices ($d = 3$), the only case for which we know the exponents exactly is the case $D = \infty$, and accordingly we write for a general critical-point exponent θ the expression

$$\theta(D) = \theta(\infty)\, R_\theta(D), \tag{11.1}$$

where

$$\theta(\infty) = \text{spherical model exponent} \tag{11.2}$$

and

$$R_\theta(D) = \text{"renormalization function" for finite } D. \tag{11.3}$$

Equation (11.1) is of course only a definition. This section will consider the numerical evidence concerning the form of the "renormalization function" $R_\theta(D)$. Accordingly, we have calculated the appropriate series for *every* value of D from 1 to 100 (this is easy to do—one needs put in only the appropriate "DO loops" in one's computer programs). When we look at the extrapolated estimates for the exponents, we find that they vary with D very smoothly and monotonically, suggesting that the function $R_\theta(D)$ should be simple in form.

The first try is a polynomial,

$$R_\theta(D) = A + BD + CD^2 + \ldots, \tag{11.4}$$

but this is clearly no good because we must have

$$R_\theta(D = \infty) = 1. \tag{11.5}$$

Accordingly, we next try the ratio of two simple polynomials, and for a first guess we try first-order polynomials,

$$R_\theta(D) = (b_\theta + D)/(c_\theta + D), \tag{11.6}$$

where the coefficients of the order-D terms must be identical because of (11.5). The reader can verify that the form (11.6) has the desired monotonicity property: just differentiate with respect to D and observe that the derivative must be nonnegative.

Surprisingly, we find that this does a remarkable job at fitting the numerical extrapolations for the entire range of D from 1 to 100. Examples of a few selected D are shown in the first few rows of Table 6.

Note that if we are prepared to believe the TF (Thermodynamic function) scaling hypothesis, we can obtain expressions for the scaling parameters a_τ, a_H for general D, and these too turn out to be of the simple bilinear form. For example, if we use our $D = 1, 2, \ldots, 100$ series to fit $\alpha(D)$ and $\gamma(D)$ accurately to the simple bilinear form,

$$\alpha(D) = (-1)\left(\frac{-2 + D}{7 + D}\right) \tag{11.7}$$

and

$$\gamma(D) = (+2)\left(\frac{4 + D}{7 + D}\right), \tag{11.8}$$

TABLE 6

Estimates for critical point exponents for a three-dimensional lattice ($d = 3$) for different values of spin dimensionality D. Shown in the last column is the prediction of the bilinear form hypothesis (Stanley and Betts 1972).

Exponent	$D = 1$ (Ising)	$D = 2$ (Planar)	$D = 3$ (Heisenberg)	$D = \infty$ (Spherical)	Bilinear Hypothesis
α	$\simeq 1/8$	$\simeq 0$	$\simeq -1/10\ldots$	-1	$-1\left(\dfrac{-2 + D}{7 + D}\right)$
β	$\simeq 5/16$	$\simeq 1/3$	$\simeq 7/20\ldots$	$1/2$	$\dfrac{1}{2}\left(\dfrac{4 + D}{7 + D}\right)$
γ	$\simeq 5/4$	$\simeq 4/3$	$\simeq 7/5\ldots$	2	$2\left(\dfrac{4 + D}{7 + D}\right)$
δ	$\simeq 5$	$\simeq 5$	$\simeq 5\ldots$	5	$5\,(1)$
Δ	$\simeq 25/16$	$\simeq 5/3$	$\simeq 7/4\ldots$	$5/2$	$\dfrac{5}{2}\left(\dfrac{4 + D}{7 + D}\right)$
a_τ	$\simeq 8/15$	$\simeq 1/2$	$\simeq 10/21\ldots$	$1/3$	$\dfrac{1}{3}\left(\dfrac{7 + D}{4 + D}\right)$
a_H	$\simeq 5/6$	$\simeq 5/6$	$\simeq 5/6\ldots$	$5/6$	$\dfrac{5}{6}\,(1)$

where $\alpha(\infty) = -1$ and $\gamma(\infty) = 2$, then we can utilize Table 3 to write

$$a_\tau(D) = \frac{1}{2 - \alpha(D)} = \frac{1}{3}\left(\frac{7 + D}{4 + D}\right) \tag{11.9}$$

and

$$a_H(D) = \frac{2 - \alpha(D) + \gamma(D)}{2[2 - \alpha(D)]} = \frac{5}{6}. \tag{11.10}$$

Since all the thermodynamic function exponents are of the form a_f/a_{path}, it follows from (11.9) and (11.10) that all exponents permit the bilinear form expansion.

Note in particular that $a_H = 5/6$ for all D, while a_τ varies from $8/15$ for $D = 1$ to $1/3$ for $D = \infty$. Kadanoff (1972) has pointed out that when the scaling parameter is less than $1/2$, the variable is strongly fluctuating, while when the parameter is greater than $1/2$, it is weakly fluctuating. Since the relationships between the variables and their thermodynamic conjugates are $a_\sigma = 1 - a_\tau$ and $a_M = 1 - a_H$, it follows that the entropy is strongly fluctuating for $D = 1$ and weakly fluctuating for $D > 2$, while the magnetization is strongly fluctuating for *all* D.

We conclude by noting that there is one encouraging fact concerning the possible validity of the bilinear form hypothesis: Wilson and Fisher (1972) have, for certain exponents, just very recently calculated the first term of the expansions in powers of $(d - 4)$ and they find that the coefficients of these expansions are indeed bilinear functions of D.

12. A PHENOMENOLOGICAL APPROACH: THE DYNAMIC CLUSTER APPROXIMATION

By now I hope I have impressed upon you the difficulty of making progress in this field, and the fact that here will be problems around for some time to come. I have described (a) the approach of building microscopic models that, while not exactly solvable, *can* be "understood" by means of numerical approximation procedures, and (b) the scaling approach which is almost a complete "stab in the dark". I shall conclude with something intermediate: an attempt at a semiphenomenological description of what is roughly going on (Matsuno and Stanley, 1972).

This approach was initially inspired by the computer simulation of the dynamics of the two-dimensional "Glauber model" (essentially an Ising model in contact with a heat bath). One sees that large clusters of correlated spins appear and "tumble", their motion characterized near T_c by a characteristic frequency that approaches zero and a characteristic spatial dimension that approaches infinity. Matsuno and I have regarded the clusters as being formally analogous to "macroscopic collective excitations" of the system, and have proceeded to find an approximate set of equations to describe the dynamics of this strongly interacting many-cluster system.

Our results produce numerical values of the critical-point exponents, and these values satisfy the relations predicted by the scaling hypotheses and the mode–mode coupling approximation. A different set of exponents is found for each "shape of cluster", as shown in Figure 15. Since there is no microscopic interaction Hamiltonian in the problem, the exponents cannot in principle depend directly upon the details of the interaction, so that, e.g., the "dynamic cluster approximation" would predict the same exponents for spins interacting with an Ising Hamiltonian as for spins interacting with a Heisenberg Hamiltonian. This is certainly a drawback of the theory, as it flies in the face of all the numerical evidence that has accumulated over the past decade. Nevertheless, it is the only approach that yields explicit information about dynamic as well as static phenomena, and even if it should be as crude as the mean field theory in giving precise numerical results, it might nevertheless be useful as a simple "first step".

	ν	α	γ	a	β
Isotropic	$\frac{2}{3}$	0	1	$\frac{1}{3}$	$\frac{1}{2}$
U.A. Weak	$\frac{2}{3}$	0	$\frac{4}{3}$	$\frac{2}{3}$	$\frac{1}{3}$
P.A. Weak	$\frac{2}{3}$	0	$\frac{5}{3}$	1	$\frac{1}{6}$
U.A. Strong	1	0	1	1	$\frac{1}{2}$
P.A. Strong	2	0	1	3	$\frac{1}{2}$

U.A. Uniaxial Anisotropy
P.A. Planar Anisotropy

Figure 15

Critical-point exponents predicted by the dynamic cluster approximation (Matsuno and Stanley, 1972) for different shapes of the cluster of correlated particles. This approach is to be regarded as purely phenomenological, and its results are not to be taken too literally.

REFERENCES

R. G. Birgeneau (1973), In: "Magnetism," G. Rado and H. Suhl, Eds., Vol. 5. Academic Press, New York [to be published].

R. Brout (1965), "Phase Transitions," W. A. Benjamin Co., New York.

M. J. Buckingham and J. D. Gunton (1969), *Phys. Rev.,* **178**. 848.

T. Chang, A. Hankey and H. E. Stanley (1972), *Phys. Rev.* (to be published).

M. J. Cooper (1968), *Phys. Rev.,* **168**, 183.

C. Domb and D. L. Hunter (1965), *Proc. Phys. Soc.,* **86**, 1147.

J. W. Essam and M. E. Fisher (1963), *J. Chem. Phys.,* **38**, 802.

M. E. Fisher (1969), *Phys. Rev.,* **180**, 594.

R. B. Griffiths (1967), *Phys. Rev.,* **158**, 176.

R. B. Griffiths (1972), In: "Cooperative Phenomena and Phase Transitions", C. Domb and M. Green, Eds., Academic Press, New York.

J. D. Gunton and M. J. Buckingham (1968), *Phys. Rev. Letters,* **20**, 143.

B. I. Halperin and P. C. Hohenberg (1967), *Phys. Rev.,* **177**, 952.

A. Hankey and H. E. Stanley (1972), *Phys. Rev.,* **B6** (Nov.).

J. T. Ho and J. D. Litster (1969), *Phys. Rev. Letters,* **22**, 603.

D. Jasnow and M. Wortis (1968), *Phys. Rev.,* **176**, 739.

B. D. Josephson (1967), *Proc. Phys. Soc.,* **92**, 276.

M. Kac (1964), *Physics Today,* **17**, No. 10, p. 40.

M. Kac (1968), In: "Brandeis Theoretical Physics Lectures", M. Chrétien, Ed., Gordon and Breach, New York.

L. P. Kadanoff (1966), *Physics,* **2**, 263.

L. P. Kadanoff (1972), In: "Critical Phenomena", M. S. Green, Ed., Academic Press, New York (*Proc. Varenna Summer School*).

J. S. Kouvel and J. B. Comly (1968), *Phys. Rev. Letters,* **20**, 1237.

M. H. Lee and H. E. Stanley (1971), *Phys. Rev.,* **B4**, 1613.

L. Liu, R. I. Joseph and H. E. Stanley (1972), *Phys. Rev.,* **B6** (Sept.).

K. Matsuno and H. E. Stanely (1972) (to be published).

N. Menyuk, K. Dwight and T. B. Reed (1971), *Phys. Rev.,* **B3**, 1689.

S. Milošević and H. E. Stanley (1972), *Phys. Rev.,* **B6**, 986, 1002.

M. A. Moore, D. Jasnow, and M. Wortis (1969), *Phys. Rev. Letters,* **23**, 861.

A. Z. Patashinskii and V. L. Pokrovskii (1966), *Sov. Phys. JETP,* **23**, 292.

G. Paul and H. E. Stanley (1971a), *Phys. Letters,* **37A**, 328.

G. Paul and H. E. Stanley (1971b), *Phys. Letters,* **37A**, 347.

G. Paul and H. E. Stanley (1972a), *Phys. Rev.,* **B5**, 2578.

G. Paul and H. E. Stanley (1972b), *Phys. Rev.,* **B5**, 3715.

G. S. Rushbrooke (1949), "Introduction to Statistical Mechanics". Oxford University Press, London and New York.

G. S. Rushbrooke (1963), *J. Chem. Phys.*, **39**, 842.

N. S. Snider (1971), *J. Chem Phys.*, **54**, 4587.

H. E. Stanley (1968a), *Phys. Rev. Letters,* **20**, 150.

H. E. Stanley (1968b), *Phys. Rev. Letters,* **20**, 589.

H. E. Stanley (1968c), *Phys. Rev.,* **176**, 718.

H. E. Stanley (1969), *Phys. Rev.* **179**, 570.

H. E. Stanley (1971), "Introduction to Phase Transitions and Critical Phenomena". Oxford University Press, London and New York. [Also translated into Russian by S. V. Vonsovskii (Mir, Moscow, 1972).]

H. E. Stanley and D. D. Betts (1972), *Phys. Rev.* (to be published).

H. E. Stanley and T. A. Kaplan (1966), *Phys. Rev. Letters,* **17**, 913.

H. E. Stanley and T. A. Kaplan (1967), *J. Appl. Phys.,* **38**, 975.

G. Stell (1968), *Phys. Rev. Letters,* **20**, 533.

G. Stell (1970), *Phys. Rev. Letters,* **24**, 1343.

V. G. Vaks and A. I. Larkin (1966), *Sov. Phys. JETP,* **22**, 678.

B. Widom (1965) *J. Chem. Phys.,* **43**, 3892, 3898.

K. Wilson and M. E. Fisher (1972), *Phys. Rev. Letters,* **28**, 240.

Connection Between Wightman and LSZ Field Theory

O. STEINMANN

Swiss Institute for Nuclear Research, Villigen (Switzerland)

Abstract

Traditionally one distinguishes between two formulations of axiomatic field theory: the Wightman formulation and the LSZ formulation. The latter is specially adapted to the needs of particle physics by stressing the importance of the S-matrix and related notions. The former is more concerned with local aspects and is therefore better suited for the discussion of particular field theory models, i.e. for getting a hold on dynamics. In the present lectures I intend to combine the two formulations by developing the LSZ theory as a special case of a Wightman theory.

In chapter II the Wightman axioms will be stated in the special form in which they are needed for the establishment of a scattering formalism. In chapter III I start discussing such a scattering formalism. Asymptotic conditions are derived which allow the introduction of the notion of particle. In chapter IV the S-matrix is defined and its properties discussed. Expressions for its matrix elements in terms of the basic fields (the so-called reduction formulae) are derived. Chapter V is devoted to a short discussion of the GLZ theorem which gives a complete characterization of an LSZ theory in terms of its retarded functions, a set of c-number functions closely related to the Green's functions of the theory.

I shall try to make my statements mathematically rigorous. Most proofs will, however, only be sketched.

I. INTRODUCTION

In talking about axiomatic field theory, one usually distinguishes between two approaches: the Wightman formulation [1–3] and the LSZ formulation [4]. (I ignore the local observables approach of Haag and Araki, which cannot properly be called a field theory.) Originally these two approaches started indeed from quite different points of view. The Wightman formalism puts the interacting fields in the center of attention. It stresses the local aspects of the theory. One never talks of particles, i.e. the formalism is somewhat removed from the experimental facts. The

LSZ formalism in its original formulation, on the other hand, put the particles in the center. It started out from a discussion of the S-matrix and its properties. The interacting fields, called "interpolating fields", entered only as auxiliary mathematical quantities needed for the discussion of causality. Thus LSZ was rather closer to experiment than Wightman, but rather farther away from any hope of understanding dynamics, which cannot be gotten hold of without a closer look at what happens locally.

As far as mathematical rigor is concerned, Wightman operated from the start on a very high level, while LSZ were satisfied with less rigor, trusting to intuition and perturbation theory whenever an existence problem came up.

Of course, the two formalisms are in no way contradictory, and I want to talk here about how they can be united. More explicitly: I shall consider the LSZ formalism as a special case of a Wightman theory, obtained from it by adding a few new assumptions that allow introducing and discussing particles and their interactions. The mathematical rigor throughout is that which one is used to from Wightman theory. What I am going to say is a concoction of ingredients taken from the papers [5–9].

II. THE AXIOMS

I assume that you are more or less familiar with the Wightman axioms and their significance. Due to lack of time, I cannot discuss their origin and justification. But I must give a list of them. Again due to lack of time, I consider only the simplest case, that of a single, scalar, real field $A(x)$. The generalization to more realistic situations is reasonably straightforward, though not entirely trivial.

Axiom 1 (quantum mechanics):
We are in the general framework of quantum mechanics: states are represented by vectors or, more generally, density matrices in a Hilbert space $\mathcal{H}$, observables and other quantities of physical interest by linear operators in $\mathcal{H}$.

Axiom 2 (relativistic invariance):
In $\mathcal{H}$ operates a continuous unitary representation $U(\Lambda, a)$ of the connected Poincaré group $\mathfrak{P}_+^\uparrow$. Here Λ stands for the homogeneous part, a for the translation part of the Poincaré transformation (Λ, a). For simplicity I will not talk about the discrete operations P and T.

Axiom 3 (spectral properties):
The representation $U(\Lambda, a)$ induces a unitary representation

$$T(a) = U(1, a) \tag{1}$$

of the 4-dimensional translation group. According to the SNAG theorem, which is a generalization of Stone's theorem, $T(a)$ can be written as

$$T(a) = \exp\left[iP_\mu\, a^\mu\right],\tag{2}$$

with P_μ being selfadjoint operators. The P_μ are the energy-momentum operators. They allow a simultaneous spectral decomposition, i.e. there exists a 4-dimensional spectral measure $dE(p)$ such that

$$P_\mu = \int p_\mu\, dE(p).\tag{3}$$

We demand that the support of $dE(p)$ lie in the closed forward cone $\bar{V}_+$:

$$\mathrm{supp}\left[dE(p)\right] \subset \bar{V}_+.\tag{4}$$

(The support of a measure is the smallest closed set outside of which the measure vanishes identically.) Condition (4) means that there are no states of negative energy in any reference frame.

We demand further that $dE(p)$ contain a discrete part at the origin $p = 0$ with a one-dimensional eigenspace. In other words: there is a normalized vector $\Omega \in \mathscr{H}$, unique up to a phase factor, with

$$P_\mu \Omega = 0.\tag{5}$$

This vector Ω is called the *vacuum*. Its uniqueness implies its invariance:

$$U(\Lambda, a)\Omega = \Omega.\tag{6}$$

Ω will sometimes also be denoted by $|0\rangle$.

Condition (4) is the spectral condition as formulated by Wightman. For going over to LSZ we must restrict the spectrum further: we demand that the spectrum be that of a particle theory. Again we restrict ourselves to the simplest case, that in which there exists only one species of stable particles, with spin 0 and mass $m > 0$. (Unfortunately, nobody knows how to formulate an LSZ theory for massless particles.) The LSZ addition to the spectral condition reads then: we demand that the vacuum point $p = 0$ be an *isolated* point of the support of $dE(p)$. Furthermore, $\mathrm{supp}\left[dE(p)\right]$ contains an *isolated* hyperboloid $p^2 = m^2$, $p_0 > 0$, and a continuum starting at $p^2 = 4m^2$. I.e. the spectrum looks like Figure 1.

To the 1-particle hyperboloid $p^2 = m^2$ belongs the eigenspace $\mathscr{H}_1$ of the mass operator $M^2 = P_0^2 - \boldsymbol{P}^2$ with eigenvalue m^2. The representation $U(\Lambda, a)$ of $\mathfrak{P}_+^\uparrow$ maps $\mathscr{H}_1$ onto itself, i.e. it induces a representation of $\mathfrak{P}_+^\uparrow$ on $\mathscr{H}_1$. We demand that this representation be irreducible and be of Wigner type $(m, 0)$, i.e. of mass m and spin 0.

Axiom 4 (field theory):

The basic quantity of the theory shall be a field $A(x)$, i.e. all the relevant physical

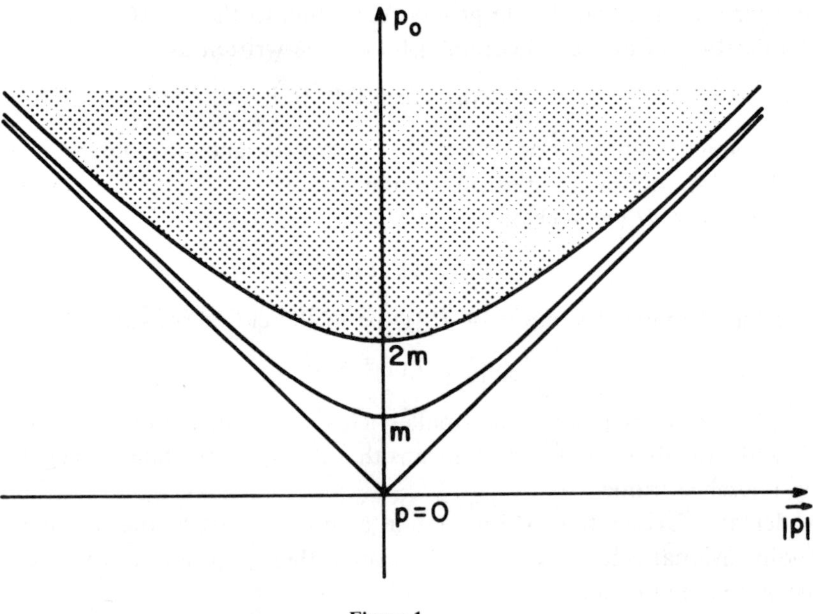

Figure 1

quantities are functions of the field operators $A(x)$, in a sense which I am not going to specify. $A(x)$ is not a function of x in the usual sense, the value of A at a point x is not defined. Rather, A is what is called a "distribution" or a "generalized function." This means that what is defined is the average

$$A(\phi) = \int dx \, A(x) \, \phi(x) \tag{7}$$

of A over a sufficiently smooth "test function" $\phi(x)$. As to what "sufficiently smooth" means we have little physical information. For simplicity I stick to Wightman's original choice, assuming that (7) exists for all ϕ from the space $\mathscr{S}$ of tempered test functions. The complex-valued function ϕ is in $\mathscr{S}$ if it is infinitely often differentiable, and all its derivatives decrease stronger than any inverse power of $\|x\|$ if x tends to infinity in any direction. However, our considerations can be applied with appropriate changes to other cases with rather more singular A's, i.e. with smaller spaces of admissible test functions. (This type of theories has achieved some popularity recently in connection with the so-called nonpolynomial Lagrangians.)

The field A shall have the following properties:

a) The operator $A(\phi)$, $\phi \in \mathscr{S}$, is defined on a dense set $D \in \mathscr{H}$ and maps D into itself. D shall be independent of ϕ and shall contain the vacuum Ω. This implies, in particular, that the vectors

$$A(\phi_1) \ldots A(\phi_n) \, \Omega$$

exist and are in D. The linear hull of all vectors of this form (for $n = 0, 1, 2, \ldots$) is called D_0 and assumed to be dense in $\mathcal{H}$. This is another way of saying that the field characterizes the theory completely.

b) The field A is real, i.e. $A(\phi)$ is hermitian if ϕ is real. This is assumed for simplicity. Otherwise we should have to consider the field A^* along with A.

c) A transforms under $\mathfrak{P}^\uparrow_+$ as a scalar:

$$A(\Lambda x + a) = U(\Lambda, a)\, A(x)\, U^*(\Lambda, a). \tag{8}$$

Actually, we shall only need covariance under translations, while the covariance properties under Lorentz transformations are inessential.

d) A satisfies local commutativity:

$$[A(x), A(y)] = 0 \quad \text{if} \quad (x - y)^2 < 0. \tag{9}$$

(I use the metric where space-like vectors have negative square.) This condition is usually justified by its connection with the causality requirement. This connection is unfortunately not as close as one would like it to be.

Again we add a specific LSZ assumption.

e) A has nonvanishing matrix elements between the vacuum and the one-particle space $\mathcal{H}_1$:

$$\langle 0 \,|\, A(x) \,|\, \mathcal{H}_1 \rangle \neq 0. \tag{10}$$

It can be shown that the matrix elements (10) differ from the corresponding elements of a free field only by a constant real factor, and (10) means that this factor is $\neq 0$. It is easy to see that multiplication of A with a nonvanishing real constant does not destroy the properties postulated until now. Hence we can assume A to be normalized such that the matrix elements (10) are equal to those of a free field.

Assumption e) is not really necessary for our purposes. It would suffice to have a polynomial in A with nonvanishing matrix elements between Ω and $\mathcal{H}_1$, and the existence of such a polynomial is guaranteed by the denseness of D_0 in $\mathcal{H}$. However, assuming (10) simplifies the formalism considerably.

This concludes our slightly expanded list of the Wightman axioms. We shall have to add yet another axiom somewhat later in the game.

III. ASYMPTOTIC CONDITIONS

As yet I have not said anything about particles. Particles are introduced with the help of the *asymptotic conditions*. The idea is based on two considerations:

1) We know how to describe *free* particles with the help of *free* fields.

2) If we look at a scattering or production event involving any number of particles, we find that sufficiently long before and after the event the particles are far apart

from one another, and their interaction can be neglected: they can be treated as free, hence can be described by free fields.

We talk, then, about particles only in the asymptotic region. In the region of interaction the notion of particle is a dubious one (remember the possibility of destruction and creation of particles!) which is best discarded.

The asymptotic conditions are the mathematical formulation of this idea. They state that the interacting field $A(x)$ tends for large positive or negative times towards two free fields A^{out} and A^{in} respectively, in a sense that will be explained presently.

But first let me briefly remind you of what a free scalar hermitian field $A^{\text{ex}}(x)$ of mass $m > 0$ looks like. It is of the form

$$A^{\text{ex}}(x) = (2\pi)^{-3/2} \int d^4 p \, e^{-ipx} \, \tilde{A}^{\text{ex}}(p),\tag{11}$$

with

$$\tilde{A}^{\text{ex}}(p) = \delta_+(p)\, \hat{A}^{\text{ex}}(p) + \delta_-(p)\, \hat{A}^{\text{ex}*}(-p),\tag{12}$$

$$\delta_+(p) = \theta(p_0)\, \delta(p^2 - m^2), \qquad \delta_-(p) = \delta_+(-p).\tag{13}$$

$\hat{A}^{\text{ex}}$ and $\hat{A}^{\text{ex}*}$ are destruction and creation operators respectively, with the commutation relations

$$[\hat{A}^{\text{ex}}(p), \hat{A}^{\text{ex}*}(q)] = 2\omega(p)\, \delta^3(p - q),$$

$$[\hat{A}^{\text{ex}}(p), \hat{A}^{\text{ex}}(q)] = 0,\tag{14}$$

with $\omega(p) = \sqrt{p^2 + m^2}$. Furthermore

$$\hat{A}^{\text{ex}}(p)\, \Omega = 0.\tag{15}$$

Strictly speaking, the value of $\hat{A}^{\text{ex}}(p)$ at a point is not defined. What is defined is the integral

$$\hat{A}^{\text{ex}}(\hat{f}) = \int \hat{A}^{\text{ex}}(p)\hat{f}^*(p)\, \frac{d^3 p}{2\omega(p)}\tag{16}$$

over a test function $\hat{f}^*$ (called "wave function"), which in this case may be any square integrable function:

$$\int \frac{d^3 p}{2\omega(p)}\, |\hat{f}(p)|^2 < \infty.\tag{17}$$

Tempered wave functions $\hat{f} \in \mathscr{S}$ are admitted by this criterion. $\mathscr{S}$ is in fact dense in the space of all admissible test functions (17), so that we can henceforth restrict ourselves to considering $\hat{f}$'s from $\mathscr{S}$.

The operators $\hat{A}^{\text{ex}}$ and $\hat{A}^{\text{ex}*}$ act in the *Fock space* $\mathscr{H}^{\text{ex}}$ of A^{ex}. $\mathscr{H}^{\text{ex}}$ is spanned by the vectors

$$|\hat{f}_1, \ldots, \hat{f}_n\rangle_{\text{ex}} = \hat{A}^{\text{ex}*}(\hat{f}_1) \ldots \hat{A}^{\text{ex}*}(\hat{f}_n)\, \Omega,\tag{18}$$

$n = 0, 1, 2, \ldots$

Two different versions of asymptotic conditions have been proved under our assumptions: the Haag–Ruelle condition and the LSZ condition. Both versions claim the existence of two free fields $A^{\text{out, in}}$ in $\mathcal{H}$ such that $A(x) \to A^{\text{out, in}}(x)$ for $x^0 \to \pm \infty$, in a sense which is different in the two versions.

A rigorous mathematical treatment cannot work with $A(x)$, which does not exist, but only with integrals of the form (7), or with wave packets of the form (16) in the free case. Since the asymptotic region $x^0 \to \pm \infty$ is defined in x-space, it is convenient to transform (16) into x-space. Define

$$f(x) = (2\pi)^{-3/2} \int d^4p \, \delta_+(p) \hat{f}(p) \, e^{-ipx}. \tag{19}$$

$f(x)$ is a solution of the Klein–Gordon equation:

$$(\Box + m^2) f(x) = 0.$$

The wave packet (16) can be written

$$\hat{A}^{\text{ex}}(\hat{f}) = i \int_{x^0 = t} d^3x \, f^*(x) \, \overset{\leftrightarrow}{\frac{\partial}{\partial x^0}} \, A^{\text{ex}}(x), \tag{20}$$

with the definition

$$f \, \overset{\leftrightarrow}{\frac{\partial}{\partial x^0}} \, g = f \frac{\partial g}{\partial x^0} - \frac{\partial f}{\partial x^0} \, g.$$

Because A^{ex} and f both solve the Klein–Gordon equation it turns out that the integral (20) does not depend on the value of $x^0 = t$ at which we integrate over $\boldsymbol{x}$.

We can try to use (20) to define "destruction operators" $A(\hat{f}, t)$ also for the interacting field A:

$$A(\hat{f}, t) = i \int_{x^0 = t} d^3x \, f^*(x) \, \overset{\leftrightarrow}{\frac{\partial}{\partial x^0}} \, A(x) =$$

$$= \int d^4p \, \frac{p_0 + \omega(\boldsymbol{p})}{2\omega(\boldsymbol{p})} \, \tilde{A}(p) \hat{f}^*(\boldsymbol{p}) \, e^{-it(p_0 - \omega(\boldsymbol{p}))} \tag{21}$$

which is time-dependent. Unfortunately, in general this integral does not exist even for $\hat{f} \in \mathcal{S}$, because $\hat{f}(\boldsymbol{p})$ is only a 3-dimensional test function, and this is not sufficient to make the 4-dimensional integral (21) converge. We must, then, use a little more sophistication. Instead of the wave functions $\hat{f}(\boldsymbol{p})$ depending only on the 3-vector $\boldsymbol{p}$ we introduce test functions $\tilde{f}(p)$ which are in $\mathcal{S}$ in all 4 components of p, i.e. they decrease strongly also in the p_0-direction. For good measure we choose them even with quite a small support in this direction: we introduce the subspace $\mathcal{G}$ of $\mathcal{S}$ of the functions $\tilde{f}(p)$ which vanish identically outside the set

$$G = \{p : 0 \leqq p^2 \leqq 4m^2, p_0 \geqq 0\}. \tag{22}$$

This restriction is important for the Haag–Ruelle condition but not for the LSZ condition. To $\tilde{f} \in \mathcal{G}$ we associate a wave function in the old sense, the restriction of $\tilde{f}$ to the mass shell:

$$\hat{f}(\boldsymbol{p}) = \tilde{f}(\omega(\boldsymbol{p}), \boldsymbol{p}). \tag{23}$$

Instead of (21) we consider

$$A(\tilde{f}, t) = \int d^4p \; e^{-itp^-} \; \tilde{f}^*(p) \, \tilde{A}(p), \tag{24}$$

with $p^- = p_0 - \omega(\boldsymbol{p})$ and $\tilde{f} \in \mathcal{G}$. If A is a free field this reduces to (16) because of the $\delta_\pm$-factors in the form (12) of a free field.

I can now formulate and discuss the asymptotic conditions.

The Haag–Ruelle Condition

Let $\tilde{f}_1, \ldots, \tilde{f}_n \in \mathcal{G}$. Consider the vector

$$\Phi(t) = A^*(\tilde{f}_1, t) \ldots A^*(\tilde{f}_n, t) \, \Omega. \tag{25}$$

$(A^*(\tilde{f}, t)$ is the adjoint of $A(\tilde{f}, t)$.) Then

$$\lim_{t \to \pm\infty} \Phi(t) = \hat{A}^{\mathrm{ex}*}(\hat{f}_1) \ldots \hat{A}^{\mathrm{ex}*}(\hat{f}_n)\Omega, \tag{26}$$

where "ex" means "out" for $t \to \infty$, "in" for $t \to -\infty$. The limit is attained in the strong topology of $\mathcal{H}$.

I cannot give the proof of (26) in detail, but I shall at least try to sketch its general outline. (Note that the asymptotic "conditions" are actually theorems, not conditions in the sense of assumptions.) The idea is to first consider the time derivative

$$\dot{\Phi}(t) = \frac{d}{dt}\Phi(t) \tag{27}$$

of the vector-valued function $\Phi(t)$ and show that its norm tends to zero at least as $|t|^{-3/2}$ for $|t| \to \infty$:

$$|\dot{\Phi}(t)| \leqq c \, |t|^{-3/2} \tag{28}$$

with c a suitable positive constant. From this we obtain that

$$\lim_{t \to \infty} \Phi(t) = \Phi(0) + \int_0^\infty dt \, \dot{\Phi}(t)$$

exists. It can then be shown that the limit is of the form claimed in (26).

An essential ingredient of the proof of estimate (28) is the *cluster property* of the truncated Wightman functions. The ordinary Wightman functions are defined as

$$W(x_1, \ldots, x_n) = \langle 0 | A(x_1) \ldots A(x_n) | 0 \rangle. \tag{29}$$

Their truncated part W^T is obtained by "taking out the vacuum as an intermediate state in a symmetric way". Less mysteriously we define W^T with the help of the *cluster expansion*

$$W(x_1, \ldots, x_n) = \sum W^T(x_1, \ldots, x_{i_\alpha}) \ldots W^T(x_{i_\zeta}, \ldots, x_{i_\omega}). \tag{30}$$

The sum in this formula extends over all partitions of $\{x_1, \ldots, x_n\}$ into any number of mutually nonoverlapping subsets. In each W^T-factor the variables x_i stand in their natural order. The expansion (30) allows a recursive calculation of W^T, starting from

$$W^T(x_1) = W(x_1) = 0, \qquad W^T(x_1, x_2) = W(x_1, x_2). \tag{31}$$

The first of these equations is the consequence of an assumption that I have not yet mentioned: that $A(x)$ have a vanishing vacuum expectation value. This can always be achieved by adding a constant c-number to $A(x)$, i.e. it is no restriction of generality.

The $W^T(x_1, \ldots, x_n)$ share all the well-known properties of the W-functions. In particular they are invariant, and they are local:

$$W^T(\ldots, x_i, x_{i+1}, \ldots) = W^T(\ldots, x_{i+1}, x_i, \ldots) \tag{32}$$

if $(x_i - x_{i+1})^2 < 0$. The spectral properties of the W^T are even nicer than those of the W. If $\tilde{W}^T(p_1, \ldots, p_n)$ is the Fourier transform of $W^T(x_1, \ldots, x_n)$, we find that

$$\tilde{W}^T(p_1, \ldots, p_n) = \delta^4(p_1 + \ldots + p_n) \, \hat{W}^T(p_2, \ldots, p_n), \tag{33}$$

and $\hat{W}^T \neq 0$ only if $p_n \in V_-^m$, $(p_n + p_{n-1}) \in V_-^m, \ldots, (p_n + \ldots + p_2) \in V_-^m$, with

$$V_-^m = \{p : p^2 \geqq m^2, p_0 < 0\}.$$

The support of W contains also the points $p_i + \ldots + p_n = 0$. This difference between W and W^T is crucial for the cluster property.

The version of the cluster property which is needed for the proof of the asymptotic condition states that $W^T(x_1, \ldots, x_n)$ tends strongly to zero if the arguments separate to infinity within a fixed t-plane. Mathematically formulated: let $\phi(x_1, \ldots, x_n) \in \mathscr{S}$, define $x + a = (x^0, \boldsymbol{x} + \boldsymbol{a})$, and form

$$\psi(\boldsymbol{a}_2, \ldots, \boldsymbol{a}_n) = \int dx_1 \ldots dx_n \, W^T(x_1, \ldots, x_n) \, \phi(x_1, x_2 + \boldsymbol{a}_2, \ldots, x_n + \boldsymbol{a}_n). \tag{34}$$

Then $\psi \in \mathscr{S}$, i.e. ψ is C^∞ and decreases strongly at ∞.

Let me just indicate why this is so in the case of the 2-point function $W(x, y)$. Due to translation invariance this function depends only on the difference $\xi = x - y$: $W(x, y) = W(\xi)$, and the expression (34) becomes

$$\psi(\boldsymbol{a}) = \int d^4\xi \, W(\xi) \, \phi(\xi + a). \tag{35}$$

It is obvious that ψ is C^∞: we can differentiate under the integral sign, and a derivative of a function in $\mathscr{S}$ is again in $\mathscr{S}$. The strong decrease at ∞ is more difficult to establish. Let me first consider the function

$$K(\xi) = W(\xi) - W(-\xi) = \langle 0 \,|\, [A(x), A(y)] \,|\, 0 \rangle. \tag{36}$$

Its support is contained in the set $\xi^2 \geqq 0$. Hence

$$\bar{\psi}(\boldsymbol{a}) = \int d^4\xi \, K(\xi)\, \phi(\xi + \boldsymbol{a}) \tag{37}$$

decreases strongly for $|\boldsymbol{a}| \to \infty$, because the values of the test function $\phi(\xi + \boldsymbol{a})$ at the points $\xi \in \operatorname{supp} K$ decrease strongly for $|\boldsymbol{a}| \to \infty$. Relation (37) can also be written

$$\bar{\psi}(\boldsymbol{a}) = \int d^4p \, \tilde{K}(p)\, \tilde{\phi}(p)\, e^{i\boldsymbol{a}p}. \tag{38}$$

The Fourier transform $\tilde{\phi}$ of ϕ is again in $\mathscr{S}$. From the spectral support of $\tilde{W}$ we obtain

$$\tilde{W}(p) = \vartheta(p_0)\, \tilde{K}(p), \tag{39}$$

where ϑ is a C^∞-function with $\vartheta(p_0) = 0$ for $p_0 \geqq m$ and $\vartheta(p_0) = 1$ for $p_0 \leqq -m$. Relation (35) becomes

$$\psi(\boldsymbol{a}) = \int d^4p \, \tilde{K}(p)\, \vartheta(p_0)\, \tilde{\phi}(p)\, e^{ip\boldsymbol{a}}. \tag{40}$$

But $\vartheta\phi \in \mathscr{S}$ if $\phi \in \mathscr{S}$, hence ψ is of the form (38), hence strongly decreasing. A similar argument can be given for general n.

By Fourier-transforming (34) with respect to $\boldsymbol{a}_i$ as well as with respect to x_i, we obtain as a corollary to the cluster property that the function

$$\tilde{\psi}(p_2, \ldots, p_n) = \int dp_{20} \ldots dp_{n0} \, \hat{W}^T(p_2, \ldots, p_n)\, \tilde{\phi}(p_2, \ldots, p_n) \tag{41}$$

is in $\mathscr{S}$ for any $\tilde{\phi} \in \mathscr{S}$.

We return to the discussion of $\Phi(t)$. From the definition (25) of Φ we find that $\dot{\Phi}$ is of the form

$$\dot{\Phi}(t) = \sum_{i=1}^{n} A^*(\tilde{f}_1, t) \ldots \dot{A}^*(\tilde{f}_i, t) \ldots A^*(\tilde{f}_n, t)\, \Omega. \tag{42}$$

Call the i-th term in this sum $\Phi_i(t)$ and consider

$$(\Phi_i(t), \Phi_j(t)) = \int \prod_1^n \left\{ dp_l \, dq_l \, e^{-itp_l^-}\, e^{-itq_l^+}\, \tilde{f}_i^*(p_l)\, \tilde{f}_i(-q_l) \right\} \times$$
$$\times\, \tilde{W}(p_1, \ldots, p_l, q_1, \ldots, q_l)\, p_i^-\, q_j^+, \tag{43}$$

with $p^{\pm} = p_0 \pm \omega(\boldsymbol{p})$.

For $\tilde{W}$ we insert the cluster expansion (30). Each term leads to a product of terms of the form

$$X^{\alpha\beta}(t) = \int \prod_1^{\alpha} \left\{ dp_l\, e^{-itp_l^-}\, \tilde{f}_l^*(p_l) \right\} \prod_1^{\beta} \left\{ dq_h\, e^{-itq_h^+}\, \tilde{f}_h(-q_h) \right\} \times$$

$$\times\, \tilde{W}^T(p_1,\ldots,p_\alpha,q_1,\ldots,q_\beta), \qquad (44)$$

where $\{p_1,\ldots,p_\alpha\}$ is any subset of $\{p_1,\ldots p_n\}$, and $\{q_1,\ldots,q_\beta\}$ of $\{q_1,\ldots,q_n\}$. If p_i or q_j are contained in these subsets, we get additional factors p_i^- or q_j^+ respectively.

The behavior of $X^{\alpha\beta}(t)$ for large t can be estimated. We distinguish two cases:

Case 1: $\alpha + \beta = 2$; at least one of the factors p_i, q_j is present. If $\alpha = 2$, $\beta = 0$ or $\alpha = 0$, $\beta = 2$, then $X^{\alpha\beta}$ vanishes because of the supports of $\tilde{f}_l$ and $\tilde{W}^T$. There remains the case $\alpha = \beta = 1$. To fix the ideas, assume that the factor q_j^+ is present:

$$X^{11} = \int dp\, dq\, e^{-itp^-}\, \tilde{f}_1^*(p)\, e^{-itq^+}\, \tilde{f}_2(-q)\, q^+\, \tilde{W}^T(p,q).$$

It can be shown that $\hat{W}^T(q)$ is of the form

$$\mathrm{const} \cdot \theta(-q_0)\, \delta(q^2 - m^2)$$

in the support of $\tilde{f}_2(-q)$. The δ-factor is annihilated by the factor q^+, hence $X^{11} = 0$.

Case 2 (remaining possibilities): For support reasons, $X^{\alpha\beta} \neq 0$ only if $\alpha \geq 1$, $\beta \geq 1$. We integrate over p_1 with the help of the δ^4 in (33) and obtain something of the form

$$X^{\alpha\beta} = \int dp_2 \ldots dp_\alpha\, dq_1 \ldots dq_\beta\, \hat{W}^T(p_2,\ldots,q_\beta)\, \tilde{f}(p_2,\ldots,q_\beta) \times$$

$$\times \prod_2^{\alpha} \exp\{it\omega(p_l)\} \prod_1^{\beta} \exp\{-it\omega(q_h)\} \exp\left\{ it\omega\left(\sum_2^{\alpha} p_i + \sum_1^{\beta} q_i \right) \right\}; \qquad (45)$$

$\tilde{f}$ is a function from $\mathscr{S}$, and this is all we need know about it. The 0-components have dropped out from the exponentials because of the factor $\delta(\sum_1^{\alpha} p_{i0} + \sum_1^{\beta} q_{j0})$ in $\tilde{W}^T$.

The integration of the product $\hat{W}^T \tilde{f}$ over p_{l0} and q_{h0} can be carried out and yields a function $w(p_2,\ldots p_\alpha, q_1,\ldots,q_\beta) \in \mathscr{S}$, because of the form (41) of the cluster property. The remaining integral

$$X^{\alpha\beta}(t) = \int \prod_2^{\alpha} d^3p_i \prod_1^{\beta} d^3q_j\, w(p_2,\ldots,q_\beta) \times$$

$$\times \exp\left[it \left\{ \sum_2^{\alpha} \omega(p_i) - \sum_1^{\beta} \omega(q_j) + \omega\left(\sum p_i + \sum q_j \right) \right\} \right]$$

can be estimated with the help of a generalized saddle point method [9], with the result

$$|X^{\alpha\beta}| \leqq C \, |t|^{-(3/2)(\alpha+\beta-2)}, \tag{46}$$

were C is a positive constant. Note that $\alpha + \beta$ is the total number of points in W^T.

We apply this result to (Φ_i, Φ_j). In the terms of the cluster expansion which contain only 2-point functions, the factors p_i^- and q_j^+ are certainly contained in 2-point functions; hence these terms vanish because of the results obtained in Case 1. In the remaining terms we have at least two 3-point functions or one 4- or more-point function, so that

$$|(\Phi_i, \Phi_j)| \leqq \text{const} \cdot |t|^{-3},$$

whence the desired relation (28) follows at once.

This proves the existence of $\lim \Phi(t)$. The form of this limit can be found by a discussion of the surviving 2-point-function-only terms in the undifferentiated product $(\Phi(t), \Phi(t))$. The result is what was stated in (26).

Before going over to the LSZ asymptotic condition, I wish to introduce a new basic assumption: asymptotic completeness. The Haag–Ruelle condition established above states that in a certain sense any reasonable state converges for $t \to \pm \infty$ towards a state with a particle interpretation. In order to really have a particle theory we must be sure that these asymptotic states contain all there is to say. No state is allowed to contain components with the behavior expected from a classical field: a local disturbance spreads, as time goes on, over the whole space and becomes diluted to unrecognizability. We exclude this possibility by introducing a new, final, axiom.

Axiom 5 (asymptotic completeness):

The Fock spaces $\mathscr{H}^{\text{in, out}}$ generated by the asymptotic fields $A^{\text{in, out}}$ are already the full Hilbert space of the theory:

$$\mathscr{H}^{\text{in}} = \mathscr{H}^{\text{out}} = \mathscr{H}. \tag{47}$$

(Note that from the asymptotic condition we can only conclude $\mathscr{H}^{\text{in, out}} \subset \mathscr{H}$.) It is a bit unpleasant that we are forced to introduce an axiom at such a late stage in the game. It would be more satisfactory if we could formulate a condition directly bearing on the interacting $A(x)$ which guarantees asymptotic completeness. But such a condition has not yet been found.

The LSZ Condition

Let $\tilde{f} \in \mathscr{G}$ and Φ^{ex} be a vector of the form

$$\Phi^{\text{ex}} = A^{\text{ex}} {}^*(\hat{f}_1) \dots A^{\text{ex}} {}^*(\hat{f}_n)\Omega, \tag{48}$$

with $\hat{f}_i \in \mathcal{S}$. Then

$$\lim_{t \to \pm \infty} A^*(\tilde{f}, t)\Phi^{\mathrm{ex}} = A^{\mathrm{ex}*}(\hat{f})\Phi^{\mathrm{ex}}, \tag{49}$$

where this time the limit must be taken in the weak topology of $\mathcal{H}$, i.e. what converges are the scalar products $(\Psi, A^*(f, t)\Phi^{\mathrm{ex}})$ with Ψ any vector in $\mathcal{H}$.

The relation (49) has only been proved from our general assumptions for *non-overlapping* Φ^{ex}, i.e. if the supports of all the $\hat{f}_i$ are disjoint. (This means that no two particles can have the same momentum.) However, in all known approximate models (e.g., perturbation theory) the LSZ condition holds on overlapping states, and it is not a deadly sin to use it on such states also in a general context. In what follows we will not have to do this.

I will not say much about the proof of the LSZ condition. Let me just pick out one important point. The first problem that presents itself in looking at the vector $A^*(\tilde{f}, t)\Phi^{\mathrm{ex}}$ is an existence problem. $A^*(\tilde{f}, t)$ is an unbounded operator, hence not defined on all of $\mathcal{H}$. By assumption it is defined on the dense set D_0 spanned by the vectors $A(\phi_1) \ldots A(\phi_n)\Omega$, but Φ^{ex} is not necessarily contained in this set. However, we can extend the definition of $A^*(\tilde{f}, t)$ (or more generally, any $A(\phi)$) to Φ^{ex} if we note that $A(\phi)$ is a closable operator. This follows from the assumed symmetry of $A(\phi)$ for real ϕ.

An operator O in $\mathcal{H}$ is called *closed* if the following is true: if $\Phi_1, \Phi_2, \ldots \in \mathcal{H}$ is a convergent sequence of vectors, all lying in the domain of definition of O,

$$\lim_{i \to \infty} \Phi_i = \Phi \tag{50}$$

such that

$$\Psi = \lim_{i \to \infty} O\,\Phi_i \tag{51}$$

exists, then O is defined on Φ and $O\,\Phi = \Psi$. O is called *closable* if it has a closed extension, i.e. if there exists a closed $\bar{O}$ which coincides with O on the latter's domain of definition.

It can be shown that Φ^{ex} is such a limit vector of a sequence in D_0, with (51) satisfied for $O = A(\phi)$, $\phi \in \mathcal{S}$, provided that Φ^{ex} is nonoverlapping [6, 7]. This is the point where the nonoverlap assumption enters the proof. (Of course, $A(\phi)$ may actually be defined on overlapping Φ^{ex}; we cannot disprove this, we can only not prove it.) The estimate for $|\dot{\Phi}(t)|$ given in the proof of the Haag–Ruelle condition can be considerably improved for nonoverlapping $\dot{\Phi}(t)$ (still meaning nonoverlapping of the $\hat{f}_i$): in this case $|\dot{\Phi}(t)|$ tends to zero more strongly than any negative power of $|t|$ for $|t| \to \infty$. By the Haag–Ruelle condition we can write Φ^{ex} as strong limit

$$\Phi^{\mathrm{ex}} = \lim_{s \to \pm \infty} \Phi(s)$$

of a vector family $\Phi(s)$ of the form (25). $\Phi(s)$ is in D_0, hence $A^*(\tilde{f}, t)\Phi(s)$ exists. We want to prove that $\lim_{s \to \pm \infty}$ of this vector exists (t is kept constant). This we do again by discussing its derivative $A^*(\tilde{f}, t)\dot{\Phi}(s)$. We find

$$\|A^*(\tilde{f}, t)\dot{\Phi}(s)\|^2 = (\dot{\Phi}(s), A(\tilde{f}, t) A^*(\tilde{f}, t)\dot{\Phi}(s)) \leqq \|\dot{\Phi}(s)\| \, \| A(\tilde{f}, t) A^*(\tilde{f}, t)\dot{\Phi}(s)\|.$$

The first factor decreases strongly for $|s| \to \infty$, because of the nonoverlap assumption. The second factor is polynomially bounded in $|s|$, due to the temperedness of the fields. (I might note that we use only that part of the temperedness assumption which survives in the more general cases considered recently: the Jaffe fields and even some nonlocalizable fields.) Hence the product decreases strongly, and

$$A^*(\tilde{f}, t)\Phi^{\mathrm{ex}} = A^*(\tilde{f}, t)\Phi(0) \pm \int_0^{\pm\infty} A^*(\tilde{f}, t)\dot{\Phi}(s)\, ds \tag{52}$$

exists.

I refrain from giving the rest of the proof of the LSZ condition. It proceeds via some reasonably simple lemmas, substantiating the plausible idea that, since

$$A^*(\tilde{f}, t)\Phi^{\mathrm{ex}} = \lim_{s \to \pm\infty} A^*(\tilde{f}, t)\Phi(s),$$

we may expect

$$\lim_{t \to \pm\infty} A^*(\tilde{f}, t)\Phi^{\mathrm{ex}} = \lim_{t \to \pm\infty} A^*(\tilde{f}, t)\Phi(t) = A^{\mathrm{ex}*}(\hat{f})\Phi^{\mathrm{ex}}, \tag{53}$$

where we have used the already proved Haag–Ruelle condition. The first equality turns out to be true in the weak topology of $\mathcal{H}$.

IV. THE S-MATRIX

We are now in a position to define the S-matrix, which is the quantity that is of main interest to theoreticians of a more phenomenological tendency. We have defined two free fields A^{in}, A^{out}, i.e. two representations of the canonical commutation relations, both with a vacuum, and any two such representations are known to be unitarily equivalent. The S-"matrix" is the unitary operator effecting this equivalence:

$$A^{\mathrm{out}}(x) = S^* A^{\mathrm{in}}(x) S. \tag{54}$$

S is of interest because it describes the probability of any scattering event occurring. The probability of an incoming n-particle state $|\hat{f}_1, \ldots, \hat{f}_n\rangle_{\mathrm{in}} = A^{\mathrm{in}*}(\hat{f}_1)\ldots A^{\mathrm{in}*}(\hat{f}_n)\Omega$ being scattered into an outgoing m-particle state $|\hat{g}_1, \ldots, \hat{g}_m\rangle_{\mathrm{out}} = A^{\mathrm{out}*}(\hat{g}_1)\ldots\Omega$ is given by the amplitude

$$_{\mathrm{out}}\langle \hat{g}_1, \ldots, \hat{g}_m | \hat{f}_1, \ldots, \hat{f}_n\rangle_{\mathrm{in}} = {}_{\mathrm{in}}\langle \hat{g}_1, \ldots, \hat{g}_m | S | \hat{f}_1, \ldots, \hat{f}_n\rangle_{\mathrm{in}}, \tag{55}$$

i.e. by a matrix element of S in a basis of $\mathcal{H}$ constructed from in-vectors.

A convenient simple expression for these matrix elements of S in terms of the field $A(x)$ is given by the *reduction formulae*. Similar formulae can also be given for the matrix elements of $A(x)$ in an in-basis. Before we can state and discuss these formulae we must introduce two new objects, the *retarded* and *time-ordered* products of fields $A(x_1), \ldots, A(x_n)$. Formally they are defined as

$$R(x_1, \ldots, x_n) = (-i)^{n-1} \sum_{(i_2 \ldots i_n)} \theta(x_1, x_{i_2}, \ldots, x_{i_n}) \times$$

$$\times [\ldots [A(x_1), A(x_{i_2})], \ldots, A(x_{i_n})], \qquad (56)$$

$$T(x_1, \ldots, x_n) = \sum_{(i_1 \ldots, i_n)} \theta(x_{i_1}, \ldots, x_{i_n}) A(x_{i_1}) \ldots A(x_{i_n}), \qquad (57)$$

with

$$\theta(x_1, \ldots, x_n) = \begin{cases} 1 & \text{if } x_1^0 > x_2^0 > \ldots > x_n^0 \\ 0 & \text{elsewhere.} \end{cases}$$

The sum extends over all permutations of the indices $(2, \ldots, n)$ in (56), of $(1, \ldots, n)$ in (57).

Unfortunately, these definitions are quite problematic, because $A(x)$ is a distribution, not a function, and the product of a distribution with a discontinuous function is not defined *a priori*. We can give more sophisticated definitions, starting from the fact that (56) and (57) make sense outside of the discontinuities of θ, where no two x_i^0 coincide. Using locality we can, in fact, reduce the trouble to the points where two x_i coincide in all 4 components. R and T are then defined as suitable continuations from the harmless points into the critical ones, such that the relevant properties (supports, invariance, etc.) are not destroyed. That this can always be done has not yet been proved. We can solve the problem by *assuming* the existence of such a continuation. This is unsatisfactory from a fundamental point of view, but quite acceptable as a working hypothesis from a more practical standpoint. Alternatively, we can work with "smooth" products instead of the "sharp" products (56), (57). They are obtained by using (56), (57) only outside some finite open neighborhood of the critical points $x_i = x_j$, and then suitably continuing into this neighborhood. Some of the nice properties of R and T must then be sacrificed, in particular the explicit covariance. This is somewhat awkward for the applications, but one can live with it. Of course, the covariance of the physical results is not destroyed by this procedure.

For simplicity I assume here the existence of the sharp products. Those of their properties that are of importance to us are:

1) covariance:

$$R(\Lambda x_i + a) = U(\Lambda, a) R(x_i) U^*(\Lambda, a) \qquad (58)$$

and the same for T;

2) symmetry: $R(x_1, \ldots, x_n)$ is symmetric in the variables $x_2, \ldots, x_n$, $T(x_1, \ldots, x_n)$ in $x_1, \ldots, x_n$;

3) support:

$$R(x_1,\ldots,x_n) = 0 \quad \text{unless} \quad (x_1 - x_i) \in \bar{V}_+ \quad \text{for} \quad i = 2,\ldots,n, \tag{59}$$

$$T(x_1,\ldots,x_n) = T(x_1,\ldots,x_m)\, T(x_{m+1},\ldots,x_n) \tag{60}$$

if $x_i^0 > x_j^0$ for all $i \in \{1,\ldots,m\}$ and all $j \in \{m+1,\ldots,n\}$.

With the help of the T-product we can formulate a new variant of the Haag–Ruelle asymptotic condition. It turns out that this condition remains valid if the ordinary product $A(x_1)\ldots A(x_n)$ in $\Phi(t)$ is replaced by the T-product. If we define

$$\Phi(t) = \int \prod_1^n \{dp_i\, \tilde{f}_i(-p_i) \exp[-itp_i^+]\}\, \tilde{T}(p_1,\ldots,p_n)\,\Omega, \tag{61}$$

then still

$$\lim_{t \to \pm\infty} \Phi(t) = A^{\mathrm{ex}*}(\hat{f}_1)\ldots A^{\mathrm{ex}*}(\hat{f}_n)\Omega. \tag{62}$$

Instead of $\tilde{T}(p_i)$ we could introduce in (61) $\tilde{\bar{T}}(p_1,\ldots,p_n) = [\tilde{T}(-p_1,\ldots,-p_n)]^*$, the Fourier transform of the anti-time-ordered product $\bar{T}$, and still (62) would hold. The proof of (62) follows the same pattern as that of the original Haag–Ruelle condition and will not be given here.

The new version is useful for the derivation of reduction formulae. Relation (61) can be transformed into x-space. We define

$$\bar{f}(x, t) = (2\pi)^{-5/2} \int d^4p\; e^{-ipx}\, e^{itp^-}\, \tilde{f}(p) \tag{63}$$

and obtain

$$\Phi(t) = \int \prod_1^n \{d^4x_i\, \bar{f}_i(x_i, t)\}\, T(x_1,\ldots,x_n)\,\Omega. \tag{64}$$

For $\bar{f}(x, t)$ the following estimate can be derived:

$$|\bar{f}(x, t)| \leqq \frac{c_{MN}(1 + |t|)^M}{(1 + |x|)^M (1 + |x^0 - t|)^N} \tag{65}$$

for any positive integers N, M, with c_{MN} a positive constant.

Similar estimates hold for all derivatives of arbitrary order of $\bar{f}$. We can fix M at a value large enough to make the x_i-integrals in (64)—and in similar expressions to be considered later—converge, while still allowing arbitrarily large N. We see that $\bar{f}(x, t)$ decreases for $|t| \to \infty$ at any fixed point x stronger than $|t|^{-(N-M)}$, with $N - M$ arbitrarily large. The region where $\bar{f}$ is large is concentrated around $x^0 \sim t$: it moves to infinity together with $|t|$.

Let us derive a reduction formula for the S-matrix element (55):

$$X = {}_{\mathrm{in}}\langle \hat{g}_1,\ldots,\hat{g}_m | S | \hat{f}_1,\ldots,\hat{f}_n \rangle_{\mathrm{in}} = {}_{\mathrm{out}}\langle \hat{g}_1,\ldots | \ldots,\hat{f}_n \rangle_{\mathrm{in}}, \tag{66}$$

with $\hat{g}_i, \hat{f}_j \in \mathscr{S}$. For simplicity we assume that all $\hat{g}_i$ are nonoverlapping with all $\hat{f}_j$:

$$\hat{g}_i(p)\hat{f}_j(p) \equiv 0$$

for all pairs i, j. This excludes processes in which a particle simply goes through without interacting at all. These processes could easily be included, at the price of having to write some more.

With the T-product variant of the HR condition we obtain

$$X = \lim_{t \to \infty} \int \prod_1^m \{dx_i \, \bar{g}_i^*(x_i, t)\} \prod_1^n \{dy_j \, \bar{f}_j(y_j, -t)\} \times$$

$$\times \langle 0 | T(x_1, \ldots, x_m) \, T(y_1, \ldots, y_n) | 0 \rangle . \qquad (67)$$

Because of the properties of $\bar{f}$ discussed above and the property (60) of the T-product, we can replace $\langle 0 | T(\ldots) T(\ldots) | 0 \rangle$ by $\langle 0 | T(x_1, \ldots, x_m, y_1, \ldots, y_n) | 0 \rangle = \tau(x_1, \ldots, y_n)$:

$$X = \lim_{t \to \infty} \int \prod_1^m \{dx_i \, \bar{g}_i^*(x_i, t)\} \prod_1^n \{dy_j \bar{f}_j(y_j, -t)\} \, \tau(x_1, \ldots, y_n). \qquad (68)$$

The difference between the two expressions under the lim in (67) and (68) tends strongly to zero for $t \to \infty$.

Consider what happens if some of the t-signs in (68) are inverted:

$$Y = \lim_{t \to \infty} \int \prod_1^m dx_i \prod_1^n dy_j \prod_1^\alpha \bar{g}_i^*(x_i, t) \prod_{\alpha+1}^m \bar{g}_i^*(x_i, -t) \times$$

$$\times \prod_1^\beta \bar{f}_j(y_j, t) \prod_{\beta+1}^n \bar{f}_j(y_j, -t) \, \tau(x_1, \ldots, y_n), \qquad (69)$$

with $\alpha < m$ or $\beta > 1$, or both. By reversing the argument leading from (67) to (68), Y becomes

$$Y = \lim_{t \to \infty} \int \ldots \times \langle 0 | T(x_1, \ldots, x_\alpha, y_1, \ldots, y_\beta) \, T(x_{\alpha+1}, \ldots, x_m, y_{\beta+1}, \ldots, y_n) | 0 \rangle .$$

In the preceding section I have given a version of the HR asymptotic condition in which only creation operators occur. However, the relation (26) holds also if some of the stars * are missing in $\Phi(t)$ and Φ^{ex}. Of course, they must be missing in the same places in the two vectors. This generalized condition gives, under our nonoverlap assumption:

$$Y = \langle 0 | A^{\text{out}*}(\hat{f}_1) \ldots A^{\text{out}*}(\hat{f}_\beta) \, A^{\text{out}}(\hat{g}_1) \ldots A^{\text{out}}(\hat{g}_\alpha) \times$$

$$\times A^{\text{in}*}(\hat{f}_{\beta+1}) \ldots A^{\text{in}*}(\hat{f}_n) \, A^{\text{in}}(\hat{g}_{\alpha+1}) \ldots A^{\text{in}}(\hat{g}_m) | 0 \rangle = 0,$$

because

$$A^{\text{in}}(\hat{g}) | 0 \rangle = 0, \qquad \langle 0 | A^{\text{out}*}(\hat{f}) = 0.$$

286   O. STEINMANN

This result permits to write X in a more complicated way:

$$
X = \lim_{t \to \infty} \int \prod dx_i \prod dy_j \prod_1^m \left[\bar{g}_i^*(x_i, t) - \bar{g}_i^*(x_i, -t) \right] \times
$$

$$
\times \prod_1^n \left[\bar{f}_j(y_j, -t) - f_j(y_j, t) \right] \tau(x_1, \ldots, y_n),
$$

and this reads in p-space

$$
X = \lim_{t \to \infty} \int \prod_1^m \left\{ dp_i \, \tilde{g}_i^*(p_i) \left[e^{-itp^-} - e^{itp^-} \right] \right\} \times
$$

$$
\times \prod_1^n \left\{ dq_j \, \tilde{f}_j(-q_j) \left[e^{itq^+} - e^{-itq^+} \right] \right\} \tilde{\tau}(p_1, \ldots, q_n). \tag{70}
$$

We define

$$
\delta_t(u) = \frac{i}{2\pi} \frac{e^{-itu} - e^{itu}}{u} \tag{71}
$$

and note

$$
\lim_{t \to \infty} \delta_t(u) = \delta(u) \tag{72}
$$

in the sense that $\int du \, f(u) \, \delta_t(u) \to f(0)$ for continuous functions $f(u)$ with compact support. With this notation, X becomes

$$
X = (-2\pi i)^{n+m} \lim_{t \to \infty} \int \prod_1^m \left\{ \frac{dp_i}{p_i^+} \, \tilde{g}_i^*(p_i) \, \delta_t(p_i^-) \right\} \times
$$

$$
\times \prod_1^n \left\{ \frac{dq_j}{-q_j^-} \, \tilde{f}_j(-q_j) \, \delta_t(q_j^+) \right\} \tilde{\tau}^{\text{amp}}(p_1, \ldots, q_n), \tag{73}
$$

with

$$
\tilde{\tau}^{\text{amp}}(p_1, \ldots, q_n) = \prod (p_i^2 - m^2) \prod (q_j^2 - m^2) \, \tilde{\tau}(p_1, \ldots, q_n), \tag{74}
$$

the "amputated" $\tilde{\tau}$-function.

Using (72) formally we obtain

$$
X = (-2\pi i)^{n+m} \int \prod \left\{ \frac{d^3 p_i}{2\omega(p_i)} \, \hat{g}_i^*(p_i) \right\} \prod \left\{ \frac{d^3 q_j}{2\omega(q_j)} \, \hat{f}_j(q_j) \right\} \times
$$

$$
\times \tilde{\tau}_{MS}^{\text{amp}}(p_1, \ldots, p_m, -q_1, \ldots, -q_n), \tag{75}
$$

with

$$
\tilde{\tau}_{MS}^{\text{amp}}(p_i, -q_j) = \tilde{\tau}^{\text{amp}}(p_i, -q_j) \big|_{p_{i0} = \omega(p_i), \, q_{j0} = \omega(q_j)} : \tag{76}
$$

the S-matrix element X is simply the mass shell restriction of $\tilde{\tau}^{\text{amp}}$ integrated over the wave functions of the states in question.

More explicitly, (76) means the following. Take $\tilde{\tau}^{\text{amp}}(\ldots)$, write it as a function of the variables p_i, p_i^-, q_j, q_j^- instead of p_i, p_{i0}, q_j, q_{j0}, and restrict p_i^-, q_j^- to the value

0. What you obtain is a distribution in p_i, q_j. At the moment we know nothing about how regular $\tilde{\tau}^{\text{amp}}$ is in the variables p_i^-, q_j^-; hence it is not yet clear what this "restriction to the values $p_i^- = 0$, $q_j^- = 0$" is. Hepp [7] has proved that $\tilde{\tau}^{\text{amp}}$ is a continuous function in p_i^-, q_j^- for nonoverlapping p_i, q_j, i.e. where no two of the 3-vectors p_i, q_j coincide. There $\tilde{\tau}_{MS}^{\text{amp}}$ is defined as the value of this continuous function in $p_i^- = q_j^- = 0$. In the critical overlapping points, (73) serves as definition of $\tilde{\tau}_{MS}^{\text{amp}}$:

$$\tilde{\tau}_{MS}^{\text{amp}}(p_1, \ldots, -q_n) = \lim_{t \to \infty} \int \prod \{ dp_i^- \, \delta_t(p_i^-) \} \prod \{ dq_j^- \, \delta_t(q_j^-) \} \, \tilde{\tau}^{\text{amp}}(\ldots). \quad (77)$$

We have proved that this limit exists, so that the definition makes sense.

Hepp's continuity proof is rather involved, and I will not talk about it. In Hepp's paper [7] you can also find a derivation of the reduction formula (75) from the LSZ asymptotic condition. This derivation is only valid for nonoverlapping states, since only for these states the LSZ condition has been proved. Historically, the LSZ derivation is the original one.

A reduction formula for the more general expression

$$X(k_1, \ldots, k_l) = {}_{\text{out}}\langle \hat{g}_1, \ldots, \hat{g}_m | \tilde{T}(k_1, \ldots, k_l) | \hat{f}_1, \ldots, \hat{f}_n \rangle_{\text{in}} \quad (77)$$

can be derived along the same lines. It reads:

$$X(k_1, \ldots, k_l) = (-2\pi i)^{n+m} \int \prod \left\{ \frac{d^3 p_i}{2\omega(p_i)} \, \hat{g}_i^*(p_i) \right\} \prod \left\{ \frac{d^3 q_j}{2\omega(q_j)} \, \hat{f}_j(q_j) \right\} \times$$

$$\times \tilde{\tau}_{MS}^{\text{amp}}(k_1, \ldots, k_l; p_1, \ldots, p_m, -q_1, \ldots, -q_n). \quad (78)$$

$\tilde{\tau}_{MS}^{\text{amp}}(\ldots; \ldots)$ is amputated and restricted to the mass shell only in the variables standing behind the semicolon.

A reduction formula can also be derived by similar methods for the matrix element

$$M = {}_{\text{in}}\langle \hat{g}_1, \ldots, \hat{g}_m | \tilde{R}(k_1, \ldots, k_l) | \hat{f}_1, \ldots, \hat{f}_n \rangle_{\text{in}}. \quad (79)$$

The special case $l = 1$ gives the matrix elements of the field operator $\tilde{A}(k)$.

In order to be sure that M exists, we must assume that at least one of the states $|\{\hat{g}_i\}\rangle_{\text{in}}, |\{\hat{f}_j\}\rangle_{\text{in}}$ is nonoverlapping. For simplicity we assume that all $\hat{g}_i$ are orthogonal to all $\hat{f}_j$. We obtain

$$M = (2\pi)^{n+m} \int \prod_1^m \left\{ \frac{d^3 p_i}{2\omega(p_i)} \, \hat{g}_i^*(p_i) \right\} \prod_1^n \left\{ \frac{d^3 q_j}{2\omega(q_j)} \, \hat{f}_j(q_j) \right\} \times$$

$$\times \tilde{r}_{MS}^{\text{amp}}(k_1, \ldots, k_l; p_1, \ldots, p_m, -q_1, \ldots, -q_n); \quad (80)$$

here $\tilde{r}_{MS}^{\text{amp}}(\ldots; \ldots)$ is defined analogously to $\tilde{\tau}_{MS}^{\text{amp}}$.

Next I wish to discuss the consequences of asymptotic completeness. With the help of this assumption we can derive sets of quadratic integral equations, the "*completeness equations,*" which are essentially off-mass shell continuations of the unitarity condition for S. They are therefore often called "unitarity equations."

We start from the algebraic identity

$$T(x_1,\ldots,x_n) + (-1)^n\, T^*(x_1,\ldots,x_n) =$$

$$= -\sum (-1)^k\, T^*(x_{i_1},\ldots,x_{i_k})\, T(x_{i_{k+1}},\ldots,x_{i_n})\,. \quad (81)$$

The sum on the right extends over all partitions of the set $\{x_i,\ldots,x_n\}$ into two nonempty subsets $\{x_{i_1},\ldots,x_{i_k}\}, \{x_{i_{k+1}},\ldots,x_{i_n}\}$.

We take the vacuum expectation value of (81) and sum on the right-hand side over a complete orthonormal set of out-states as intermediate states. Such a complete set is formed by the vectors

$$|i_1,\ldots,i_l\rangle = c_{i_1\ldots i_l}\, A^{\mathrm{out}}{}^*(\hat{f}_{i_1})\ldots A^{\mathrm{out}}{}^*(\hat{f}_{i_l})\,\Omega\,, \quad (82)$$

where the wave functions f_i are chosen from a basis of the L^2-space defined by

$$\int \frac{d^3 p}{2\omega(p)}\, |\hat{f}(p)|^2 < \infty\,.$$

$c_{i_1\ldots i_l}$ is a normalization constant.

The matrix elements $\langle 0|T^*|i_1,\ldots,i_l\rangle$ and $\langle i_1,\ldots,i_l|T|0\rangle$ occurring on the r.h.s. can be evaluated with the reduction formula (78), and the sum over the $|i_1,\ldots,i_l\rangle$ with a fixed l can be carried out with the help of the completeness relation in L^2:

$$\sum_\alpha \hat{f}_\alpha{}^*(p)\hat{f}_\alpha(q) = 2\omega(p)\,\delta^3(p - q)\,.$$

The result is

$$\tau(x_1,\ldots,x_n) + (-1)^n\, \tau^*(x_1,\ldots,x_n) =$$

$$= -\sum_{l=0}^{\infty} \frac{i^l}{l!} \sum_{(i_h)(i_j)} \int \prod_1^l \{du_i\, dv_i\, \Delta_+(u_i - v_i)\} \times$$

$$\times\, \tau^{\mathrm{amp}}{}^*(x_{i_1},\ldots,x_{i_k}; u_1,\ldots,u_l)\, \tau^{\mathrm{amp}}(x_{i_{k+1}},\ldots,x_{i_n}; v_1,\ldots,v_l)\,. \quad (83)$$

Here τ^{amp} stands for

$$\tau^{\mathrm{amp}}(x_1,\ldots,x_k; u_1,\ldots,u_l) = \prod_1^l K_{u_i}\tau(x_1,\ldots;\ldots,u_l)\,, \quad (84)$$

$$K_u = -\,\square_u - m^2\,.$$

A similar equation, called GLZ-equation (for Glaser, Lehmann, Zimmermann [4]), can be derived from the identity

$$R(x, y, x_1,\ldots,x_n) - R(y, x, x_1,\ldots,x_n) =$$

$$= -i\sum [R(x, x_{i_1},\ldots,x_{i_k}), R(y, x_{i_{k+1}},\ldots,x_{i_n})] \quad (85)$$

with the help of the reduction formula (80). The sum in (85) extends over all partitions of $\{x_1, \ldots, x_n\}$ into two complementary subsets, where this time the empty subset is admitted. We obtain:

$$r(x, y, x_1, \ldots, x_n) - r(y, x, x_1, \ldots, x_n) =$$

$$= -i\left[\ \sum_{l=1}^{\infty} \frac{i^l}{l!} \sum_{(i_j)(i_h)} \int \prod_1^l \{du_i\, dv_i\, \Delta_+(u_i - v_i)\} \times\right.$$

$$\left.\times\ r^{\mathrm{amp}}(x, x_{i_1}, \ldots, x_{i_k}; u_1, \ldots, u_l)\, r^{\mathrm{amp}}(y, x_{i_{k+1}}, \ldots, x_{i_n}; v_1, \ldots, v_l) - (x \leftrightarrow y)\ \right]. \quad (86)$$

V. THE GLZ THEOREM

The GLZ theorem is a LSZ counterpart of Wightman's reconstruction theorem. The latter states that a Wightman field theory can be completely characterized by its Wightman functions and details the conditions under which a set of such functions determines a field. Similarly, the GLZ theorem states that a LSZ theory can be characterized by its retarded functions and gives necessary and sufficient conditions which these functions must satisfy in order to determine a field.

These necessary and sufficient conditions are the following.

The $r(x_1, \ldots, x_n)$ must be tempered distributions which are invariant under $\mathfrak{P}^{\uparrow}_+$, symmetric in the variables $x_2, \ldots, x_n$, have their support contained in the set $(x_1 - x_i) \in \bar{V}_+, i = 2, \ldots, n$, and satisfy the completeness equations (86).

Unfortunately, this is not quite the full story yet. The last assumption, fulfillment of the completeness equations, needs some elaboration.

If we transform (86) into p-space, we find that the r.h.s. is a sum over integrals of the form

$$I = \int dk_i \ldots dk_l \prod_1^l \delta_+(k_i)\, \tilde{r}^{\mathrm{amp}}(p_1, \ldots, p_\alpha; -k_1, \ldots, -k_l) \times$$

$$\times\ \tilde{r}^{\mathrm{amp}}(p_{\alpha+1}, \ldots, p_n; k_1, \ldots, k_l). \quad (87)$$

The integrand is a product of 3-distributions in k_i, and such a product is in general not defined. In order to guarantee its existence we must make some assumptions about the behavior of $\tilde{r}^{\mathrm{amp}}(\ldots; \pm k_1, \ldots, \pm k_l)$ in the neighborhood of the k_i-mass-shell (only this region is relevant, because of the δ_+-factors). Necessary and sufficient conditions are known, but they look extremely complicated, and I will not state them (see ref. [8]). Instead, let me state a simpler condition which is sufficient, but presumably not necessary, and which is acceptable from a practical point of view, i.e. it does not seem to be too restrictive.

Consider $\tilde{r}^{\mathrm{amp}}(p_1, \ldots, p_n; k_1, \ldots, k_\alpha, -k'_1, \ldots, -k'_\beta)$. We restrict this function to a neighborhood of the mass shell in k_i, k'_j by multiplying it with functions $\chi(k_i)$, $\chi(k'_j)$ with the following properties: χ is C^∞, $\chi(k) \equiv 1$ on the mass shell $k_0 = \omega(\boldsymbol{k})$; $\chi \equiv 0$ outside of $0 < k^2 < 4m^2$, $k_0 > 0$. We write the k_i and k'_j dependence of this product as dependence on the variables k_i^-, $\boldsymbol{k}_i$, k'^-_j, $\boldsymbol{k}'_j$. Our assumption is: there exists a linear space $\mathcal{T}$ of test functions $\psi(\boldsymbol{k}'_1, \ldots, \boldsymbol{k}'_\beta)$, $\beta = 1, 2, \ldots$, which contains $\mathcal{S}$ and is dense in the L^2-space defined by

$$\int \prod_1^\beta \frac{d^3 \boldsymbol{k}'_i}{2\omega(\boldsymbol{k}'_i)} \, | \psi(\ldots, \boldsymbol{k}'_i, \ldots) |^2 < \infty,$$

such that, for $\phi(p_1, \ldots, p_n) \in \mathcal{S}$, the function

$$f(k_1^-, \ldots, k_\alpha^-, \boldsymbol{k}_1, \ldots, \boldsymbol{k}_\alpha, k'^-_1, \ldots, k'^-_\beta) = \int \prod dp_i \, \tilde{\phi}(\ldots, p_i, \ldots) \times$$

$$\times \prod d^3 \boldsymbol{k}'_j \, \psi(\boldsymbol{k}'_1, \ldots \boldsymbol{k}'_\beta) \, \tilde{r}^{\mathrm{amp}}(p_i; k_h, -k'_j) \prod \chi(k_h) \prod \chi(k'_j) \quad (88)$$

is continuous in k_h^-, k'^-_j and $\in \mathcal{T}$ in $\boldsymbol{k}_1, \ldots, \boldsymbol{k}_\alpha$.

It is easy to see that this condition implies the existence of the integral I. As to the space $\mathcal{T}$, I can only say that in perturbation theory a possible and convenient choice is $\mathcal{T} = H_\varepsilon$: the space of Hölder-continuous functions of index ε, $0 < \varepsilon < \frac{1}{2}$ with strong decrease at infinity [10].

For proving the GLZ theorem one needs, actually, a slightly more complicated condition allowing for a dependence of ψ on the nonintegrated variables k_h. I refrain from stating the condition in its full complexity. Its essential nucleus is what I have said.

Under the conditions enumerated above, the r are the retarded functions of a field $A(x)$ satisfying Axioms 1–5. The field A is explicitly given by the "Haag expansion":

$$A(x) = A^{\mathrm{in}}(x) + \sum_{l=2}^\infty \frac{1}{l!} \int du_1 \ldots du_l \, r^{\mathrm{amp}}(x; u_1, \ldots u_l) \times \, :A^{\mathrm{in}}(u_1) \ldots A^{\mathrm{in}}(u_l):, \quad (89)$$

with A^{in} a free field of mass m. This expansion converges strongly on all states in the Fock space $\mathcal{H}^{\mathrm{in}}$ of A^{in} with finitely many particles with nonoverlapping wave functions from $\mathcal{S}$. These states are total in $\mathcal{H}^{\mathrm{in}}$.

Let me briefly indicate how (89) is proved. Consider, as a simple example,

$$A(x)\Omega = A^{\mathrm{in}}(x)\Omega + \sum_{l=2}^\infty \frac{1}{l!} \int \prod du_i \, r^{\mathrm{amp}}(x; u_1, \ldots) \times \, :\prod A^{\mathrm{in}}(u_i): \Omega.$$

Under our assumptions on r^{amp} this vector clearly exists if the l-series is broken off after a finite number of terms. In order to show convergence of the series we must prove the existence of the 2-point function $\langle 0 | A(x) A(y) | 0 \rangle$. We introduce the Haag series (89) for the two A-factors. We obtain products $: \ldots : : \ldots :$ of two

Wick products, which can be reduced to a sum of Wick products with the help of Wick's theorem. Because $\langle 0| :\ldots: |0\rangle = 0$, only the completely reduced terms remain (i.e. the completely paired terms). The result is a sum which is simply the positive frequency part of the right-hand side of the completeness equation (86), for $n = 0$; hence the sum converges. This proves the existence of $A(x)$. The proof that (89) converges on more complicated states than Ω follows the same lines but is combinatorially quite involved.

In the same way it can be shown that the generalized Haag series

$$R(x_1, \ldots, x_n) = r(x_1, \ldots, x_n) + \sum_{l=1}^{\infty} \frac{1}{l!} \iint du_1 \ldots du_l \times$$

$$\times r^{\mathrm{amp}}(x_1, \ldots, x_n; u_1, \ldots, u_l) : \prod_1^l A^{\mathrm{in}}(u_i): \qquad (90)$$

converges on the same states. Moreover, the R defined by (90) are the retarded products of $A(x)$. In particular:

$$[A(x), A(y)] = R(x, y) - R(y, x) = \sum_{l=0}^{\infty} \frac{1}{l!} \int du_1 \ldots du_l \times$$

$$\times [r^{\mathrm{amp}}(x, y; \ldots u_i \ldots) - r^{\mathrm{amp}}(y, x; \ldots u_i \ldots)] : A^{\mathrm{in}}(u_1) \ldots A^{\mathrm{in}}(u_l): , \qquad (91)$$

whence the locality of A follows because of the assumed support of r. Covariance follows from the invariance of r. The spectral properties are clearly satisfied, because the Hilbert space of the theory is simply the Fock space $\mathscr{H}^{\mathrm{in}}$ of A^{in}, which has a particle spectrum. In short: A satisfies all our axioms.

REFERENCES

[1] A. S. Wightman, *Phys. Rev.,* **101**, 860 (1956).

[2] R. F. Streater and A. S. Wightman, "PCT, Spin and Statistics, and All That". Benjamin, New York (1964).

[3] R. Jost, "The General Theory of Quantized Fields." Am. Math. Soc., Providence (1965).

[4] H. Lehmann, K. Symanzik, and W. Zimmermann, *Nuovo Cim.,* **1**, 205 (1955); *ibid.,* **6**, 319 (1957);
V. Glaser, H. Lehmann, and W. Zimmermann, *Nuovo Cim.,* **6**, 1122 (1957).

[5] D. Ruelle, *Helv. phys. acta,* **35**, 147 (1962).

[6] K. Hepp, *Helv. phys. acta,* **37**, 639 (1964).

[7] K. Hepp, *Comm. Math. Phys.,* **1**, 95 (1965).

[8] O. Steinmann, *Comm. Math. Phys.,* **10**, 245 (1968).

[9] O. Steinmann, *Comm. Math. Phys.,* **18**, 179 (1970).

[10] O. Steinmann, "Perturbation Expansions in Axiomatic Field Theory", *Lecture Notes in Physics,* Springer, Berlin (1971).

STATISTICAL MECHANICS AND FIELD THEORY
R. N. Sen and C. Weil, Editors

Unstable Particles in Field Theory

O. STEINMANN

Swiss Institute for Nuclear Research, Villigen (Switzerland)

Abstract

We discuss a field-theoretical description of long-lived unstable particles, assuming that these particles are associated with poles on the second sheet of the Green's functions of the underlying fields. The spatio-temporal evolution of scattering events of stable particles is discussed under certain smoothness assumptions for the Green's functions. It is shown that second-sheet poles near the real axis induce geometrical features in these processes, which can be interpreted as manifestations of unstable particles in intermediate stages. An S-matrix for processes involving these unstable particles is defined in an approximate sense and its properties discussed. In particular, S is unitary in a suitable approximation.

The problem I want to discuss is this: how can quantum field theory describe unstable particles and their interactions, i.e. their production and decay and scattering events involving them?

Let me first remind you of how this problem is traditionally tackled for stable particles. In this case we start from two observations.

A) The relation field $\leftrightarrow$ particle is well understood in the free case.

B) An interacting state with a finite number of particles (a scattering state) will, for sufficiently large positive or negative times, be indistinguishable from a suitable free state, because the particles will be so far apart from each other that their interaction is negligible. The mathematical expressions of this fact are the asymptotic conditions, which can be proved from the generally accepted postulates of relativistic quantum field theory [1].

We then take the point of view that it is reasonable to talk of particles in the two asymptotic regions, where the situation can be described with the help of free fields. In the interaction region we talk only about fields, the notion of particles not being very appropriate for describing what is going on there. Remember that the particle number is not conserved: the particles lose their individuality in the region of interaction, and this makes them of doubtful value.

The asymptotic region is here simply defined by the requirement that $|t|$ be "sufficiently large", and it is left open what this exactly means.

We see at once that such an approach does not work for unstable particles, since they are no longer around in the truly asymptotic region. The corresponding mathematical problem becomes apparent if we look at the assumptions under which the asymptotic conditions have been proved. They are the usual Wightman axioms plus an additional condition on the energy-momentum spectrum: the spectrum must contain an isolated hyperboloid in $p^2 = m^2 > 0$ corresponding to the 1-particle states. This condition is not satisfied in the unstable case, because the 1-particle states lie in the continuum formed by their decay products. Hence no asymptotic condition can be derived.

Nevertheless, the ideas and notions developed on the basis of the asymptotic conditions, notably the notion of the S-matrix, are, in practice, constantly used for unstable particles, with good success at least as far as relatively long-lived particles are concerned. I shall only deal with these "metastable" particles, which can be roughly defined as having mean lives which allow them to travel over macroscopic distances between production and decay. I wish to propose an approach for treating these particles which allows one to define an S-matrix for them in an approximate sense. I shall not attempt to associate a field with an unstable particle, but shall work only with the fields connected with the stable particles of the theory (the decay products of the unstable objects).

I propose to achieve my object by having a closer look at the spatio-temporal development of scattering events. In other words, I intend to treat a particle as what the experimentalist considers to be a particle: an object that sticks together, i.e. is permanently concentrated in a small volume and, as long as it is left alone, travels along a straight path without changing its characteristic aspects. Remember that the asymptotic particle definition mentioned before relies on what is known in the free case, and there the notion of particle is based on p-space properties: a particle is a lump of energy and momentum (a "quantum") about whose localization properties nothing is said. That these two ideas about what a particle is have anything to do with each other is not clear *a priori* and, indeed, is true only under certain favorable conditions.

How do we define the location of a particle? Attempts to define a "localization operator" as quantum-mechanical observable from natural looking abstract requirements have failed to yield any satisfying results. We must, therefore, look a bit closer at how the experimentalist solves this problem. He does it with the help of detectors, and so we must find a field-theoretical description of a detector.

I consider the registration of a particle by a detector as an operation in the sense of Haag and Kastler [2]. A state, represented by a vector Φ in the Hilbert space $\mathscr{H}$ of field theory [1], is an ensemble of identical systems, prepared in a certain well-defined way. We can think, e.g., of an in-state with two stable particles of mass $m > 0$, with prescribed wave functions, one prepared from an accelerator beam with slits, magnets, monitoring counters, etc., the other one sitting in a target. We check all the systems in the ensemble with a counter (as example of a detector) in a fixed location. The systems that have triggered the counter form a new ensemble Φ', the other systems are not considered further. This procedure is an operation in the sense of Haag and Kastler. It can be represented by an operator C in $\mathcal{H}$:

$$\Phi' = C\Phi, \tag{1}$$

such that

$$\frac{|\Phi'|^2}{|\Phi|^2} = \frac{|C\Phi|^2}{|\Phi|^2} \tag{2}$$

is the probability that a system in Φ triggers the counter. Obviously, in order for this statement to make sense, we must demand

$$\|C\| \leqq 1. \tag{3}$$

If $T(a)$ is the unitary representation of the translation group which we have in $\mathcal{H}$ according to Wightman's axioms, then

$$C(a) = T(a)\,CT^*(a) \tag{4}$$

represents the same counter as C, translated by the 4-vector a. If we choose a frame of reference such that its origin is contained in the region of localization of C, then $C(x)$ is located around the point x. (We think of a counter as localized in a bounded 4-dimensional region: we assume that it is switched on for a short time only.) By saying that C is localized in a bounded neighborhood $\mathcal{N}$ of the origin I mean the following: there is an operator C' in the local algebra of $\mathcal{N}$, generated by the field operators at the points in $\mathcal{N}$, such that $\|C - C'\|$ is so small as to be negligible compared with the experimental inaccuracy. We cannot demand $C = C'$, because we also want to have $C\Omega = 0$, Ω the vacuum state: the vacuum shall not be counted. But, according to a theorem by Reeh and Schlieder, no strictly local operator $\neq 0$ annihilates the vacuum. The diameter of the localization region $\mathcal{N}$ will be called d_1.

We can form the Fourier transform $\tilde{C}(p)$ of the operator-valued function $C(x)$. The argument p is simply the 4-momentum transferred from particle to counter in the counting process. We want that the counting disturb the particle as little as possible, i.e. we demand that $\|\tilde{C}(p)\| \neq 0$ only for $\|p\| < d_2 \ll m$. m is the mass of the particle to be observed, $\|p\|$ is the Euclidean length of p, d_2 is a positive constant characterizing the counter. Note that d_1 and d_2 cannot be made arbitrarily small simultaneously. They satisfy some sort of uncertainty relation, i.e. their product cannot be smaller than some finite number depending on the experimental accuracy used in the definition of the local approximant C'.

Let us consider a simple model: an LSZ field theory of a single scalar field $A(x)$, describing one type of stable particles with mass $m > 0$. Let Φ be any *in*-state. We can subject it to an arbitrary number of counters $C(x_1), \ldots, C(x_n)$. The state of the systems that have triggered all the counters is given by

$$\Phi' = T[C(x_1)\ldots C(x_n)]\Phi. \tag{5}$$

The time-ordering operator T has been introduced because we do not know what it means to subject Φ first to one counter and then to another one located in a region anterior to the first one. We assume that the localization regions of the various counters do not overlap, that they are, in fact, separated by macroscopic distances (macroscopic means $\gg 1/m$). The probability that all counters are triggered is

$$\mathscr{P}(x_1, \ldots, x_n) = \frac{\left|T[C(x_1)\ldots C(x_n)]\Phi\right|^2}{|\Phi|^2}. \tag{6}$$

Take $\Phi = |\mathbf{p}, \mathbf{q}\rangle_{\text{in}}$, a 2-particle state with momenta p, q, on the mass shell, such that $4m^2 < (p + q)^2 < 9m^2$, i.e. in the elastic region. Then

$$|\Phi|^2\mathscr{P} = {}_{\text{in}}\langle p, q | T^*[C(x_1)\ldots] T[\ldots] | p, q\rangle_{\text{in}}. \tag{7}$$

We sum over a complete intermediary set of *out*-states, assuming asymptotic completeness of the stable particles. Only the 2-particle states contribute, because of momentum conservation. The matrix elements ${}_{\text{out}}\langle p', q' | T[\ldots] | p, q\rangle_{\text{in}}$ can be calculated with the help of the reduction formulae. The result, omitting some inessential numerical factors, is:

$$\int \prod_{h=1}^{n} dP_h \exp\left\{i\sum_h P_h x_h\right\} \tilde{\tau}(P_1, \ldots, P_n; -p', -q', p, q), \tag{8}$$

where $\tilde{\tau}(\,.\,.\,;\,.\,.)$ is the Fourier transform of the time-ordered function of n C-"fields" $C(x_i)$ and 4 A-fields, amputated with respect to the latter.

The asymptotic form of such a Fourier integral for $\|x_i - x_j\| \to \infty$ is determined by the singularities of $\tilde{\tau}$, the worst singularities contributing the asymptotically leading terms. Now, we know that certain singularities must be present in $\tilde{\tau}$ for physical reasons. In particular, we must have the well-known 1-particle poles and less virulent singularities at the 2-particle-, 3-particle thresholds etc. The contribution of the poles to $\mathscr{P}(x_1, \ldots, x_n)$ can be calculated, and it can be shown that these contributions dominate $\mathscr{P}$ completely at the physically realizable $x_i - x_j$, provided that $\tilde{\tau}$ is sufficiently smooth outside the physical singularities.

Let us, then, make the following smoothness assumption: $\tilde{\tau}(\ldots)$ has the physically necessary singularities, but is smooth outside of them. Smoothness means here not only infinite differentiability but also the absence of strong oscillations and sharp kinks. Under this condition we can evaluate $\mathscr{P}(x_1, \ldots, x_n)$ in an excellent approximation [3], and we find that $\mathscr{P}$ is measurably different from zero only if the counters are in the following geometrical situation in 4-space: There is a row of counters on a straight line in direction p, another row on a line in direction q. These two lines meet approximately at a point, and from this point issue two rows of counters in the directions p', q'. All the counters belong to one of these 4 rows. Clearly, we are observing the scattering of two particles. Moreover, the dependence of $\mathscr{P}$ on the momenta p, q, p', q', is given by the factor $|\tilde{\tau}(p, q, -p', -q')|^2$, $\tilde{\tau}$ the amputated Green's function of 4 A-fields, and this is the good old S-matrix element squared, i.e. our procedure has provided us with an alternative definition of the S-matrix.

The advantage of this formalism is that it can take care of unstable particles. We relax the smoothness requirement for $\tilde{\tau}(p, q, -p', -q')$ by adding a second-sheet pole

$$\frac{c(p, .., -q')}{(p + q)^2 - M^2 + iM\Gamma} \tag{9}$$

with $\Gamma \ll d_2$, $4m^2 < M^2 < 9m^2$, to the list of singularities. If we choose p, q, such that $(p + q)^2 \cong M^2$, then our results on $\mathscr{P}(x_1, .., x_n)$ are modified: $\mathscr{P}$ is measurably different from zero only if all the counters are arrayed along five straight lines, as shown in Figure 1.

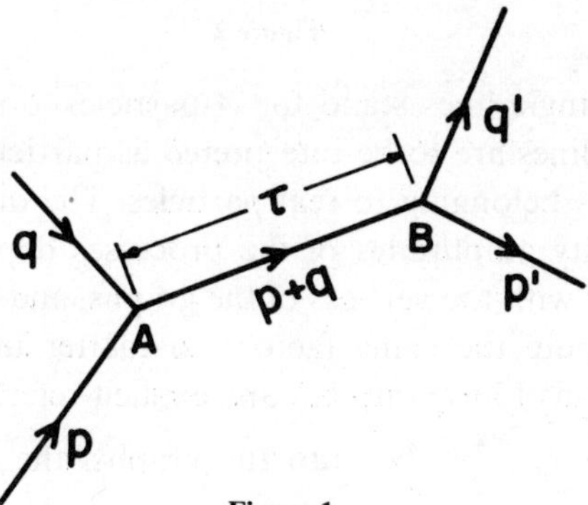

Figure 1

The dependence of $\mathscr{P}$ on the Minkowski-length τ of the middle segment is essentially given by a factor $\exp(-\Gamma\tau)$. The interpretation of this situation is clear: two stable particles meet in A and form an unstable particle of mass M travelling in the direction $p + q$ and decaying in B into two stable particles with momenta p', q'. The decay is exponential, with decay-constant Γ. The momentum dependence of $\mathscr{P}$ is given by the residue c of the pole (9): $\mathscr{P} \propto |c(p, ..., -q')|^2$. If our particle interpretation of the arrangement of Figure 1 is to make sense, then the production in A and the decay in B must be independent events. This means that c factorizes:

$$c(p, q, -p', -q') = \alpha(p, q)\,\alpha(-p', -q'). \tag{10}$$

In our simple model the two factors are values of the same function α at different arguments. We assume that α is Lorentz-invariant. This implies that the unstable particle, called "B-particle" or simply "B", has spin zero.

                    O. STEINMANN

The above considerations establish the presence of a B-particle, as far as the process $2A \to 2A$ is concerned. In order to have the right to talk about a B-particle *per se*, we must show that all other possible processes involving B's also occur, for instance the processes shown in Figure 2.

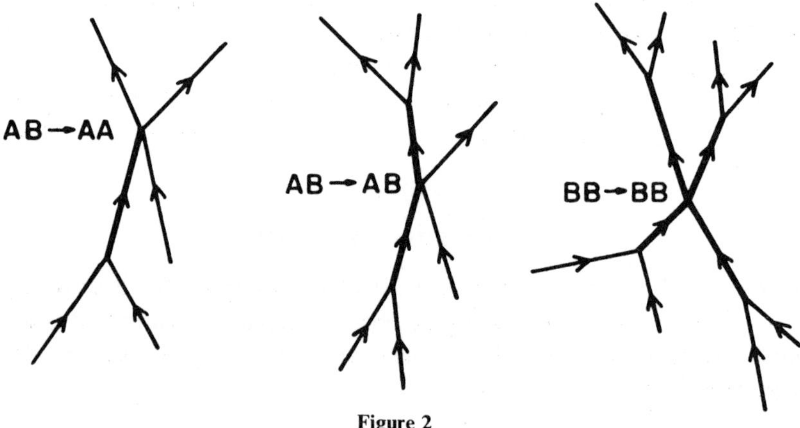

Figure 2

In these diagrams single lines stand for A-particles, double lines for B-particles. Remember that these lines are to be interpreted as particle tracks, for instance in a bubble chamber, i.e. as belonging to real particles. The diagrams are not Feynman graphs! The probability amplitudes of the processes depicted should be products of functions associated with the vertices of the graphs, and a particular type of vertex should always contribute the same factor, no matter in what process it occurs.

In these considerations I have omitted any explicit reference to the counters. They give insertions of the type into the graphs, the wiggly lines standing for counters. The corresponding vertex functions factor out and are not interesting in the present context. I shall therefore forget about the counters from now on.

The desired structure as mentioned above is indeed present, because the following can be proved. Let $\tilde{\tau}(p_1, \ldots, p_n)$ be the amputated Green's functions ($=$ time-ordered functions) in p-space of the field A. They satisfy the completeness equations ($=$ unitarity equations), which are of the form*

$$\tilde{\tau}(p_1, \ldots, p_n) - \tilde{\tau}^*(p_1, \ldots, p_n) =$$

$$= -i \sum_{L,R} \sum_{l=0}^{\infty} \frac{(2\pi)^{2l}}{l!} \int \prod_{1}^{l} \{dk_i \, \delta_+(k_i)\} \, \tilde{\tau}^*(P_L, k_1, \ldots, k_l) \, \tilde{\tau}(P_R, -k_1, \ldots, -k_l). \quad (11)$$

The symbols P_L, P_R, denote two complementary subsets of the set $\{p_1, \ldots, p_n\}$. δ_+ is the δ-function of the positive mass shell. Assume that the $\tilde{\tau}$ have the axiomatic

---

* We use a definition of $\tau(x_1, \ldots, x_n)$ which differs from the one given in ref. [1] by a factor $(-i)^{n-1}$.

analyticity [4] and are, furthermore, analytically continuable in the variable $p_i + p_j$ through the 2-particle cut $4m^2 < (p_i + p_j)^2 < 9m^2$ as far as $(p_i + p_j)^2 = M^2 - iM\Gamma$. Assume that the 4-point function $\tilde{\tau}(p_1, .., p_4)$ contains a pole on the second sheet with factorizing residue:

$$\frac{\alpha(p_1, p_2)\, \alpha(p_3, p_4)}{(p_1 + p_2)^2 - M^2 + iM\Gamma} . \tag{12}$$

Then it can be shown that the higher functions $\tilde{\tau}(p_1, \ldots, p_n)$ contain the poles

$$\frac{\alpha(p_1, .., p_m)\, \alpha(p_{m+1}, .., p_n)}{\left(\sum_1^m p_i \right)^2 - M^2 + iM\Gamma}, \tag{13}$$

with factorizing residues, in any partial sum $\sum_1^m p_i$ of the arguments. (Since $\tilde{\tau}$ is totally symmetric, I have taken the liberty to number the variables of this partial set consecutively.) The functions α are the same in all cases, i.e. they depend only on the number m of arguments occurring in them, not on the total number n of points in the original function. Pole products in different partial p_i-sums occur in the expected combinations, so that we obtain a consistent particle interpretation of all processes involving B's.

I cannot give the proof of this statement here (for a proof see ref. [5]). Let me just point out the main idea. We write down the completeness equation (11) for any n. On the right-hand side occurs the 4-point function $\tilde{\tau}(p_1, p_2, -k_1, -k_2)$ which contains a B-pole in $p_1 + p_2$ by assumption. Such a pole must then also be present on the left-hand side, in $\tilde{\tau}(p_1, .., p_n)$. We conclude further that any $l = 2$ term on the right: $\tilde{\tau}(P_R, -k_1, -k_2)$, contains a pole in $k_1 + k_2 = \sum_{P_R} p_i$ (remember momentum conservation!), which then again must be present on the left, and so on. Comparison of the residues on both sides shows the correct factorization properties.

The vertex factors α can be interpreted as S-matrix elements by definition, so that no asymptotic conditions are needed. More exactly: the S-matrix element for the process with m incoming A-particles of momenta $p_1, .., p_m$, n outgoing A-particles of momenta $q_1, \ldots, q_n$, and one outgoing B-particle with momentum $\sum p_i - \sum q_i$, is defined to be $\alpha(p_1, .., p_m, -q_1, .., -q_n)$, etc. If more than one B is involved in a process, then the B-momenta occur explicitly as arguments of the corresponding α— these new α's coming in as residue factors of products of poles in several partial sums.

More useful information can be obtained from the completeness equations in the lowest orders of perturbation theory. We have assumed Γ to be small, i.e. B to be long-lived. This means that the decay interaction is weak. Let us therefore introduce a small coupling constant g for this interaction, and let us assume $\Gamma = \mathcal{O}(g^2)$, $\alpha(..) = \mathcal{O}(g)$. The A among themselves may interact strongly, also the B among themselves.

We expand $\tilde{\tau}$ in powers of g:

$$\tilde{\tau}(p_1,\ldots,p_n) = \sum_{\sigma=0}^{\infty} g^{\sigma}\tilde{\tau}_{\sigma}(p_1,\ldots,p_n), \tag{14}$$

and analogously for α and Γ. $\tilde{\tau}_0$ need not be the $\tilde{\tau}$ of a free field; it may belong to a self-interacting field $A_0(x)$. We assume, however, that $\tilde{\tau}_0$ satisfies the original smoothness conditions without any second sheet poles or other singularities which are not physically necessary. $\tilde{\tau}_1$ is still maximally smooth, but $\tilde{\tau}_2(p_1,\ldots,p_n)$ is supposed to contain the poles

$$\frac{\alpha_1(P_L)\alpha_1(P_R)}{P_L^2 - M^2 + i\varepsilon} \tag{15}$$

with P_L, P_R, any two non-empty complementary subsets of $\{p_1,\ldots,p_n\}$. The denominator in (15) results from the expansion

$$\frac{1}{Q^2 - M^2 + iM\Gamma} = \frac{1}{Q^2 - M^2 + i\varepsilon}\left\{1 - \frac{i\Gamma_2 g^2}{Q^2 - M^2 + i\varepsilon}\right\} + \mathcal{O}(g^3) \tag{16}$$

which holds in the sense of distributions, i.e. after integration over a test function in Q, whose support is allowed to contain the dangerous points where $Q^2 = M^2$.

Take the completeness equation (11) for $n = 4$, insert the perturbation expansion (14), and pick out the terms of order g^2. It is important to note that terms of this order are contributed on the r.h.s. by the expression

$$g^4 \int dk_1 dk_2 \tilde{\tau}_2^*(p_1, p_2, k_1, k_2)\,\tilde{\tau}_2(p_3, p_4, -k_1, -k_2)\,\delta_+(k_1)\,\delta_+(k_2) \tag{17}$$

in spite of its 4th-order look. This is so because the $\tilde{\tau}_2^*$-factor contains the pole $[(p_1 + p_2)^2 - M^2 - i\varepsilon]^{-1}$, the $\tilde{\tau}_2$-factor the pole $[(p_1 + p_2)^2 - M^2 + i\varepsilon]^{-1}$ (note $p_3 + p_4 = -p_1 - p_2$ because of momentum conservation), and the product of these two poles is no distribution. If we, however, remember the origin of the $i\varepsilon$-poles as approximations to $iM\Gamma$-poles and use

$$\frac{1}{Q^2 - M^2 - iM\Gamma}\,\frac{1}{Q^2 - M^2 + iM\Gamma} =$$

$$= \frac{1}{2iM\Gamma}\left\{\frac{1}{Q^2 - M^2 - iM\Gamma} - \frac{1}{Q^2 - M^2 + iM\Gamma}\right\} =$$

$$= \frac{1}{2iM\Gamma_2 g^2}\left\{\frac{1}{Q^2 - M^2 - i\varepsilon} - \frac{1}{Q^2 - M^2 + i\varepsilon}\right\} + \mathcal{O}(g^0), \tag{18}$$

we find that

$$g^4\left[(p_1 + p_2)^2 - M^2 - i\varepsilon\right]^{-1}\left[(p_1 + p_2)^2 - M^2 + i\varepsilon\right]^{-1} =$$

$$= \frac{g^2}{2iM\Gamma_2}\left\{\left[(p_1 + p_2)^2 - M^2 - i\varepsilon\right]^{-1} - \left[(p_1 + p_2)^2 - M^2 + i\varepsilon\right]^{-1}\right\}. \tag{19}$$

This exists, but is of order g^2, not g^4! Taking due account of this and comparing residues on both sides of the completeness equation, we find

$$4M\Gamma_2 = (2\pi)^5\, I\,|t_1|^2\,, \tag{20}$$

I being the phase space volume for the process $B \to 2A$ and t_1 the value of $\alpha_1(p_1, p_2)$ on the mass shell $p_1^2 = p_2^2 = m^2$, $(p_1 + p_2)^2 = M^2$. This is, of course, an old result, which says that the attenuation of the B-beam given by the factor $\exp(-\Gamma\tau)$ is compatible with the decay probability given by $|t|$, at least in the lowest non-vanishing order of perturbation theory.

The appearance of g^2-terms in contributions that are apparently of order g^4 occurs, of course, also in the completeness equations with any number n of points. Apart from giving equation (20), this has another pleasant consequence.

We define the S-matrix elements for processes involving A-particles only in the usual way, as the mass shell restrictions of the Green's functions $\tilde\tau(\dots)$, those of processes involving B's as the residue factors α, again restricted to the mass shell of both the A- and the B-variables. Then this S-matrix satisfies, in the lowest orders of perturbation theory, unitarity equations in which both A's and B's occur as particles in the intermediate states. More explicitly: consider the Fock space of the particles A and B together, the latter treated as stable. Let $|a, b\rangle$ be a state with a A-particles and b B-particles, wave functions not specified. Let $\langle a, b|S|a', b'\rangle_\sigma$ be the terms of order g^σ in the S-matrix elements as specified above. Then the unitarity equations

$$\sum_{\tau=0}^{\sigma}\ \sum_{(a',b')} \langle a, b|S^*|a', b'\rangle_\tau \langle a', b'|S|a'', b''\rangle_{\sigma-\tau} = \delta_{0\sigma}\langle a, b|a'', b''\rangle \tag{21}$$

hold up to order $\sigma = 3$ if $a \neq 0$ or $a'' \neq 0$, to order $\sigma = 1$ if $a = a'' = 0$. Here it is assumed that the B's interact strongly among themselves, i.e. that $\langle b|S|b'\rangle_0 \neq \langle b|b'\rangle$. Otherwise (21) holds to order $\sigma = 3$ in all cases.

This result is not self-evident: the equations (21) are derived by perturbation expansion of the exact completeness equations for the A-field Green's functions, and these equations contain only a summation over A-states. The B-terms come in as mentioned above, as contributions from terms of seemingly higher order. But these exceptional terms are exactly of the right form to mimic unitarity expressions only in low orders.

This result justifies a method which is often used in the discussion of unstable particles (Weisskopf–Wigner method) and which uses an expansion of a state with respect to a seemingly overcomplete basis containing both a B-state *and* its $2A$ decay product. We have shown that this method is legitimate in an approximate way, namely in the lowest orders of perturbation theory.

A detailed account of the considerations I have sketched here can be found in [3, 5].

REFERENCES

[1] O. Steinmann, Lectures on "Connection between Wightman and LSZ field theory", this volume.

[2] R. Haag and D. Kastler, *J. Math. Phys.*, **5**, 848 (1964).

[3] O. Steinmann, *Comm. Math. Phys.*, **7**, 112 (1968)

[4] H. Epstein, in: "Axiomatic Field Theory", ed. by M. Chrétien and S. Deser. Gordon and Breach, New York (1966).

[5] O. Steinmann, *Helv. phys. acta,* **44**, 618 (1971).

STATISTICAL MECHANICS AND FIELD THEORY
R. N. Sen and C. Weil, Editors

The Notion of Ergodicity in Solvable Models

A. VERBEURE

Instituut voor Theoretische Natuurkunde, Universiteit Leuven (Belgium)

Abstract

Conditions are given for ergodicity for Fermi lattice systems. The result is compared with ergodicity for classical systems.

I. INTRODUCTION

In this note we want to discuss the property of ergodicity in quantum systems. We study bilinear Hamiltonians in the Fermi field operators. In particular we are interested in the Fermi lattice system (F.L.S.).

Let $\mathbb{Z}$ be the set of all positive and negative integers and $\{\phi_m\}_{m \in \mathbb{Z}}$ an orthonormal basis in H, a real vector space equipped with a real scalar product $s: (\psi, \phi) \to s(\psi \mid \phi)$. Let $\mathfrak{A}$ be the C^*-algebra generated by the Hermitian Fermi fields $\{B(\phi_i)\}_{i \in \mathbb{Z}}$; B is a real linear map of H into $\mathfrak{A}$, satisfying

$$[B(\psi), B(\phi)]_+ = 2s(\psi \mid \phi)\,\mathbf{1},$$

where $\mathbf{1}$ is the unit element of $\mathfrak{A}$. For any finite subset Λ of $\mathbb{Z}$ consider the C^*-sub-algebra $\mathfrak{A}(\Lambda)$ of $\mathfrak{A}$ generated by $\{B_k \equiv B(\phi_k) \mid k \in \Lambda\}$; $\mathfrak{A}$ is the norm closure of $\cup_{\Lambda \subset \mathbb{Z}} \mathfrak{A}(\Lambda)$ and can be considered as the algebra of quasi-local observables of the lattice $\mathbb{Z}$.

Let $\sigma: a \in \mathbb{Z} \to \sigma_a$ be a mapping of the group $\mathbb{Z}$ of lattice translations into the *-automorphisms σ_a of $\mathfrak{A}$ defined by $\sigma_a B_k = B_{k+a} = B(U_a \phi_k)$, where U_a is an orthogonal operator on H. We consider also the time translations induced by the following locally defined bilinear Hamiltonians: for each finite subset $\Lambda \subset \mathbb{Z}$, the Hamiltonian for region Λ is given by

$$H_\Lambda = \sum_{i,j \in \Lambda} v_{ij} B_i B_j, \tag{1}$$

where the v_{ij} are c-numbers such that

(i) $v_{ij} = \bar{v}_{ji}$

(ii) $\sum_{i \in \mathbb{Z}} |v_{ij}| < \infty$

(iii) $v_{i+k, j+k} = v_{ij}$ for any $i, j \in \mathbb{Z}$; $k \in S_n$

where $S_n = \{l \in \mathbb{Z} \,|\, l = na, a \in \mathbb{Z}\}$.

Before going on we point out that the linear XY-model is a special case given by the following Hamiltonian:

$$H_\Lambda = i \sum_{k \in \Lambda} (- J_x B_{2k+1} B_{2k+2} + J_y B_{2k} B_{2k+3} + \mu h B_{2k} B_{2k+1}). \tag{2}$$

This Hamiltonian is invariant under S_2.

In the following we adopt the definition of ergodicity given essentially in [1, 2]. We give conditions in order that a system be ergodic in its equilibrium state. The XY-model satisfies this condition. Finally, we discuss the notion of ergodicity in quantum and classical systems.

Part of this note is based on joint work with R. Lima, CNRS, Marseille (France).

II. EQUILIBRIUM-STATE ERGODICITY

Define α_t^Λ by $\alpha_t^\Lambda(x) = e^{iH_\Lambda t} x e^{-iH_\Lambda t}$ for $x \in \mathfrak{A}$, $t \in R$ and H_Λ as in (1). If $\{\Lambda_n\}_n$ is a sequence of finite subsets of $\mathbb{Z}$ tending to infinity in the sense that it contains every finite subset, then for each $t \in R$, the sequence $\{\alpha_t^{\Lambda_n}\}_n$ tends strongly to a *-automorphism α_t of $\mathfrak{A}$, defining a strongly continuous one-parameter group of automorphisms $\{\alpha_t \,|\, t \in R\}$ (see [2]).

An easy verification shows that for $k \in \mathbb{Z}$

$$\alpha_t^\Lambda(B_k) = B_k + \frac{it}{1!}[H_\Lambda, B_k] + \frac{(it)^2}{2!}[H_\Lambda, [H_\Lambda, B_k]] + \ldots = B(e^{D_\Lambda t} \phi_k),$$

where D_Λ is an operator on H, defined by

$$s(\phi_k | D_\Lambda \phi_l) = i(v_{kl} - v_{lk}) \quad \text{for} \quad k, l \in \Lambda$$

$$= 0 \qquad \text{for} \quad k \text{ or } l \notin \Lambda.$$

Using the existence of the limit $\Lambda \to \infty$, it is easy to verify that D_Λ tends strongly to an operator D on a dense domain of H consisting of finite linear combinations of the basis vectors. The operator D is defined by

$$s(\phi_k | D \phi_l) = i(v_{kl} - v_{lk}) \qquad k, l \in \mathbb{Z}. \tag{3}$$

Furthermore,

$$s(\phi_k|D\phi_l) = - s(D\phi_k|\phi_l)$$

and

$$U_a D = D U_a \quad \text{for all} \quad a \in S_n.$$

We summarize:

Proposition 1:

For the F.L.S. the set of automorphisms α_t^Λ; $\Lambda \subset \mathbb{Z}$, $t \in \mathbb{R}$, tend in the limit $\Lambda \to \infty$ to the group of automorphisms $\{\alpha_t | t \in \mathbb{R}\}$ such that

$$\alpha_t(B(\phi_k)) = B(e^{Dt}\phi_k),$$

where D is an antisymmetric, translation-invariant operator on (H, s) defined in (3).

As definition of equilibrium state for a certain evolution (Hamiltonian) we take the state satisfying the Kubo–Martin–Schwinger (KMS) boundary condition with respect to that evolution. For our F.L.S. the existence of an equilibrium state is guaranteed by the following proposition, the proof of which can be extracted from [3].

Proposition 2:

For the evolution $t \to \alpha_t : \alpha_t(B(\phi)) = B(e^{Zt}\phi)$ there exists a unique quasi-free state [4] ω_A on $\mathfrak{A}$, satisfying the KMS boundary condition with respect to α_t. The operator A is given by

$$A = J \frac{e^X - e^{-X}}{e^X + e^{-X}}$$

$$D = J|D| \text{ (polar decomposition of } D)$$

$$X = \beta|D|, \beta = \frac{1}{kT}.$$

Now we define the notion of ergodicity of a state. Let $(\pi_A, \mathscr{H}_A, \Omega_A)$ be the GNS-triplet induced by the state ω_A.

Definition:

The state ω_A is called ergodic if for each quasi-local observable $x \in \pi_A(\mathfrak{A})''$ there exists a time mean $\mathfrak{M}_{(t)}(x)$ and a lattice mean $\mathfrak{M}_{(a)}(x)$ such that $\mathfrak{M}_{(t)}(x) = \mathfrak{M}_{(a)}(x)$.

Consider the operator $D = J|D|$ defined before. The operator J is a complex structure for the real space (H, s) and we can consider the complex Hilbert space H_J; this is the vector space H equipped with the scalar product $(\psi|\phi) = s(\psi|\phi) + is(J\psi|\phi)$; $\psi, \phi \in H$. For all notions about the spectrum of D we refer to the complex Hilbert space H_J.

Lemma 1:

(i) Let U_a, $a \in S_n$ be the operator defining the space translations. Then

$$\lim_{|a| \to \infty} s(U_a \psi | \phi) = 0 \quad \text{for all} \quad \psi, \phi \in H.$$

(ii) Let the spectrum of D be purely continuous. Then

$$\lim_{t \to \infty} s(e^{Dt} \psi | \phi) = 0 \quad \text{for all} \quad \psi, \phi \in H.$$

Proof:

To prove (i), note that

$$s(U_a \phi_k | \phi_l) = s(\phi_{k+a} | \phi_l) = 0 \quad \text{for all} \quad a \in \mathbb{Z},$$

such that $k + a > l$. The rest is trivial.

Further, by a theorem of Wiener [5], the spectrum of D is purely continuous if and only if

$$\lim_{T \to \infty} \frac{1}{2T} \int_{-T}^{T} |(e^{Dt} \psi | \phi)|^2 dt = 0$$

for all $\psi, \phi \in H_J$. Now

$$|(e^{Dt} \psi | \phi)|^2 = |s(e^{Dt} \psi | \phi)|^2 + |s(e^{Dt} J \psi | \phi)|^2,$$

hence for t tending to infinity $(e^{Dt} \psi | \phi)$ tends to zero for all $\psi, \phi \in H_J$ if and only if $s(e^{Dt} \psi | \phi)$ tends to zero for all $\psi, \phi \in H$. QED.

Lemma 2:

Let ω be any state on $\mathfrak{A}$; ζ_n the *-automorphism of $\mathfrak{A}$:

$$\zeta_n = \sigma_a \quad \text{if} \quad n = a \in S_n \quad \text{(lattice translations)}$$
$$\zeta_n = \alpha_t \quad \text{if} \quad n = t \in \mathbb{R} \quad \text{(time translations; prop. 1)}$$

If the spectrum of D is purely continuous, then

$$M_u(|\omega([\zeta_u(x), y])|) = 0, \quad x, y \in \mathfrak{A}.$$

M_u is defined by

$$M_u(\mathcal{F}_u) = \lim_{a \to \infty} \frac{1}{2a} \sum_{n=-a}^{a} \mathcal{F}_n \qquad \text{in the discrete case,}$$

and

$$M_u(\mathcal{F}_u) = \lim_{T \to \infty} \frac{1}{2T} \int_{-T}^{T} \mathcal{F}_t \, dt \qquad \text{in the continuous case.}$$

Proof can be found in [6, §3].

Theorem 1:

Let $t \to \alpha_t(B(\psi)) = B(e^{Dt}\psi)$ be the evolution of the F.L.S. such that the spectrum of the operator D is purely continuous; then the corresponding equilibrium KMS state, determined in proposition 2, is ergodic, and *vice versa*.

Proof:

Consider the faithful vector state ω_{Ω_A} on $\pi_A(\mathfrak{A})$ such that

$$\omega_{\Omega_A}(x) = (\Omega_A | x \Omega_A), \qquad x \in \pi_A(\mathfrak{A}).$$

This state is invariant for lattice and time translations. Hence there exist representations $U_A(t)$, $V_A(a)$; $t \in \mathbb{R}$, $a \in S_n$ of the groups of automorphisms of time and lattice translations such that

$$\pi_A(\alpha_t(x)) = U_A(t) \pi_A(x) U_A(-t)$$

$$\pi_A(\sigma_a(x)) = V_A(a) \pi_A(x) V_A(-a)$$

$$U_A(t)\Omega_A = V_A(a)\Omega_A = \Omega_A \quad \text{for all} \quad a \in S_n, t \in \mathbb{R}.$$

By Lemma 2, $\pi_A(\mathfrak{A})$ is a concrete C^*-algebra on $\mathscr{H}_A$ which is weakly asymptotically abelian as well for the lattice as for the time translations. Hence by [1] each element $x \in \pi_A(\mathfrak{A})''$ has a time mean

$$\mathfrak{M}_{(t)}(x) = E_{(t)} x E_{(t)}$$

and a lattice mean

$$\mathfrak{M}_{(a)}(x) = E_{(a)} x E_{(a)},$$

where $E_{(t)}$ ($E_{(a)}$) is the orthogonal projection operator on the set

$$\{\Omega \in \mathscr{H}_A | U_A(t)\Omega = \Omega, \quad t \in \mathbb{R}\}$$

$$(\{\Omega \in \mathscr{H}_A | V_A(a)\Omega = \Omega, \quad a \in S_n\}).$$

Now the spectrum of D is purely continuous if and only if Ω_A is the unique time-invariant vector [7, Lemma 3.2]. Also, by [3, Lemma 2.3.9] and [9, Theorem 2] Ω_A is the unique lattice-invariant vector; hence $E_{(a)} = E_{(t)} = E_A$, where E_A is the projection operator on Ω_A. Hence for all $x \in \pi_A(\mathfrak{A})''$

$$\mathfrak{M}_{(t)}(x) = \mathfrak{M}_{(a)}(x) = E_A x E_A$$

if and only if the spectrum of D is purely continuous. QED.

Now we derive a sufficient condition on the Hamiltonian (1), and hence on the coefficient v_{ij}, in order that this equilibrium state be ergodic.

First suppose that $[D, U_a]_- = 0$ for all $a \in Z$. Note that $a_i = D_{0i}$, $i \in \mathbb{N}^+$ (positive integers); using the translation invariance and the antisymmetry of D,

$$D\phi_m = \sum_{n>0}^{\infty} a_n \phi_{-n+m} - \sum_{n>0}^{\infty} a_n \phi_{n+m} \,.$$

Realize the complex space H_J as the Hilbert space of square integrable functions o' the unit circle $\mathbb{C}_1$; an orthonormal basis is given by the functions

$$\phi_m : z \in \mathbb{C}_1 \rightarrow \frac{1}{\sqrt{2\pi}} \, z^m \,;$$

then

$$(D\phi_m)(z) = g(z)\,\phi_m(z)$$

$$g(z) = f(\bar{z}) - f(z)$$

$$f(z) = \sum_{n>0}^{\infty} a_n z^n \,.$$

Hence the operator D is realized as a multiplication operator.

If now, in general, $[D, U_a]_- = 0$ for all $a \in S_n$, then it is possible to realize D^n as a multiplication operator: e.g., in the linear XY-model (2) the corresponding operator D is invariant for S_2 and

$$(D^2\phi)(z) = h(z)\,\phi(z)$$

$$h(z) = -g(z)\,g(\bar{z})$$

$$g(z) = J_x \bar{z} + \mu h z + J_y z^3 \,.$$

Now the spectrum of D^n or of D is purely continuous if the function $g(z)$ is not constant on a set of measure different from zero on $\mathbb{C}_1$. Otherwise the characteristic function of that set should be an eigenfunction of D^n. For the linear XY-model the function

$$-g(z) = J_x^2 + J_y^2 + (\mu h)^2 + \mu h(J_x + J_y)(x^{-2} + z^2) + J_x J_y(z^{-4} + z^4)$$

is constant only if $J_x = J_y = 0$.

Now using Theorem 1, we can formulate:

Theorem 2:

Let $t \rightarrow \alpha_t(B(\psi)) = B(e^{Dt}\psi)$, $\psi \in H$ be the evolution induced by the bilinear Hamiltonians H_Λ (see (1)); then the corresponding equilibrium KMS state is ergodic if the function $g(z)$, $z \in \mathbb{C}_1$ constructed above is not constant on a nontrivial subset of $\mathbb{C}_1$.

III. DISCUSSION

We studied systems interacting with bilinear Hamiltonians in the Fermi field operators. Assuming that for quantum systems the notion of ergodicity of a system in a particular state is given by the equality of time mean to space mean for each observable, we give in Theorem 1 a necessary and sufficient condition for ergodicity. In Theorem 2 we give a sufficient condition, in a more explicit way, directly in terms of the coefficients v_{ij} appearing in the Hamiltonian.

Now we can see in which sense our notion of ergodicity can be compared with the notion of metric transitivity for classical systems. For classical systems, an abstract dynamical system is given by the triplet (M, μ, ϕ_t): M is a measure space, μ a measure and ϕ_t, $t \in \mathbb{R}$ a group of automorphisms of (M, μ) [10]. The system is ergodic if the metric space has no ϕ-invariant subsets. For our bilinear systems the *-automorphism α_t leaves invariant the different particle sectors, e.g. the n-particle sectors

$$\alpha_t(B(\psi_1) \ldots B(\psi_n)) = B(e^{Dt}\psi_1) \ldots B(e^{Dt}\psi_n).$$

This feature is due to bilinear Hamiltonians. Now look at the one-particle subspace

$$\alpha_t(B(\psi)) = B(e^{Dt}\psi).$$

Ergodicity of our system is equivalent to the condition that the spectrum of the operator D be purely continuous. This is equivalent to the fact that the test function space H has no finite-dimensional invariant subsets under the one-parameter group $\{U_t = e^{Dt} | t \in \mathbb{R}\}$ induced by the time evolution automorphism group, except eigenvalue zero. This clarifies a connection with the property of metric transitivity for classical systems.

REFERENCES

[1] S. Doplicher, R. V. Kadison, D. Kastler, and D. W. Robinson, *Comm. Math. Phys.*, **6**, 101 (1967).

[2] D. Ruelle, "Statistical Mechanics". W. A. Benjamin, New York (1969).

[3] F. Rocca, M. Sirugue, and D. Testard, *Comm. Math. Phys.*, **13**, 317 (1969).

[4] E. Balslev, J. Manuceau, and A. Verbeure, *Comm. Math. Phys.*, **8**, 315 (1968).

[5] A. Zygmund, "Trigonometric Series", Vol II. Cambridge (1959).

[6] R. H. Hermann and M. Takesaki, "States and Automorphism Group of Operator Algebras" (preprint).

[7] D. Testard, *Ann. Inst. Henri Poincaré*, **12**, 329 (1970).

[8] J. Manuceau, F. Rocca, and D. Testard, *Comm. Math. Phys.*, **12**, 43 (1969).

[9] D. Kastler and D. W. Robinson, *Comm. Math. Phys.*, **3**, 151 (1966).

[10] V. I. Arnold and A. Avez, "Problèmes Ergodiques de la Mécanique Classique". Gauthier-Villars, Paris (1967).

Some General Properties of Thermodynamic States in an Algebraic Approach

MARINUS WINNINK

Institute for Theoretical Physics, University of Groningen (The Netherlands)

Abstract

The local normality and the separating character of thermodynamic limit states and some properties of K.M.S. states are discussed in an algebraic setting. It is shown that a K.M.S. state, being separating, is locally normal. A C^*-algebra such that (i) thermodynamic limit states are states on it, and (ii) time translations induce a one-parameter automorphism group on it, is constructed.

Contents

INTRODUCTION

In these lectures we are going to be concerned with some general properties of states describing thermodynamic systems. Everything we shall discuss will be in terms of an algebraic setting. For a quantum system in a box of finite volume V we shall adopt Mackey's scheme [1]. In this scheme all observables are generated by elementary observables admitting only yes or no answers, the so-called questions.

The questions are in one-to-one correspondence with the projections on an infinite dimensional separable Hilbert space $\mathfrak{H}_V$. In this scheme a bounded observable, via von Neumann's spectral theorem, is a bounded selfadjoint operator on $\mathfrak{H}_V$. To every pair consisting of an observable A and a state α, a probability distribution is given on the (possibly) measured values of A. Let $\langle A \rangle_\alpha$ denote the expectation value computed with this probability distribution, i.e. $\langle A \rangle_\alpha$ is the expectation value of the observable A, when the system is in the state α. It follows from a theorem of Gleason [2] that $\langle A \rangle_\alpha$ is of the form:

$$\langle A \rangle_\alpha = \mathrm{Tr}\,(\rho_\alpha A), \tag{1}$$

where ρ_α is a positive trace-class operator on $\mathfrak{H}_V$ with $\mathrm{Tr}\,\rho_\alpha = 1$.

Whenever we speak of a state of a system confined to a box V in space, we shall have in mind the extension of $\mathrm{Tr}(\rho_\alpha)$ to all elements of the von Neumann algebra generated by the questions, i.e. all bounded operators $(B(\mathfrak{H}_V))$ on $\mathfrak{H}_V$.

There is only one reasonable description of the dynamics in Mackey's scheme, namely as a group of transformations of the states that preserves convex combinations of states and is continuous in t. This then is equivalent to the existence of the Hamiltonian as an unbounded observable, i.e. an unbounded selfadjoint operator on $\mathfrak{H}_V$, whose spectral projections trivially are questions.

For thermodynamic systems the total energy is not acceptable as an observable (it is infinite) or anything near to an observable like in Mackey's scheme. Therefore the full Mackey scheme is not useful for the description of thermodynamic systems.

For thermodynamic systems we assume that the observables generate the quasi-local C^*-algebra $\mathfrak{A}$, constructed from local algebras $\mathfrak{A}(V)$. More precisely, we assume that $\mathfrak{A}$ is the C^*-inductive limit [3] of local von Neumann algebras $A(V)$, where $A(V)$ is isomorphic to $B(\mathfrak{H}_V)$. In particular, there exist isomorphisms ϕ_V that map $A(V)$ onto $\mathfrak{A}(V)$ such that

$$\mathfrak{A} = \overline{\bigcup_{\substack{V \subset R^3 \\ V\,\text{finite}}} \mathfrak{A}(V)}^{\,n},$$

where $\overline{(.)}^{\,n}$ means the closure in the norm on the C^*-inductive limit of $A(V)$'s.

A thermodynamic state ω is a positive linear functional on $\mathfrak{A}$. From the construction of $\mathfrak{A}$ we see that the kinematical structure for a subsystem in a finite volume V of the thermodynamic system is described by essentially the same algebra as in the Mackey scheme.

We shall now want to know to what extent a state ω, in its restriction to $\mathfrak{A}(V)$ is consistent with Mackey's scheme. [Consistency with Mackey's scheme means that $(\omega \circ \phi_V \circ \alpha_V)(A) = \mathrm{Tr}\,\rho_{(\omega \circ \phi_V \circ \alpha_V)}(A)$, where α_V is the isomorphism that maps $B(\mathfrak{H}_V)$ onto $A(V)$].

The dynamics of a thermodynamic system is, under certain conditions to be specified later, to be thought of as a group of transformations on the states of $\mathfrak{A}$. In certain cases this group of transformations in the state space of $\mathfrak{A}$ reduces by transposition to a set of automorphisms of $\mathfrak{A}$. [In the last section we shall investigate the possible structures of the dynamics more precisely.] Even in the cases where one would have a group of automorphisms for the dynamics, the dynamical behavior of a local subsystem of a thermodynamic system is in general not in accordance with the dynamical structure of the Mackey scheme. In particular there can be a change of particle-number in V that is governed not solely by the properties of the subsystem in V, but also by the part of the system outside V.

Whenever a state of a thermodynamic system is locally consistent with Mackey's scheme as discussed above, it is said to be *locally normal*.

Local normality is a property that is true for many thermodynamic states; in fact, it holds for all states that are obtained by a limit procedure, i.e. thermodynamic limit states. Furthermore, it is true for all states that are separating [4]. [Let ω be a state of $\mathfrak{A}$. From this state we can, by the Gelfand–Segal construction, find a representation π_ω with a cyclic vector Ω_ω on a Hilbert space $\mathfrak{H}_\omega$. ω is said to be separating whenever Ω_ω is separating for the von Neumann algebra $\pi_\omega(\mathfrak{A})''$].

We shall therefore investigate to what extent thermodynamic states are separating. As we shall see, thermodynamic limit states at a temperature different from zero are separating under very weak conditions. Of course, the fact that they are separating is not very important for their local normality, since they are locally normal due to the very fact that they are limit states. Separating states however have many other nice structure properties (in relation to Tomita's theorem), as we shall see later on.

Regardless of how a thermodynamic state is obtained, whenever it satisfies the K.M.S. condition, it is separating and therefore locally normal [4].

We shall also see that a K.M.S.-like condition for a state already implies that the state is separating (cf. Section 1b).

1. LOCAL NORMALITY AND SEPARATING CHARACTER OF THERMODYNAMIC STATES

1a. Local Normality of Thermodynamic Limit States

Let ω_{V_n} be a Gibbs state on $\mathfrak{A}(V)$; then it is consistent with the Mackey scheme.

$\omega(A)$ is a thermodynamic limit state if we can find an absorbing sequence of finite volumes $\{V_n\}$ (i.e. $V_n \subset V_{n+1}, \forall n$) such that eventually every finite volume V is, from a certain n onwards, contained in $V_{n+p}, p \geqq 0$, and if, furthermore

$$\forall A \in \bigcup_{V \subset R^3} \mathfrak{A}(V), \qquad \omega_{V_n}(A) \to \omega(A).$$

One easily shows that $\omega(A)$, although primarily defined on $\mathfrak{A}_L (\equiv \bigcup_{V \subset R^3} \mathfrak{A}(V))$, can be extended to a positive linear functional on $\mathfrak{A}$.

A state ρ on a von Neumann algebra (or rather a W^*-algebra) $\mathfrak{M}$ is normal in the mathematical sense whenever the map

$$\mathfrak{M} \xrightarrow{\omega} \mathbb{C} \text{ (the complex numbers)}$$

is $\sigma(\mathfrak{M}, \mathfrak{M}_*)$ continuous, where $\mathfrak{M}_*$ is the predual of $\mathfrak{M}$. Furthermore, any isomorphism ϕ between two W^*-algebras, $\mathfrak{M}_1$ and $\mathfrak{M}_2$ say, is normal in the sense that it is continuous when $\mathfrak{M}_i$ are equipped with their $\sigma(\mathfrak{M}_i, \mathfrak{M}_{i_*})$ topologies.

Whenever we consider $B(\mathfrak{H}_V)$ the normal states are precisely the ones given by density operators such as (1). Therefore, since the mapping

$$B(\mathfrak{H}_V) \to A(V) \to \mathfrak{A}(V)$$

is composed of isomorphisms only, we have that ω_{V_n} can be considered as normal on $B(\mathfrak{H}_V)$. Stated differently, a normal state on $B(\mathfrak{H}_V)$ can be considered as a normal state on $\mathfrak{A}(V)$.

Sakai's theorem: Let ρ_n be the sequence of normal states on a W^*-algebra that converges pointwise to a limit ρ. Then ρ is normal [5].

The fact that $\omega_{V_n}(A)$ for A fixed in $\mathfrak{A}(V)$ stays normal on $\mathfrak{A}(V)$ when we let n grow is due to isotony, in the sense that

$$\mathfrak{A}(V) \subset \mathfrak{A}(V_n) \subset \mathfrak{A}(V_{n+1}),$$

where the injections are by supposition isomorphisms (cf. [3]).

From this we conclude that a thermodynamic limit state is locally normal.

1b. The Separating Character of Thermodynamic Limit States

Let us now see to what extent thermodynamic limit states are separating.
ω_V is a Gibbs state for a system enclosed in a box V, i.e.

$$\omega_V(A) = \frac{\text{Tr}(\exp - \beta(H'_V - \mu N_V)\,A)}{\text{Tr}\,\exp - \beta(H'_V - \mu N_V)},$$

where β, H'_V, μ, N_V are respectively the inverse temperature, the Hamiltonian, the chemical potential and the particle number operator.

Define a new Hamiltonian $H_V = H'_V - \mu N_V$. We then have

$$\omega_V(A) = \frac{\text{Tr}\,(e^{-\beta H_V}\,A)}{\text{Tr}\,e^{-\beta H_V}}.$$

We know that the unitary group $U_t = \exp iH_V t$ gives a group of automorphisms of $B(\mathfrak{H}_V)$ in the following sense:

$$A \to \alpha_t^V(A) = \exp iH_V t \cdot A \cdot \exp -iH_V t.$$

The following statements hold:

i. $\omega_V(A\alpha_t^V(B))$ and $\omega_V(\alpha_t^V(B)\,A)$ are continuous functions in t.

ii. $\int \omega_V(A\alpha_t^V(B))\,f(t - i\beta)\,dt = \int \omega_V(\alpha_t^V(B)\,A)\,f(t)\,dt,$

where $\hat{f}$ (the Fourier transform of f) belongs to the Schwarz class D.

For a proof of these statements see [6, 7].

[We stick to $B(\mathfrak{H}_V)$, but by transport of structure one can of course discuss everything in terms of $\mathfrak{A}(V)$].

We shall now suppose that there exists an absorbing sequence of volumes V_n such that for every A, B, C fixed in $\mathfrak{A}_L$ there exist:

$$\lim_{n\to\infty} \omega_{V_n}(A\alpha_t^{V_n}(B)), \qquad \left[\; \lim_{n\to\infty} \omega_{V_n}(\alpha_t^{V_n}(B)\,A)\; \right] \tag{Ia}$$

$$\lim_{n\to\infty} \omega_{V_n}(A\alpha_t^{V_n}(B)\,C). \tag{Ib}$$

Definition:

$$\lim_{n\to\infty} \omega_{V_n}(A\alpha_t^{V_n}(B)) = F(A, B, t);$$

$$\lim_{n\to\infty} \omega_{V_n}(A\alpha_t^{V_n}(B)\,C) = F(A, B, C, t);$$

$$\left[\; \lim_{n\to\infty} \omega_{V_n}(\alpha_t^{V_n}(B)\,A) = F(B, A, -t)\; \right].$$

We first observe that due to the Lebesgue dominated convergence theorem,

$$\int F(A, B, t)\,f(t - i\beta)\,dt = \int F(B, A, -t)\,f(t)\,dt. \tag{2}$$

If we had chosen $B = 1_V,$* then $\omega_{V_n}(A) \to F(A, 1, t) = F(A, 1, 0)$, i.e., $F(A, 1, 0)$ is the state ω we encountered before.

Let us construct from ω the Hilbert space $\mathfrak{H}_\omega$ and the representation π_ω on it with the cyclic vector Ω_ω. One now sees easily that**

$$\left| F(B^*, A, C, t) \right| \leqq \|A\| \sqrt{\omega(B^*B)\,\omega(C^*C)}, \qquad B, A, C \in \mathfrak{A}_L. \tag{3}$$

Furthermore, since $\mathfrak{A}_L$ is norm-dense in $\mathfrak{A}$, one extends (3) to all $B, A, C \in \mathfrak{A}$.

One can rewrite (3) in the following way:

$$\left| F(B^*, A, C, t) \right| \leqq \|A\|\, \|\pi_\omega(B)\,\Omega_\omega\|\, \|\pi_\omega(C)\,\Omega_\omega\|.$$

---

* 1_V is the unit operator in $\mathfrak{A}(V)$.

** This follows from the following observations:

$$\left| F(B^*, A, C, t) \right| \leqq \left| \omega_{V_n}(B^*\alpha_t^{V_n}(A)\,C) \right| + \varepsilon$$

for n sufficiently large. Furthermore,

$$\left| \omega_{V_n}(B^*\alpha_t^{V_n}(A)\,C) \right|^2 \leqq \omega_{V_n}(B^*B)\,\omega_{V_n}(C^*C)\,\|A\|^2$$

and

$$\lim_{n\to\infty} \omega_{V_n}(B^*B) = \omega(B^*B), \qquad \lim_{n\to\infty} \omega_{V_n}(C^*C) = \omega(C^*C).$$

F equals zero on all B or C in the left kernel I_l of ω $[I_l = \{A \in \mathfrak{A} : \omega(A^* A) = 0\}]$, and we have that $F(B^*, A, C, t)$ is a sesquilinear form on a dense set of vectors in $\mathfrak{H}_\omega$, which by the continuity can be extended to all $\psi, \phi \in \mathfrak{H}_\omega$. Therefore by the Riesz representation theorem there exists a uniquely determined bounded operator A_t on $\mathfrak{H}_\omega$ such that

$$F(B^*, A, C, t) = (\pi_\omega(B)\, \Omega_\omega, A_t \pi_\omega(C)\, \Omega_\omega), \tag{4}$$

where $\|A_t\| \leq \|A\|$.

If we had chosen $C = 1_V$ in the above construction then we would have had instead of (3):

$$F(B^*, A, t) \leqq \sqrt{\omega(B^*B)\, \omega(C^*C)}, \qquad A, B \in \mathfrak{A}_L.$$

By the same series of arguments we gave above, we establish the existence of a bounded operator U_t such that

$$F(B^*, A, t) = (\pi_\omega(B)\, \Omega_\omega, U_t \pi_\omega(A)\, \Omega_\omega). \tag{5}$$

From (5) and (4) we deduce by taking $C = 1_V$ in (4):

$$A_t \Omega_\omega = U_t \pi_\omega(A)\, \Omega_\omega. \tag{6}$$

U_t has the following properties:

$$U_0 = 1, \qquad \|U_t\| = 1,$$

$$U_t^* = U_{-t}, \qquad U_t \Omega_\omega = \Omega_\omega.$$

From our assumptions we cannot yet conclude that

$$U_{t_1 + t_2} = U_{t_1} U_{t_2},$$

i.e. the family of operators does not *a priori* form a group of operators.

Let us now gather all the information we have and plug it into equation (2). We then have

$$\int (\Omega_\omega, \pi(A)\, U_t\, \pi(B)\, \Omega_\omega)\, f(t - i\beta)\, dt = \int (\Omega_\omega, \pi(B)\, U_{-t}\, \pi(A)\, \Omega_\omega)\, f(t)\, dt. \tag{2'}$$

We know from [7, 9] that $\mathfrak{H}_\omega$ is separable, namely from the fact that it is locally normal. Therefore the unit ball of the von Neumann algebra $\pi_\omega(\mathfrak{A})''$ is metrizable in the strong topology. We can therefore find sequences

$$\pi(A_n^*) \xrightarrow{s} A^* \in \pi_\omega(\mathfrak{A})'',$$

$$\pi(A_n) \xrightarrow{s} A \in \pi_\omega(\mathfrak{A})''.$$

By the Lebesgue dominance theorem we can extend this equation to

$$\int (\Omega_\omega, A\, U_t\, B\Omega_\omega)\, f(t - i\beta)\, dt = \int (\Omega_\omega, B\, U_{-t}\, A\Omega_\omega)\, f(t)\, dt, \quad \forall\, A, B \in \pi_\omega(\mathfrak{A})''. \tag{2''}$$

Let us now assume that $\lim_{n \to \infty} \omega_{V_n}(A\alpha_t^{V_n}B)$ is continuous on

$$t \in [-\varepsilon, \varepsilon], \qquad \text{where } \varepsilon \text{ is arbitrary positive}, \quad A, B \in \mathfrak{A}_L. \qquad \text{(II)}$$

Then also $(\Omega_\omega, AU_t B\Omega_\omega)$ is continuous $\forall\, A, B \in \pi_\omega(\mathfrak{A})''$.

Indeed, as before, we find that due to the separability of $\mathfrak{H}_\omega$, $\pi(A_n)^* \to A^*$ and $\pi(A_n) \to A$ simultaneously in the strong operator topology, and hence

$$\left|(\Omega_\omega, \pi(A_n)\, U_t \pi(B)\, \Omega_\omega) - (\Omega_\omega, AU_t \pi(B)\, \Omega_\omega)\right| \leq \left\|(\pi(A_n)^* - A^*)\, \Omega_\omega\right\| \left\|\pi(B)\, \Omega_\omega\right\| < \eta$$

for n sufficiently large.

In particular,

$$\sup_{t \in [-\varepsilon, \varepsilon]} \left|(\Omega_\omega, \pi(A_n)\, U_t \pi(B)\, \Omega_\omega) - (\Omega_\omega, AU_t \pi(B)\, \Omega_\omega)\right| < \eta$$

for n sufficiently large.

Therefore, $(\Omega_\omega, AU_t \pi(B)\, \Omega_\omega) = \lim_n (\Omega_\omega, \pi(A_n)\, U_t \pi(B)\, \Omega_\omega)$ uniformly in t on $[-\varepsilon, \varepsilon]$, which shows that $(\Omega_\omega, AU_t \pi(B)\, \Omega_\omega)$ is a continuous function of t on $[-\varepsilon, \varepsilon]$. The same trick can be performed for $\pi(B_n) \to B, \pi(B_n^*) \to B^*$, which shows that $(\Omega_\omega, AU_t B\Omega_\omega)$ is continuous for $t \in [-\varepsilon, \varepsilon]$.

Suppose now that $B\Omega_\omega = 0$. Then the left-hand side of equation (2'') is zero and hence the right-hand side is zero, i.e.

$$\int (B^*\Omega_\omega, U_{-t} A\Omega_\omega)\, f(t)\, dt = 0, \qquad \forall\, A \in \pi_\omega(\mathfrak{A})'', \quad \hat{f} \in D.$$

From this it follows that

$$\int (B^*\Omega_\omega, U_{-t} A\Omega_\omega)\, f(t)\, dt = 0, \qquad \forall\, A \in \pi_\omega(\mathfrak{A})'', \quad f \in S,$$

and therefore $(B^*\Omega_\omega, U_{-t} A\Omega) = 0$ almost everywhere.

Since we assumed that this function is continuous on $[-\varepsilon, \varepsilon]$ and since $U_0 = 1$, we have that

$$(B^*\Omega_\omega, A\Omega_\omega) = 0, \qquad \forall\, A \in \pi_\omega(\mathfrak{A})'',$$

and therefore we conclude that $B^*\Omega_\omega = 0$.

Thus, under the assumption of continuity, we have proved that

$$B\Omega_\omega = 0 \leftrightarrow B^*\Omega_\omega = 0, \qquad B \in \pi_\omega(\mathfrak{A})''.$$

The left ideal $\mathscr{I}_l = \{B \in \pi_\omega(\mathfrak{A})'' : (\Omega_\omega, B^*B\Omega_\omega) = 0\}$ is therefore equal to

$$\mathscr{I}_r = \{B \in \pi_\omega(\mathfrak{A})'' : (\Omega_\omega, BB^*\Omega_\omega) = 0\}.$$

Hence $\mathscr{I}_l = \mathscr{I}_r$, i.e. $\mathscr{I}_l$ is a two-sided ideal, which we denote by $\mathscr{I}$. Let $B \in \mathscr{I}$; then

$$\left|(\Omega_\omega, A^*CB\Omega_\omega)\right|^2 \leq (\Omega_\omega, A^*A\Omega_\omega)(\Omega_\omega, (CB)^*(CB)\Omega_\omega),$$

$C \in \mathscr{I}$, and by the very fact that $\mathscr{I}$ is a two-sided ideal in $\pi_\omega(\mathfrak{A})''$ we have that $CB \in \mathscr{I}$, and hence $\omega((CB)^*(CB)) = 0$.

We have thus proved that for $B \in \mathscr{I}$

$$(A\Omega_\omega, BC\Omega_\omega) = 0, \qquad \forall\, A, B \in \pi_\omega(\mathfrak{A})''.$$

But this implies that $B = 0$.

Under the assumptions we made, we have:

$$B\Omega_\omega = 0 \to B = 0, \qquad B \in \pi_\omega(\mathfrak{A})''.$$

Hence Ω_ω is separating for $\pi_\omega(\mathfrak{A})''$, and thus ω is separating.

Remark: *

To arrive at equation (2″) and the continuity of $(\Omega, AU_t B\Omega)$ from assumption (II), we explicitly used the fact that ω, as a thermodynamic limit state, is locally normal (Sakai's theorem, see above) and therefore gives rise to a representation on a separable Hilbert space. One can also prove the same results along the following lines: $(\Omega_\omega, \pi(A)\, U_t \pi(B)\, \Omega)$ is Lebesgue-measurable as a function of t, because it is a pointwise limit in t of continuous functions in t, $\forall\, A, B \in \mathfrak{A}_L$. Since $\mathfrak{A}_L$ is norm-dense in $\mathfrak{A}$ and $\{\pi(\mathfrak{A})\,\Omega_\omega\}$ is dense in $\mathfrak{H}_\omega$, it follows that $\{\pi(\mathfrak{A}_L)\,\Omega_\omega\}$ is dense in $\mathfrak{H}_\omega$. Therefore $(\phi, U_t\psi)$, $\phi, \psi \in \mathfrak{H}_\omega$, is a pointwise limit of a sequence of Lebesgue-measurable functions and therefore it is at least measurable.

We then see easily that

$$\int (\phi, U_t\psi) f(t - i\beta)\, dt \quad \text{and} \quad \int (\phi, U_{-t}\psi) f(t)\, dt$$

are sesquilinear forms on $\mathfrak{H}_\omega$ defining bounded linear operators U_{f_β} and $U_{\check{f}}$ respectively, with the properties

$$\int (\phi, U_t\psi) f(t - i\beta)\, dt = (\phi, U_{f_\beta}\psi),$$

$$\int (\phi, U_{-t}\psi) f(t)\, dt = (\phi, U_{\check{f}}\psi).$$

The l.h.s. and the r.h.s. of (2′) can be rewritten in terms of the operators U_{f_β} and $U_{\check{f}}$ respectively as:

$$\int (\Omega_\omega, \pi(A)\, U_t \pi(B)\, \Omega_\omega)\, f(t - i\beta)\, dt = (\Omega_\omega, \pi(A)\, U_{f_\beta}\, \pi(B)\, \Omega_\omega) =$$

$$= \int (\Omega_\omega, \pi(B)\, U_{-t}\, \pi(A)\, \Omega_\omega) f(t)\, dt = (\Omega_\omega, \pi(A)\, U_{\check{f}}\pi(B)\, \Omega_\omega).$$

We then have that for $A, B \in \pi_\omega(\mathfrak{A})''$

$$\left|(\Omega_\omega, AU_{f_\beta} B\Omega_\omega) - (\Omega_\omega, BU_{\check{f}}A\Omega)\right| \leqq \left|(\Omega_\omega, AU_{f_\beta} B\Omega_\omega) - (\Omega_\omega, \pi(C)\, U_{f_\beta} B\Omega_\omega)\right| +$$

$$+ \left|(\Omega_\omega, \pi(C)\, U_{f_\beta} B\Omega_\omega) - (\Omega_\omega, \pi(C)\, U_{f_\beta} \pi(D)\, \Omega_\omega)\right| +$$

$$+ \left|(\Omega_\omega, \pi(C)\, U_{f_\beta} \pi(D)\, \Omega_\omega) - (\Omega_\omega, \pi(D)\, U_{\check{f}}\pi(C)\, \Omega_\omega)\right| +$$

* A discussion on this point with Michel Sirugue is gratefully acknowledged.

$$+ \left|(\Omega_\omega, \pi(D)\, U_{f}\pi(C)\,\Omega_\omega) - (\Omega_\omega, \pi(D)\, U_{f}A\Omega_\omega)\right| +$$

$$+ \left|(\Omega_\omega, \pi(D)\, U_{f}A\Omega_\omega) - (\Omega_\omega, BU_{f}A\Omega_\omega)\right| \leqq$$

$$\leqq \left\|(\pi(C^*) - A^*)\,\Omega_\omega\right\| \cdot \left\|B\Omega_\omega\right\| \cdot \smallint|f_\beta|\,dt +$$

$$+ \left\|\pi(C^*)\,\Omega\right\| \cdot \left\|(B - \pi(D))\,\Omega_\omega\right\| \cdot \smallint|f_\beta|\,dt + 0 \ (\text{due to } 2') +$$

$$+ \left\|\pi(D)\,\Omega\right\| \cdot \left\|(\pi(C) - A)\,\Omega_\omega\right\| \cdot \smallint|f|\,dt + \left\|(\pi(D^*) - B^*)\,\Omega_\omega\right\| \cdot \left\|A\Omega_\omega\right\| \cdot \smallint|f|\,dt.$$

Since we can always choose $\pi(C), \pi(D)$ such that

$$\left\|(\pi(C) - A)\,\Omega_\omega\right\| < \varepsilon, \quad \left\|(\pi(C^*) - A^*)\,\Omega_\omega\right\| < \varepsilon, \quad \left\|(\pi(D) - B)\,\Omega_\omega\right\| < \varepsilon,$$

and

$$\left\|(\pi(D^*) - B^*)\,\Omega_\omega\right\| < \varepsilon,$$

we find that

$$\left|(\Omega_\omega, AU_{f_\beta}B\Omega_\omega) - (\Omega_\omega, BU_{f}A\Omega_\omega)\right| < \eta$$

η arbitrary. Hence

$$(\Omega_\omega, AU_{f_\beta}B\Omega_\omega) = (\Omega_\omega, BU_{f}A\Omega_\omega).$$

But this is (2″) $\forall\, A, B \in \pi_\omega(\mathfrak{A})''$.

Concerning the continuity argument one finds that there always exists $\pi(C)$, given A and η, such that

$$\left|(\Omega_\omega, \pi(C)\, U_t\pi(B)\,\Omega_\omega) - (\Omega_\omega, AU_t\pi(B)\,\Omega_\omega)\right| \leqq \left\|(A^* - \pi(C^*))\,\Omega_\omega\right\| \cdot \left\|\pi(B)\,\Omega_\omega\right\| < \eta,$$

and hence

$$\sup_{t \in [-\varepsilon, \varepsilon]} \left|(\Omega_\omega, \pi(C)\, U_t\pi(B)\,\Omega_\omega) - (\Omega_\omega, AU_t\pi(B)\,\Omega_\omega)\right| < \eta.$$

The argument then proceeds in the same way as before, and one finds from assumption (II) that $(\Omega_\omega, AU_tB\Omega_\omega)$ is continuous on $[-\varepsilon, \varepsilon]$.

Let us now adopt the following set of assumptions (cf. also [11]) instead of the assumption (I):

The existence of

$$\lim_m \lim_n \omega_{V_n}(A\alpha_t^{V_m}(B)\,C), \qquad \lim_n \omega_{V_n}(A\alpha_t^{V_n}(B)\,C), \qquad \lim_m \lim_n \omega_{V_n}(A\alpha_{t_1}^{V_n}(B)\,\alpha_{t_2}^{V_m}(C))$$

and

$$\lim_n \omega^{V_n}(A\alpha_{t_1}^{V_n}(B)\,\alpha_{t_2}^{V_n}(C)), \quad A, B, C \in \mathfrak{A}_L \tag{IIIa}$$

with the following properties:

$$\lim_m \lim_n \omega_{V_n}(A\alpha_t^{V_m}(B)\,C) = \lim_{n \to \infty} \omega_{V_n}(A\alpha_t^{V_n}(B)\,C), \tag{IIIb}$$

$$\lim_m \lim_n \omega_{V_n}(A\alpha_{t_1}^{V_n}(B)\,\alpha_{t_2}^{V_m}(C)) = \lim_{n \to \infty} \omega_{V_n}(A\alpha_{t_1}^{V_n}(B)\,\alpha_{t_2}^{V_n}(C)). \tag{IIIc}$$

Clearly, this set of assumptions covers assumptions (I). Therefore every conclusion drawn from (I) still holds.

From (IIIb) we have:

$$(\pi(A^*)\,\Omega_\omega, B_t\pi(C)\,\Omega_\omega) = \lim_m \,(\pi(A^*)\,\Omega_\omega, \pi(\alpha_t^{V_m}(B))\,\pi(C)\,\Omega_\omega),$$

which implies

$$(\phi, B_t\psi) = \lim_m \,(\phi, \pi(\alpha_t^{V_m}(B))\,\psi),$$

i.e.

$$B_t = \text{weak-lim}\ \pi(\alpha_t^{V_m}(B)). \tag{7}$$

Proof :

$$\left|(\phi, B_t\psi) - (\phi, \pi(\alpha_t^{V_m}(B))\psi)\right| \leq \left|(\phi - \psi_A, B_t\psi)\right| + \left|(\psi_A, B_t\psi) - (\psi_A, B_t\psi_C)\right| +$$

$$+ \left|(\psi_A, B_t\psi_C) - (\psi_A, \pi(\alpha_t^{V_m}(B))\,\psi_C)\right| + \left|(\psi_A, \pi(\alpha_t^{V_m}(B))\,\psi_C) - (\phi, \pi(\alpha_t^{V_m}(B))\,\psi_C)\right| +$$

$$+ \left|(\phi, \pi(\alpha_t^{V_m}(B)\,\psi_C) - (\phi, \pi(\alpha_t^{V_m}(B))\,\psi)\right| \to 0.$$

From (6) and (7) we infer:

$$U_t\pi(B)\,\Omega_\omega = B_t\Omega_\omega = \text{weak-lim}\ \pi(\alpha_t^{V_m}(B))\,\Omega_\omega, \tag{8}$$

where the weak limit is meant as a weak limit of vectors in $\mathfrak{H}_\omega$.

If we now choose $A = 1_V$, $B = C^*$, and $t_1 = t_2 = t$ in (IIIb), we have:

$$\lim_m \lim_n \omega_{V_n}(\alpha_t^{V_n}(C^*)\,\alpha_t^{V_m}(C)) = \lim_n \omega_{V_n}(\alpha_t^{V_n}(C^*)\,\alpha_t^{V_n}(C)) = \lim_n \omega_{V_n}(C^*C). \tag{9}$$

The last member of (9) is due to the invariance of the Gibbs state.

Rewriting (9) with the help of (8) we find that

$$\lim_m \lim_n \omega_{V_n}(\alpha_t^{V_n}(C^*)\,\alpha_t^{V_m}(C)) = \lim_m \,(U_t\pi(C)\,\Omega_\omega, \pi(\alpha_t^{V_m}(C))\,\Omega_\omega) \overset{(8)}{=\!=}$$

$$\overset{(8)}{=\!=} \|U_t\pi(C)\,\Omega_\omega\|^2 \overset{(9)}{=\!=} \|\pi(C)\,\Omega_\omega\|^2,$$

i.e. U_t is unitary.

Using the unitarity of the operators U_t we obtain instead of (8):

$$\|U_t\pi(B)\,\Omega_\omega - \pi(\alpha_t^{V_n}(B))\,\Omega_\omega\|^2 =$$

$$= \|U_t\pi(B)\,\Omega_\omega\|^2 + \|\pi(\alpha_t^{V_n}(B))\,\Omega_\omega\|^2 - (U_t\pi(B)\,\Omega_\omega, \pi(\alpha_t^{V_n}(B))\,\Omega_\omega) -$$

$$- (\pi(\alpha_t^{V_n}(B))\,\Omega_\omega, U_t\pi(B)\,\Omega_\omega) \to 0,$$

i.e.

$$U_t\pi(B)\,\Omega_\omega = s - \lim_n \pi(\alpha_t^{V_n}(B))\,\Omega_\omega.$$

We can now prove the group property for the operators $\{U_t\}$:

$$U_{t_1 + t_2} = U_{t_1} U_{t_2}. \tag{*}$$

Indeed:

$$\lim_{n \to \infty} \omega_{V_n}(B^* \alpha_{t_1 + t_2}^{V_n}(C)) = (\pi(B)\,\Omega_\omega,\, U_{t_1 + t_2}\pi(C)\,\Omega_\omega). \tag{10}$$

The left-hand side of (10) can be rewritten, due to invariance and assumption (IIIa), as:

$$\lim_{n \to \infty} \omega_{V_n}(\alpha_{-t_1}^{V_n}(B^*)\,\alpha_{t_2}^{V_n}(C)) = \lim_m \lim_n \omega_{V_n}(\alpha_{-t_1}^{V_n}(B^*)\,\alpha_{t_2}^{V_m}(C)). \tag{11}$$

Therefore, since the right-hand side of (11) converges,

$$(U_{-t_1}\pi(B)\,\Omega_\omega,\, U_{t_2}\pi(C)\,\Omega_\omega) = (\pi(B)\,\Omega_\omega,\, U_{t_1} U_{t_2}\pi(C)\,\Omega_\omega). \tag{12}$$

From (10) and (12) we conclude that (*) holds.

We noticed that assumptions (IIIa, b, c) implied assumptions (I). They also imply assumption (II). Indeed, from (IIIa, b, c) we just found that the family of operators form an abelian group of unitaries, with the property that $(\phi, U_t\psi)$ is measurable in t. [The latter fact followed already from assumption (I)]. If we now use the fact that $\mathfrak{H}_\omega$ is separable, we conclude that $(\phi, U_t\psi)$ is a continuous function of t [10].

Because on the basis of assumptions (I) and (II) we already showed that ω is separating, we conclude that ω is separating also on the basis of (IIIa, b, c).

We claim that

$$A_t = U_t\pi(A)\,U_{-t}. \tag{13}$$

Since $A_t \in \pi_\omega(\mathfrak{A})''$ (cf. (7)), the right-hand side maps a dense set in $\pi_\omega(\mathfrak{A})''$, namely $\{\pi_\omega(A),\, A \in \mathfrak{A}_L\}$, into another dense set in $\pi_\omega(\mathfrak{A})''$, namely $\{A_t,\, A \in \mathfrak{A}_L\}$. We can therefore extend (13) to a $*$-automorphism of $\pi_\omega(\mathfrak{A})''$.

To prove (13) it suffices to prove that

$$\lim_{n \to \infty} \omega_{V_n}(\alpha_{-t}^{V_n}(B^*)\, A\, \alpha_{-t}^{V_n}(C)) = \lim_m \lim_n \omega_{V_n}(\alpha_{-t}^{V_n}(B^*)\, A\, \alpha_{-t}^{V_m}(C)). \tag{14}$$

The above holds because we can make the following series of estimates:

$$\left|\omega_{V_n}(\alpha_{-t}^{V_n}(B^*)\, A\, \alpha_{-t}^{V_n}(C)) - \omega_{V_n}(\alpha_{-t}^{V_n}(B^*)\, A\, \alpha_{-t}^{V_m}(C)\right|^2 =$$

$$= \left|\omega_{V_n}(\alpha_{-t}^{V_n}(B^*)\, A\, [\alpha_{-t}^{V_n}(C) - \alpha_{-t}^{V_m}(C)])\right|^2 \leq$$

$$\leqq \omega_{V_n}(\alpha_{-t}^{V_n}(B^*)\,\alpha_{-t}^{V_n}(B))\,\omega_{V_n}(\{A[\alpha_{-t}^{V_n}(C) - \alpha_{-t}^{V_m}(C)]\}^* \{A[\alpha_{-t}^{V_n}(C) - \alpha_{-t}^{V_m}(C)]\}) \leqq$$

$$\leqq \omega_{V_n}(B^*B)\,\omega_{V_n}([\alpha_{-t}^{V_n}(C) - \alpha_{-t}^{V_m}(C)]^* [\alpha_{-t}^{V_n}(C) - \alpha_{-t}^{V_m}(C)]) \cdot \|A\|^2 \leqq$$

$$\leqq \|A\|^2 \omega_{V_n}(B^*B)\,\{\omega_{V_n}(C^*C) - \omega_{V_n}(\alpha_{-t}^{V_n}(C)^*\alpha_{-t}^{V_m}(C)) - \omega_{V_n}(\alpha_{-t}^{V_m}(C)^*\alpha_{-t}^{V_n}(C)) +$$

$$+ \omega_{V_n}(\alpha_{-t}^{V_m}(C)^*\alpha_{-t}^{V_m}(C))\}.$$

$\lim_m \lim_n (\text{r.h.s.}) = 0$, using (7)....(12). This means that (14) holds, which in turn means that

$$\lim_n \omega_{V_n}(B^*\alpha_t^{V_n}(A)\, C) = \lim_m \lim_n \omega_{V_n}(\alpha_{-t}^{V_n}(B^*)\, A\, \alpha_{-t}^{V_m}(C))$$

and this implies (13). Q.E.D.

1c. Local Normality of Separating Thermodynamic States

In this chapter we want to investigate to what extent separating states on $\mathfrak{A}$ are locally normal. In the previous chapters we discussed thermodynamic limit states. Here we just discuss thermodynamic states and we do not bother whether they are or are not obtained by taking the thermodynamic limit of Gibbs states.

We then have the following theorem [4]:

Every separating state on a quasi-local algebra constructed from local algebras that are properly infinite σ-finite W^*-algebras (or finite factors) is locally normal.

For the proof we refer the reader to [4].

If now we had somehow weakened Mackey's scheme such that the local algebras were as in the theorem above, and if we could argue that the only interesting states, by a Gleason-like theorem, would be just the normal states on the local algebras, then by the theorem above every separating state would be locally normal. One proves also [4] that every locally normal state on a countably generated funnel [i.e. $\mathfrak{A}$ is generated by a countable number of $\mathfrak{A}(V)$'s (as in the previous sections), where every $\mathfrak{A}(V)$ is properly infinite with a separable predual (hence $\mathfrak{A}(V)$ is σ-finite)] gives rise to a representation on a separable Hilbert space. Moreover, in this case the converse of the statement is also true. This generalizes the separability theorem shown in [8], [9], and [12] where only factors are admitted for the local algebras.

Let us now turn to the K.M.S. states on $\mathfrak{A}$, where $\mathfrak{A}$ is again as in Sections 1a and 1b.

1d. Properties of K.M.S. States on $\mathfrak{A}$

A K.M.S. state ω on $\mathfrak{A}$ is a state on $\mathfrak{A}$ which, with respect to an automorphism $A \in \mathfrak{A} \to \alpha_t(A) \in \mathfrak{A}$, has the following properties:

 i. $\omega(A\alpha_t(B))$ is continuous in t.

 ii. $\int \omega(A\alpha_t(B)) f(t - i\beta)\, dt = \int \omega(\alpha_t(B)\, A) f(t)\, dt, \qquad \hat{f} \in D.$ (15)

$A \to \alpha_t(A)$ is then interpreted as the time-evolution automorphism.

One does not know whether for general systems the time evolution is represented in terms of a one-parameter automorphism group of $\mathfrak{A}$ (For a discussion of what one can actually expect, compare Section 2). As is pointed out in [6], and verified for lattice systems with interactions that satisfy reasonable criteria [13, 14], one has indeed that time evolution is described by an automorphism group if $\alpha_t^{V_n}(A) - \alpha_t^{V_m}(A)$ is a Cauchy sequence in norm for every $A \in \mathfrak{A}_L$ and t. In the case of the lattice gas as discussed in [13], one verifies that this condition holds and furthermore that $\|\alpha_t(A) - A\| \xrightarrow[t \to 0]{} 0$, i.e. the automorphism group is strongly continuous. Condition i above is then trivially satisfied.

One proves from i and ii that

$$\omega(\alpha_t(A)) = \omega(A)$$

and hence one can rewrite ii as

$$\int (\Omega_\omega, \pi_\omega(A)\, U_t \pi_\omega(B)\, \Omega_\omega)\, f\,(t - i\beta)\, dt = \int (\Omega_\omega, \pi_\omega(B)\, U_{-t}\pi_\omega(A)\, \Omega_\omega)\, f\,(t)\, dt. \qquad (16)$$

It is well known that this equation holds for all A and B in $\pi_\omega(\mathfrak{A})''$, i.e.

$$\int (\Omega_\omega, A U_t B \Omega_\omega)\, f\,(t - i\beta)\, dt = \int (\Omega_\omega, B U_{-t} A \Omega_\omega)\, f\,(t)\, dt,$$

$$A, B \in \pi_\omega(\mathfrak{A})'', \qquad \hat{f} \in D.$$

A proof follows straightforwardly from the remark made in Section 1 b.

Again by the same arguments as in Section 1b, we find that ω is separating.

One can also show the converse (Tomita's theorem [15, 16]).

Given a von Neumann algebra $\mathfrak{M}$ on a Hilbert space $\mathfrak{H}$ with a cyclic and separating vector Ω, there exists a one-parameter abelian group of unitaries V_t on $\mathfrak{H}$ with the properties that

i. V_t is strongly continuous on $\mathfrak{H}$, and

ii. $\int (\Omega, A V_t B \Omega)\, f\,(t - i)\, dt = \int (\Omega, B V_{-t} A \Omega)\, f\,(t)\, dt, \qquad \hat{f} \in D, \qquad A, B \in \mathfrak{M}.$ (17)

V_t is the so-called modular automorphism group of $\mathfrak{M}$ [16]. [By a scale transformation one can rewrite the equation with $i\beta$ instead of i].

If we substitute $\pi_\omega(\mathfrak{A})''$ for $\mathfrak{M}$ and Ω_ω for Ω, then we have the following: $V_t = U_t$ [16, 17] or, stated differently, the group of automorphisms of $\mathfrak{M}$ that appears in Tomita's theorem is unique. In particular it shows that the group of operators in Section 1b following from assumptions (III) generates the same automorphism of $\pi_\omega(\mathfrak{A})''$ as the modular automorphism [16, 17, 7]. If $\mathfrak{A}$ were simple then this would imply that in the case of a K.M.S. state on $\mathfrak{A}$, α_t is unique (cf. [18, appendix] for a direct proof).

From (16) (and hence from (15)) follows that on $\mathfrak{H}$ there exists a conjugation J with the properties

$$J\Omega = \Omega,$$

$$JV_tJ = V_t, \qquad (18)$$

$$J\mathfrak{M}J = \mathfrak{M}'.$$

The latter expression tells us that $\mathfrak{M}$ is as large as its commutant $\mathfrak{M}'$. This fact was already manifest in the Bose gas treated in [19].

The structure properties in (18) are common to all separating states and therefore hold for the ones discussed in Section 1b. In particular, if we have conditions (III) then the thermodynamic-limit state satisfies (17), and $U_t = V_t$.

As pointed out above, (15) implies (17). The converse however is not true in general, i.e. Tomita's theorem is a typical theorem on von Neumann algebras and therefore cannot always be related to an automorphism of an arbitrarily given C^*-algebra that is weakly dense in $\mathfrak{M}$.

The analysis in Section 1b suggests that under the assumptions (IIIa, b, c) time translation may be described as a representation-dependent automorphism of the von Neumann algebra $\pi_\omega(\mathfrak{A})''$. In Section 2 we shall come back to this.

We want to conclude this section with a few remarks on decomposition theorems in a compact set of K.M.S. states at a given temperature.

Let K_β be the set of K.M.S. states at a given temperature β. We assume that K_β is compact (if $\| \alpha_t(A) - A \| \xrightarrow[t \to 0]{} 0$ were true, one could show that K_β is compact). K_β is not always compact; for a counterexample see [4].

However, it does follow that K_β is a simplex. This is due to the fact that for every $\omega \in K_\beta$, the center of $\pi_\omega(\mathfrak{A})''$ is pointwise invariant. In particular, K_β is a simplex of locally normal states [4].

One can also show the following: Let G be any group that can be represented in the automorphism group of $\mathfrak{A}$, i.e. $g \in G \to \tau_g \in \mathrm{aut}\ \mathfrak{A}$. Let $I_g(\mathfrak{A})$ be the set of states invariant under τ_g. Then

$$I_G(\mathfrak{A}) \cap K_\beta$$

is a simplex [4].

The fact that this is true follows again from the fact that K_β is a simplex. Roughly speaking: it is the K.M.S. condition which provides constraints on the other properties of the states. For a further discussion of the constraints imposed on nondynamical properties of thermodynamic states by the K.M.S. condition, see [18].

Suppose that the action of G is G-abelian on $\mathfrak{A}$ [20]. Then we know that $I_G(\mathfrak{A})$ is also a simplex. The above statement then means that the intersection of the simplex $I_G(\mathfrak{A})$ with the simplex K_β is again a simplex [4].

For an algebra $\mathfrak{A}$ which is generated by a countable number of local algebras each of which has a separable predual, the abovementioned simplexes K_β and $I_G(\mathfrak{A}) \cap K_\beta$, as compact sets of locally normal states, are metrizable [4], which means [21] that $\omega \in K_\beta$ can be written in a unique way as

$$\omega = \int d\mu(\phi)\, \phi,$$

where ϕ is an extremal K.M.S. state, i.e. a primary K.M.S. state. Moreover, $\omega \in K_\beta \cap I_G(\mathfrak{A})$ admits a unique decomposition

$$\omega = \int d\nu(\phi)\, \phi,$$

where ϕ is an extremal element in $K_\beta \cap I_G(\mathfrak{A})$.

For general information on K.M.S. states we refer to [6, 7, 16, 17].

2. TIME TRANSLATIONS AS A GROUP OF AUTOMORPHISMS OF A SUITABLY CHOSEN ALGEBRA*

In the preceding section we studied, amongst other things, the conditions which permit time translation to be described as a one-parameter group of automorphisms of the quasi-local algebra $\mathfrak{A}$. With the conditions formulated in the assumptions (III) (Section 1b), we saw that time translations are described by a group of implementable automorphisms of at least the von Neumann algebra $\pi_\omega(\mathfrak{U})''$ obtained from a thermodynamic limit state. In this section we intend to construct a C^*-algebra $(\mathfrak{A}_4)$ with the following properties:

i. Time translations induce a one-parameter automorphism group.

ii. Thermodynamic limit states are states on $\mathfrak{A}_4$.

We shall first construct the algebra $\mathfrak{A}_4$ (For more details see [22]).

Denote by $\mathfrak{A}_F$ the free *-algebra with identity E generated by monomials in (A_i, t_i) and E, where $A_i \in \mathfrak{A}_L$. The general element of $\mathfrak{A}_F$ is of the form

$$\mathbb{P}(A_i, t_i) = \lambda_0 E + \sum_i^n \lambda_i (A_i, t_i) + \sum_{i_1, i_2}^{n_1, n_2} \lambda_{i_1, i_2}(A_{i_1}, t_{i_1})(A_{i_2}, t_{i_2}) + \ldots +$$

$$+ \sum_{i_1, \ldots, i_k}^{n_1, \ldots, n_k} \lambda_{i_1 \ldots i_k}(A_{i_1}, t_{i_1}) \ldots (A_{i_k}, t_{i_k}).$$

$\mathbb{P}(A_i, t_i)^*$ is given by

$$\mathbb{P}(A_i, t_i)^* = \bar\lambda_0 E + \sum_i^n \bar\lambda_i (A_i, t_i) + \sum_{i_1, i_2}^{n_1, n_2} \overline{\lambda_{i_1, i_2}}(A_{i_2}^*, t_{i_2})(A_{i_1}^*, t_{i_1}) + \ldots +$$

$$+ \sum_{i_1 \ldots i_k}^{n_1 \ldots n_k} \overline{\lambda_{i_1 \ldots i_k}}(A_{i_k}^*, t_{i_k}) \ldots (A_{i_1}^*, t_{i_1}).$$

We introduce on $\mathfrak{A}_F$ the following seminorm:

$$\|\mathbb{P}(A_i, t_i)\|_1 = |\lambda_0| + \sum_i |\lambda_i| \cdot \|A_i\| + \sum_{i_1, i_2}^{n_1, n_2} |\lambda_{i_1, i_2}| \cdot \|A_{i_1}\| \cdot \|A_{i_2}\| + \ldots +$$

$$+ \sum_{i_1 \ldots i_k}^{n_1 \ldots n_k} |\lambda_{i_1 \ldots i_k}| \cdot \|A_{i_1}\| \cdot \|A_{i_2}\| \ldots \|A_{i_k}\|.$$

This seminorm obviously vanishes on all elements (A_i, t_i) for which $\lambda_0 = 0$ and for which every other monomial contains an element $(0, t_{i_j})$. These elements form a two-sided *-ideal in $\mathfrak{A}_F$, which we denote by K_0.

---

* This section is an account of work done in collaboration with Michel Sirugue.

Consider now the map $\mathfrak{A}_F$ into $\mathfrak{A}_F/K_0$. $\|\cdot\|_1$ defines a norm on $\mathfrak{A}_F/K_0$. In fact one shows easily that the closure of $\mathfrak{A}_F/K_0$ in this norm is a Banach *-algebra. The norm on the closure of $\mathfrak{A}_F/K_0 \equiv \mathfrak{A}_1$ is not a C^*-norm [22].

Let ω_n be any state on $\mathfrak{A}(V_n)$ (not necessarily a Gibbs state) and consider

$$\omega_n(\alpha^n_{t_{i_1}}(A_{i_1})\dots\alpha^n_{t_{i_k}}(A_{i_k})).$$

Suppose that for $n \to \infty$ the limit of this object exists.

Definition:

$$\lim_{n \to \infty} \omega_n(\alpha^n_{t_{i_1}}(A_{i_1})\dots\alpha^n_{t_{i_k}}(A_{i_k})) = \omega((A_{i_1}, t_{i_1})\dots(A_{i_k}, t_{i_k})).$$

ω can be extended to a linear form on $\mathfrak{A}_F$ by linear extension and the convention that $\omega(E) = 1$.

ω is clearly vanishing on K_0. Furthermore, ω is continuous in the norm $\|\cdot\|_1$ on $\mathfrak{A}_F/K_0$. Indeed,

$$|\omega(A_i, t_i)| \leq |\omega_n(\alpha^n_{t_i}(A_i))| + \varepsilon,$$

$$\leq \|A\| + \varepsilon,$$

for n sufficiently large.

Hence $|\omega(A_i, t_i)| \leq \|A_i\|$ and also

$$|\omega(\mathbb{P}(A_i, t_i))| \leq \|\mathbb{P}(A_i, t_i)\|_1.$$

ω can therefore be extended to a continuous linear functional of $\mathfrak{A}_1$.

One also sees that ω is positive on $\mathfrak{A}_1$, by first observing that it is positive on $\mathfrak{A}_F/K_0$ and then by extending it to $\mathfrak{A}_1$.

Consider now the images in $\mathfrak{A}_1$ of the elements of the form

$$(AB, t) - (A, t)(B, t)$$

$$E - (I, t)$$

$$(\alpha A + \beta B, t) - \alpha(A, t) - \beta(B, t)$$

in $\mathfrak{A}_F$. Let K_1 denote the smallest closed two-sided *-ideal generated by these elements in $\mathfrak{A}_1$.

One easily sees that ω vanishes on K_1. One can therefore extend ω to a positive linear functional on the Banach *-algebra $\mathfrak{A}_1/K_1 \equiv \mathfrak{A}_2$.

Consider now the set of generalized zeros in $\mathfrak{A}_2$, i.e. $\{x \in \mathfrak{A}_2 : f(x^*x) = 0 \;\forall$ positive linear functionals on $\mathfrak{A}_2\}$. Denote it by N. Then we know that N is a two-sided *-ideal and is closed because $\mathfrak{A}_2$ is a Banach *-algebra. Defining now $\mathfrak{A}_3 = \mathfrak{A}_2/N$, we have the following structure:

    i. $\mathfrak{A}_3$ and $\mathfrak{A}_2$ have the same positive linear functionals [23].

    ii. $\mathfrak{A}_3$ has a C^*-norm [23] (the minimal regular norm).

Since ω is positive on $\mathfrak{A}_2$ and vanishes on N (like any positive functional on $\mathfrak{A}_2$), it can be regarded as a positive linear functional on $\mathfrak{A}_3$ and, by extension, as a positive linear functional on the C^*-closure of $\mathfrak{A}_3$ which we shall call $\mathfrak{A}_4$. (This extension can of course be made by the very definition of the minimal regular norm [23]). What we said above can be represented by the following schematic sequence of mappings:

$$\mathfrak{A}_F \to \mathfrak{A}_F/K_0 \to \mathfrak{A}_1 \to \mathfrak{A}_1/K_1 \equiv \mathfrak{A}_2 \to \mathfrak{A}_2/N = \mathfrak{A}_3 \to \mathfrak{A}_4. \tag{19}$$

$\mathfrak{A}_1/K_1$ and $\mathfrak{A}_2/N$ are equipped with the infinum norm inherited from $\mathfrak{A}_1$ and $\mathfrak{A}_2$ respectively.

We now want to discuss time evolution and show that it can be represented as a one-parameter group of *-automorphisms of $\mathfrak{A}_4$.

Indeed, let Γ_t denote the map of $\mathfrak{A}_F$ onto itself defined by the following equation:

$$\Gamma_t\left[\sum \lambda_{i_1\ldots i_k}(A_{i_1}, t_{i_1})\ldots(A_{i_k}, t_{i_k})\right] = \sum \lambda_{i_1\ldots i_k}(A_{i_1}, t_{i_1} + t)\ldots(A_{i_k}, t_{i_k} + t).$$

Γ_t leaves K_0 invariant, preserves algebraic structure, is semi-norm-preserving on $\mathfrak{A}_F$ and therefore norm- and *- preserving on $\mathfrak{A}_1$ and can therefore be extended to the closure of $\mathfrak{A}_F/K_0$, i.e. to $\mathfrak{A}_1$.

The extension of Γ_t to $\mathfrak{A}_1$ leaves K_1 invariant and hence we have a *-automorphism of $\mathfrak{A}_2$. (For more details we refer to [22]). One also verifies that N is an invariant two-sided ideal under the action of the image of Γ_t on $\mathfrak{A}_2$.

The C^*-norm on $\mathfrak{A}_2/N$ is the minimal regular norm, which can be expressed in terms of the positive linear functionals f with $f(e) = 1$ (e is the unit in $\mathfrak{A}_2/N$) in the following way:

$$x \in \mathfrak{A}_2/N \to \|x\| = \sup_{\substack{f \geq 0 \\ f(e) = 1}} \sqrt{f(x^*x)}.$$

Moreover, $\mathfrak{A}_2$ and $\mathfrak{A}_2/N$ have the same positive linear functionals. It follows that the image of Γ_t defines a map of $\mathfrak{A}_2/N$ onto itself that preserves the C^*-norm and the algebraic structure and therefore can be extended to a *-automorphism of $\mathfrak{A}_4$ [22], denoted by α_t. Consider

$$\omega_n(\mathbb{P}(\alpha_{t_i}^n(A_i))\,\alpha_t^n(\mathbb{P}(\alpha_{t_k}^n(B_k)))) = \omega_n(\mathbb{P}(\alpha_{t_i}^n(A_i))\,\mathbb{P}(\alpha_{t_k+t}^n(B_k))).$$

Let us assume that ω_n is locally a Gibbs state. Then, if the limit of the above-mentioned expression exists, we arrive at the following equation:

$$\int \omega(\mathbb{P}(A, t_i))\,\alpha_t(\mathbb{P}(B_k, t_k)))\,f(t - i\beta)\,dt = \int \omega(\alpha_t(\mathbb{P}(B_k, t_k))\,\mathbb{P}(A_i, t_i))f\,dt,$$

$$\hat{f} \in D. \tag{20}$$

$[\mathbb{P}(\alpha_{t_i}^n(A_i))$ is the same polynomial in $\alpha_{t_i}^n(A_i)$ as $\mathbb{P}(A_i, t_i)$ is in $(A_i, t_i)]$.

We assume that the integrands on both sides are continuous for $t \in [-\varepsilon, \varepsilon]$ [cf. assumptions (I) and (II) in Section 1b].

To reduce our notation to a minimum, we denote the image of $\mathbb{P}(A_i, t_i)$ in $\mathfrak{A}_4$ again by $\mathbb{P}(A_i, t_i)$, and the state ω on $\mathfrak{A}_F$, considered as a state on $\mathfrak{A}_4$, again by ω.

From ω we can construct a representation π of $\mathfrak{A}_4$, which acts on a Hilbert space $\mathfrak{H}_\omega$. The elements of π that are representatives of the images $\mathbb{P}(A_i, 0)$ by (19) in $\mathfrak{A}_4$ are denoted by π_0. The C^*-algebra generated by π_0 is denoted by $C^*(\pi_0)$. $C^*(\pi)$ denotes the C^*-algebra generated by the representatives (on $\mathfrak{H}_\omega$) of the images of $\mathfrak{A}_F$ (via (19)) in $\mathfrak{A}_4$. Evidently we have $C^*(\pi) = \pi(\mathfrak{A}_4)$.

Equation (20) is primarily defined on a dense subset in $\mathfrak{A}_4$, namely the images of $\mathfrak{A}_F$ in $\mathfrak{A}_4$, but can of course be extended to all of $\mathfrak{A}_4$. Since continuity was assumed, we have invariance of ω for α_t, hence implementation of α_t.

Furthermore, the cyclic vector Ω in $\mathfrak{H}_\omega$ is separating for π'' [cf. Section 1b]. U_t is the group of unitaries that implement α_t on $\mathfrak{H}_\omega$. Furthermore, $U_t \pi(\mathbb{P}(A_i, t_i)) U_{-t} = {} $ $= \pi(\mathbb{P}(A_i, t_i + t))$. $\mathfrak{H}_0 = \overline{\pi_0 \Omega}$ (i.e. the subspace generated by applying π_0 on the cyclic vector Ω).

We then have the following lemma:

$$U_t \mathfrak{H}_0 = \mathfrak{H}_0 \text{ is equivalent to the cyclicity of } \Omega_\omega \text{ for } \pi_0''.$$

Proof:

The only nontrivial part is that $U_t \mathfrak{H}_0 = \mathfrak{H}_0$ implies the cyclicity of Ω for π_0''. This however follows directly from the structure of the representation, the fact that $U_t \pi(\mathbb{P}(A_i, t_i)) U_{-t} = \pi(\mathbb{P}(A_i, t_i + t))$ and that $P_{\mathfrak{H}_0}$ (the projection operator onto $\mathfrak{H}_0$) commutes with U_t.

Let us now recall some definitions and a theorem due to Kadison [24].

A convex set S of states on a C^*-algebra $\mathfrak{A}$ is *full* if $\phi(A) \geqq 0, \forall\, \phi \in S$ implies $A \geqq 0, A \in \mathfrak{A}$.

A *unimorphism* v_t of a full set of states S on a C^*-algebra onto itself relative to the weak *-structure is an affine map of S onto itself with the property that for any $B \in \mathfrak{A}$ and any t, there exists $A_1 \ldots A_n \in \mathfrak{A}$ such that if $|\rho(A_i) - \tau(A_i)| < 1, \rho, \tau \in S$ then $|(v_t \rho)(B) - (v_t \tau)(B)| < 1$. A unimorphism is continuous in t if $t \to (v_t \rho)(A)$ is a continuous function of t for each $\rho \in S$ and $A \in \mathfrak{A}$.

Theorem:

A mapping v of a full family of states of a C^*-algebra $\mathfrak{A}$ is a unimorphism *iff* there exists a C^*-automorphism α of $\mathfrak{A}$ such that $(v\rho)(A) = \rho(\alpha(A))$ for each ρ in S, $A \in \mathfrak{A}$.

This theorem is a specialized version of [24, corollary 4.7].

Recall that a C^*-automorphism α of $\mathfrak{A}$ is a linear *-preserving mapping of $\mathfrak{A}$ onto $\mathfrak{A}$ such that $\alpha(A^2) = \alpha(A)^2$ for each $A \in \mathfrak{A}$ (equivalently $\alpha(AB + BA) = \alpha(A)\alpha(B) + {} $ $+ \alpha(B)\alpha(A)$).

Let B be a von Neumann algebra on a space $\mathfrak{H}$ and $\mathfrak{A}$ a C^*-algebra that is weakly dense in B. Let S be a full set of states on B (clearly, S is then also a full set of states of $\mathfrak{A}$). We shall call a *W^*-unimorphism v relative to B* a unimorphism of S that gives rise to a C^*-automorphism of B but not of $\mathfrak{A}$. A C^*-*unimorphism* v *relative to* $\mathfrak{A}$ of S is a unimorphism that gives rise to a C^*-automorphism of $\mathfrak{A}$.

Suppose v is a W^*-unimorphism, relative to B, of S. Since by definition and the theorem v is not a unimorphism of S relative to $\mathfrak{A}$, there exists at least one $B \in \mathfrak{A}$ such that, for every n-tuple $A_1 \ldots A_n$, there exists at least one pair $\rho, \tau \in S$ such that $|\rho(A_i) - \tau(A_i)| < 1$ and $|(v\rho)(B) - (v\tau)(B)| \geqq 1$.

Consider the dense set $\{\pi(\mathbb{P}(A_i, t_i))\,\Omega\}$ in $\mathfrak{H}_\omega$. From this dense set we construct the following full set of states S on $C^*(\pi_0)$ and π_0'':

$$S = \left\{ \phi : \phi = \sum_{i=1}^{N} \lambda_i \omega_{\pi(\mathbb{P}(A_i, t_i))\Omega}, \ \lambda_i > 0, \ \sum_{i=1}^{N} \lambda_i = 1 \right\}.$$

U_t maps $\{\pi(\mathbb{P}(A_i, t_i))\,\Omega\}$ onto itself, and its transpose on S is denoted by v_t.

We now distinguish between the following cases:

Case I:

v_t is a C^*-unimorphism of S relative to $C^*(\pi_0)$. Then clearly $(v_t\phi)(\pi(\mathbb{P}(A_k, 0))) =$
$= \phi(U_t\pi(\mathbb{P}(A_k, 0)\,U_{-t}))$ and for $C \in C^*(\pi_0)$, $C \to U_t C U_{-t}$ generates a *-*automorphism* of $C^*(\pi_0)$. Because $\Omega_\omega = U_t\Omega_\omega$, we see that $U_t\mathfrak{H}_0 = \mathfrak{H}_0$ and therefore Ω_ω is cyclic for $C^*(\pi_0)$ by the lemma. The fact that we have an automorphism and not merely a C^*-automorphism is shown as follows:

Let α be the C^*-automorphism that emerges from Kadison's theorem. Then $\phi(U_t\pi(\mathbb{P}(A_k, 0))\,U_{-t}) = \phi(\alpha_t(\pi(\mathbb{P}(A_k, 0)))), \ \phi \in S$; in particular,

$$(\psi, U_t\pi(\mathbb{P}(A_k, 0))\,U_{-t}\psi) = (\psi, \alpha_t(\pi(\mathbb{P}(A_k, 0)))\,\psi)$$

for ψ in a dense set in $\mathfrak{H}$, namely S, and hence for all $\chi \in \mathfrak{H}_\omega$ we have

$$(\chi, \alpha_t(\pi(\mathbb{P}(A_k, 0)))\,\chi) = (\chi, U_t(\pi(\mathbb{P}(A_k, 0)))\,U_{-t}\chi),$$

from which follows that

$$\alpha_t(\pi(\mathbb{P}(A_k, 0))) = U_t(\pi(\mathbb{P}(A_k, 0)))\,U_{-t}.$$

We now have the following two equations:

$$\int (\Omega, AU_tB\Omega)\,f_\beta dt = \int (\Omega, BU_{-t}A\Omega)\,f\,dt, \tag{21}$$

$A, B \in \pi_0''$, where $U_t A U_{-t} \in \pi_0''$.

Our starting point was

$$\int (\Omega, AU_tB\Omega)\,f_\beta dt = \int (\Omega, BU_{-t}A\Omega)\,f\,dt, \tag{22}$$

$A, B \in \pi''$, $U_t A U_{-t} \in \pi''$.

From (21) we have $J_1 T_1 A\Omega = A^*\Omega$, where J_1 is the usual involution that follows from K.M.S. and $T_1 = \exp - (\beta/2)\,H$, where H is the generator of U_t, $A \in \pi_0''$(cf. [7]).

Furthermore, $J_1\pi_0''J_1 = \pi_0'$.

From (22) we have $J_2 T_2 A\Omega = A^*\Omega$, for $A \in \pi''$ and in particular for $A \in \pi_0''$.

Since $T_2 = T_1$ and Ω is cyclic for π_0'', we find from $J_1 A\Omega = J_2 A\Omega$, $A \in \pi_0''$ that $J_1 = J_2$. We know that $\pi_0'' \subset \pi''$. Suppose now that

$$B \in \pi'', \qquad B \notin \pi_0'',$$

$$J_2 B J_2 \in \pi' \subseteq \pi_0'.$$

Since $J_1 B J_1 = J_2 B J_2$, we have that $B = J_1 (J_1 B J_1) J_1 \in \pi_0''$, which contradicts our assumption. *Hence we conclude that $\pi_0'' = \pi''$.*

One can show [22] that $\mathfrak{A}_L$ is continuously represented in $\mathfrak{A}_4$ by the successive maps in (19), preceded by the map

$$A \in \mathfrak{A}_L \rightarrow (A, 0). \tag{23}$$

Denote these maps preceded by (23) by ϕ_0. Then

$$\|\pi(\phi_0(A))\| \leq \|\phi_0(A)\| \leq \|A\|,$$

where the last inequality follows from the previous remark. From this it follows that π can be extended to all of $\mathfrak{A}$. Let the extension be $\tilde{\pi}$. The quasi-local algebra we have in mind is the one where every $\mathfrak{A}(V)$ is a type-I_∞ factor. This algebra can be shown to be simple [12]. [If we had chosen the algebra for the quantum lattice systems then we would also have a simple algebra].

Therefore $\|\tilde{\pi}(A)\| = \|A\| \ \forall A \in \mathfrak{A}$. This means of course that the automorphism of $C^*(\pi_0)$ induced by v_t is an automorphism of $\mathfrak{A}$.

Case II:

v_t is a W^*-unimorphism of S relative to π_0''. By exactly the same reasoning we arrive at an automorphism of π_0'', namely $A \rightarrow U_t A U_{-t}$, whose transpose on S is v_t. Again, $\pi_0'' = \pi''$. However $A \rightarrow U_t A U_{-t}$ is not an automorphism of $C^*(\pi_0)$, and hence not of $\mathfrak{A}$.

Case III:

v_t is neither a C^*-unimorphism of S relative to $C^*(\pi_0)$ nor a W^*-unimorphism of S relative to π_0'', but nevertheless U_t leaves $\mathfrak{H}_0$ invariant.

From the lemma we know that Ω is cyclic for π_0''. Equation (21) still holds, but $\{U_t\}$ no longer generates a group of automorphisms of π_0''. However, by Tomita's theorem (Ω is cyclic and separating for π_0'') we can find a group of unitaries V_t that generates an automorphism of π_0'' and satisfies:

$$\int (\Omega, A V_t B\Omega) f_\beta dt = \int (\Omega, B V_{-t} A\Omega) f \, dt, \qquad A, B \in \pi_0'', \quad \hat{f} \in D.$$

Case IV:

v_t is as in Case III, and furthermore we assume that $U_t \mathfrak{H}_0 \neq \mathfrak{H}_0$.

From the lemma we conclude again that Ω is not cyclic for π_0''. $P_{\mathfrak{H}_0} \in \pi_0'$.

Consider now the isomorphism

$$\pi_0'' \leftrightarrow [\pi_0'']_{P_{\mathfrak{H}_0}} \quad \text{(i.e. the restriction of } \pi_0'' \text{ to } \mathfrak{H}_0\text{)}.$$

The fact that this is not merely a homomorphism but an isomorphism stems from the fact that Ω is separating for π''.

Indeed let A belong to π_0'' and $A_{P_{\mathfrak{H}_0}} = 0$. Then $A_{P_{\mathfrak{H}_0}} \Omega = 0$, hence $P_{\mathfrak{H}_0} A\Omega = = 0 \Rightarrow A\Omega = 0 \Rightarrow A = 0$.

Consider now $A, B \in \pi_0''$:

$$\int (\Omega, A_{P_{\mathfrak{H}_0}} U_t B_{P_{\mathfrak{H}_0}} \Omega) f_\beta dt = \int (\Omega, A P_{\mathfrak{H}_0} U_t P_{\mathfrak{H}_0} B\Omega) f \, dt =$$

$$= \int (\Omega, A U_t B\Omega) f \, dt = \int (\Omega, B U_{-t} A\Omega) f \, dt =$$

$$= \int (\Omega, B_{P_{\mathfrak{H}_0}} U_{-t} A_{P_{\mathfrak{H}_0}} \Omega) f \, dt. \tag{24}$$

Since Ω is separating and cyclic for $[\pi_0'']_{P_{\mathfrak{H}_0}}$ in $\mathfrak{H}_0$, we can find by Tomita's theorem a group of unitaries V_t that generates a group of automorphisms of $[\pi_0'']_{P_{\mathfrak{H}_0}}$ and satisfies the following equation:

$$\int (\Omega, A_{P_{\mathfrak{H}_0}} V_t B_{P_{\mathfrak{H}_0}} \Omega) f_\beta dt = \int (\Omega, B_{P_{\mathfrak{H}_0}} V_{-t} A_{P_{\mathfrak{H}_0}} \Omega) f \, dt.$$

The equation (24) can be rewritten in the following form:

$$\int (\Omega, A_{P_{\mathfrak{H}_0}} [U_t]_{P_{\mathfrak{H}_0}} B_{P_{\mathfrak{H}_0}} \Omega) f_\beta dt = \int (\Omega, B_{P_{\mathfrak{H}_0}} [U_{-t}]_{P_{\mathfrak{H}_0}} A_{P_{\mathfrak{H}_0}} \Omega) f \, dt.$$

Since we have assumed that $U_t \mathfrak{H}_0 \neq \mathfrak{H}_0$, and therefore $[U_t, P_{\mathfrak{H}_0}] \neq 0$, we have by [22 (proposition 3.12)] that $[U_t]_{P_{\mathfrak{H}_0}}$ is not a group [i.e. $[U_{t_1}]_{P_{\mathfrak{H}_0}} [U_{t_2}]_{P_{\mathfrak{H}_0}} \neq \neq [U_{t_1+t_2}]_{P_{\mathfrak{H}_0}}$]. The transition from the family of operators $[U_t]_{P_{\mathfrak{H}_0}}$ to the group V_t is essentially the same process as we have in the preceding case.

We want to conclude with some remarks:

In Cases I and II, the whole structure as we have described it could be obtained in a less cumbersome way by just considering the usual algebra of quasi-local observables and looking at thermodynamic limits as states on this algebra. This follows from the fact that the cyclic representation given by the thermodynamic limit state is then unitarily equivalent to the representation π we considered restricted to $\mathbb{P}(A_i, 0)$. In order to account for the fact that time translation induces a group of automorphisms of the representation of the quasi-local algebra (Case I) or its generated von Neumann algebra (Cases I and II), one has to have an extra condition on the way the thermodynamic limit is taken. Indeed, a prototype of Case I is the quantum lattice gas as treated by Robinson where, together with the existence of the thermodynamic states, the local time-translation automorphisms α_t^V tend strongly to a limit automorphism α_t of the algebra of quasi-local observables, i.e.

$$\|\alpha_t^V(A) - \alpha_t(A)\| \xrightarrow[V \to \infty]{} 0.$$

A situation where Case I or Case II is obtained is discussed in [11]. The assumptions (III) in Section 1b are sufficient conditions for v_t to have, in the above language, a C^*-unimorphism relative to π_0''.

However we do not see, at this stage, whether indeed one has Case II or Case I in this situation.

If, as in the case discussed by Ruskai, one would be able to exclude Case I, the possibilities left are clearly Cases II, III and IV.

REFERENCES

[1] G. W. Mackey, "Mathematical Foundations of Quantum Mechanics," Chapter 2, Benjamin, New York (1963).

[2] A. M. Gleason, *J. Math. and Mech.*, **6**, No. 6, 885 (1957).

[3] S. Sakai, "C^*- and W^*-Algebras," Section 1, 23. Springer Verlag, Berlin (1971).

[4] M. Takesaki and M. Winnink, "Local Normality in Quantum Statistical Mechanics" (to be published).

[5] S. Sakai, *Proc. Jap. Acad.*, **33**, 439, proposition 1 (1957).

[6] R. Haag, N. M. Hugenholtz, and M. Winnink, *Comm. Math. Phys.*, **5**, 215 (1967).

[7] M. Winnink, "Cargèse Lectures in Physics", Ed. D. Kastler, Vol.4, pp. 235–255. Gordon and Breach, New York (1969).

[8] D. Ruelle, *J. Math. Phys.*, **8**, 1657 (1967).

[9] N. M. Hugenholtz and J. D. Wieringa, *Comm. Math. Phys.*, **11**, 183 (1969).

[10] E. Hille and R. S. Phillips, "Functional Analysis and Semi-Groups", Vol. XXXI. Amer. Math. Soc. Coll. Publ. (1957).

[11] D. Dubin and G. L. Sewell, *J. Math. Phys.*, **11**, No. 10, 2990 (1970).

[12] R. Haag, R. V. Kadison, and D. Kastler, *Comm. Math. Phys.*, **16**, 81 (1970).

[13] D. W. Robinson, *Comm. Math. Phys.*, **7**, 337 (1968).

[14] H. J. Brascamp, *Comm. Math. Phys.*, **18**, 82 (1970).

[15] M. Tomita, "Quasi-Standard von Neumann Algebras", Kyushu University (preprint, 1967).

[16] M. Takesaki, "Tomita's Theory of Modular Hilbert Algebras and Its Applications", Springer, Berlin (1970).

[17] M. Winnink, "An Application of C^*-Algebras to Quantum-Statistical Mechanics" (Thesis, Groningen, 1968).

[18] M. Sirugue and M. Winnink, *Comm. Math. Phys.*, **19**, 161 (1970).

[19] H. Araki and E. J. Woods, *J. Math. Phys.*, **4**, 637 (1963).

[20] D. Ruelle and O. Lanford, *J. Math. Phys.*, **8**, 1460 (1967).

[21] See for instance:

R. Phelps, "Lectures on Choquet's Theorem". Van Nostrand (1966).

G. Choquet, "Lectures on Analysis", Ed. J. Marsden, T. Lance, and S. Gelbart, Benjamin, New York (1969).

[22] M. Sirugue and M. Winnink, "Translations dans le temps comme groupe d' *-automorphismes", Marseille (preprint 71/p. 338, July 1971).

[23] M. A. Naimark, "Normed Rings", Noordhoff, Groningen (1964).

[24] R. V. Kadison, *Topology*, **3**, Suppl. 2, 177–198 (1965).

[25] M. B. Ruskai, *Comm. Math. Phys.*, **20**, 193 (1971).